国家自然科学基金面上项目（编号：52078178）
河北省高等学校人文社会科学研究项目（编号：BJ2020[illegible]62）

Building Environment and Service System

建筑环境及服务系统

侯国英　魏　绅　著

图书在版编目(CIP)数据

建筑环境及服务系统 = Building Environment and Service System：英文 / 侯国英，魏绅著. -- 天津：天津大学出版社，2023.12
ISBN 978-7-5618-7421-9

Ⅰ. ①建… Ⅱ. ①侯… ②魏… Ⅲ. ①建筑工程－环境管理－高等学校－教材－英文 Ⅳ. ①TU-023

中国国家版本馆CIP数据核字（2023）第042516号

JIANZHU HUANJING JI FUWU XITONG

出版发行 天津大学出版社
地　　址 天津市卫津路92号天津大学内（邮编：300072）
电　　话 发行部：022-27403647
网　　址 www.tjupress.com.cn
印　　刷 北京虎彩文化传播有限公司
经　　销 全国各地新华书店
开　　本 787mm×1092mm　1/16
印　　张 20
字　　数 669千
版　　次 2023年12月第1版
印　　次 2023年12月第1次
定　　价 68.00元

Contents

Chapter 1

Introduction

1.1 The Link between Architecture and Environment

Since the emergence of architecture, it has been inseparable from the environment. In ancient times when the level of productivity was very low, people chose to build their houses in places rich in natural resources. Architecture is alive and symbiotic with the environment. The life of architecture exists not only in its physical cycle, but also in its symbiotic environment. It is because any building does not exist in isolation, and it is a whole consisting of the environmental range in which it is located. Architecture exists in various natural and man-made environments. It is an artificial space separated from the natural environment by artificial materials. Architecture and environment are closely related, and the construction of a building means the interconnection between the building and its environment, as well as the integration between the natural environment system and the building environment system.

One of the basic functions of architecture is to create a comfortable indoor thermal environment. Separated from outdoor space, a comfortable indoor environment with proper temperature, humidity, radiation and airflow can be created by adjusting or utilizing outdoor climate resources, and using building envelope to filter unfavorable factors; it is beneficial to people's work and life. How to create a comfortable thermal environment? First of all, it is necessary to have a complete understanding or more accurate description of the environment, understand what kind of indoor environment people and production process need, and master the characteristics and influencing factors of the formation of indoor and outdoor environments, based on the elaboration of the main environmental impact factors, such as solar radiation, climate and other external interference factors of the outdoor environment and the internal interference factors such as heat, humidity and air pollutants of the indoor environment. Secondly, based on the fully understanding of the environment, it is necessary to apply knowledge of heat transfer and thermal design principles, use building materials and structures rationally, and master the basic principles and methods of designing and controlling the indoor environment.

Climate is the external condition of the building thermal environment, which depends on the changing characteristics and combination of various elements. Various climate elements interrelate, interact and affect the architectural design. Climate factors related to architectural design include solar radiation, air temperature, atmospheric humidity, wind direction and speed, lighting, condensation and precipitation. Solar radiation is the main heat source of building, which brings heat to the exterior wall and brings direct sunlight to the interior space; air temperature is the calculation basis for thermal design of building insulation, heat-resistance, heating, ventilation and air conditioning; wind direction and speed affect the layout of buildings and the natural ventilation organization of buildings; condensation and precipitation affect architectural modeling and drainage, as well as dew coagulation on the surface of building envelope, internal condensation and insulation material setting, etc. While climate is stable for a long time, human civilization shows diversified characteristics due to climate conditions.

The architectural design is to master the climate characteristics. According to the solar radiation, temperature, wind and precipitation conditions, architects should use technical methods to carry out thermal insulation, heat insulation, ventilation and moisture-proof design from the aspects of building site selection, lighting, shading, heat preservation, heat insulation , heat storage and heat collection, heating and cooling, ventilation, wind and moisture-protection, so as to form a good indoor thermal environment.

The climatic conditions and the external environment where the building is located can directly affect the indoor environment through building envelope. The envelope is the physical boundary between the interior and exterior of the building. Original buildings are aimed at sheltering, heat insulation, shading and ventilation. With the development of technology, the function and role of building envelope have become more and more complete, followed by a more comfortable indoor thermal environment. Today, the performance of traditional envelopes is constantly improving, while new walls, windows, doors, roofs and floors are unique in terms of materials and construction. On the one hand, advances in materials and construction technology have improved the thermal insulation and air tightness of the envelope, improving thermal comfort and meeting energy-saving requirements; on the other hand, more updated content is added to the envelope, not only to make it meet traditional functional requirements, but also to make it become a place to use renewable resources. In order to meet the requirements of solar energy collection, storage and distribution, solar collectors are integrated into the roofs, walls, doors and windows, while interlayers and pipes are laid on the roofs, walls, doors, windows and floors for energy transmission, storage, redistribution and utilization.

The problem of the building thermal environment is ultimately an energy problem. Before the fire for heating, original shelters were energy-free but uncomfortable. After the industrial revolution, the application of equipment technology such as artificial cooling, heating and ventilation exchanged high energy consumption for high comfort, creating unprecedentedly suitable living conditions, but at the same time, the building gradually lost its sensitivity to the climate environment, thus aggravating environmental pollution, global warming and other problems, and making building and climate turn into a relatively disharmonious state.

With the increasingly serious problem of global warming, countries around the world have gradually reached an agreement on the goal of building energy conservation. People gradually realize that one of the main reasons that affect climate warming is the carbon dioxide produced by energy consumption in the process of building use. In order to protect the environment, measures must be taken in the building. No matter how rapid the development of technology is, the natural environment should not be ignored when constructing buildings. On the basis of people-oriented development, always put the first priority on providing a natural, harmonious and healthy living environment for human beings. In line with the laws of natural development, follow the ecological principles, and blend the building and the natural environment. Under the premise of protecting and respecting nature, the three separately existing individuals of environment, people and building fully integrate and play their respective roles to achieve green and sustainable development.

1.2 The Link between Architecture and Human

Since ancient times, the relationship between human and nature has been developing and evolving, which is a continuous process. With the rapid development of urban construction, people are suffering from the consequences of excessive consumption of nature. The relationship between human and nature has changed from human attempting to control nature to human attempting to coexist harmoniously with nature. As the material carrier of human settlement activities, architecture is also evolving to meet people's spiritual needs as well as material needs.

As a complex organic whole, the human body is constantly exchanging material and energy with the external environment, which is the fundamental factor for human to survive. People feel the external environment and its changes through sensory organs, regulate human physiological activities and heat exchange, and radiate the heat generated by metabolism to avoid temperature changes and endangering survival. The maintenance of normal body temperature is the result of the dynamic balance of heat production and heat dissipation, which is achieved within the thermal comfort range. Thermal comfort will be affected by physiology, physics and psychology. Physiology refers to the body's temperature, blood pressure, skin humidity, skin temperature, sweating degree, etc.; physical factors refer to the heat balance relationship, which is more inclined to the influence of environmental comfort on human senses; psychological factors refer to the subjective impression of the receiver to distinguish comfort level from cold and hot. It can be seen that thermal comfort is essentially a pleasant feeling caused psychologically by a series of activities of the human nervous system, and the thermal balance between people and the environment is a necessary condition for thermal comfort.

The health of human beings not only depends on their physical condition , but also benefits from a healthy living environment, which is equally important. Creating a comfortable thermal environment for people's daily life is the basic function of architecture. With the constant advances in air conditioning and heating technology based on the consumption of non-renewable energy, the design of the thermal environment has shifted from the traditional insulation of buildings to a mechanical approach that relies on the consumption of more energy, but abusing energy blindly goes against the very nature of climate regulation. The use of air conditioning and other mechanical equipment for ventilation not only consumes a lot of energy, but also creates a closed air conditioning environment where dust and germs can easily accumulate, causing indoor air pollution and the spread of disease, which is harmful to human health. The long-term use of artificial lighting will also lead to a decline in indoor hygiene conditions, having a negative impact on human physiology.

In response to this tendency, architects play an important role of coordinating, integrating and working closely with HVAC engineers, focusing on the building's shape coefficient and orientation. They also focus on the physical parameters such as thermal insulation, thermal performance, thermal resistance and thermal storage coefficient of materials, on the natural ventilation and lighting characteristics of different planes and sections, and on the functions of the components in terms of ventilation, lighting and shading. The design of the building is to ensure the comfort and health of the building in the most efficient way by making the best possible use of the natural ventilation and lighting characteristics of different planes and sections, as well as the functions of the components in terms of ventilation, lighting and shading. The architectural design is to solve the thermal environment problem of the building by using the natural conditions as much as possible, and propose the corresponding architectural methods to respect the climate, so as to create a more comfortable and healthy space environment that meets the requirements of modern society, and realize a truly humanized building.

Building is an important carrier of social energy consumption, and people are the main body of energy consumption inside buildings. In recent years, the quality and level of people's daily life have been significantly improved, and the demand for various types of energy is also rising, so the overall energy consumption is getting higher and higher. Human behaviors, mainly including personnel activities, one's surroundings and operation management, are the important factor affecting the level of building energy consumption. On the premise that the enclosure structure, outdoor weather and equipment system form are determined, occupants' regulation and control of various energy-using devices inside the building will completely determine the overall energy consumption of the building. Building energy consumption is closely related to people's living habits, so it is very important to find the control rules of personnel for indoor equipment in buildings. Therefore, sorting out human adaptive behaviors, such as air conditioning behavior, window opening behavior or the use of other cooling and heating facilities, will help to adjust indoor environmental comfort and reduce building energy consumption.

1.3 The Link between Architecture and Technique

The 21st century is an era of information, and more importantly, it is an era of ecological civilization. Human beings use high and new technology to explore the sustainable development model of survival, production and living environment. Nowadays, with the continuous development of computer technology, network technology, communication technology and control technology, the public facilities in the building can be intelligent, and the intelligent building is the main station of the future information highway. Strengthening the combination of architecture with the development of modern science and technology and social needs, and using energy consumption simulation as an aid, are important means for building scheme design, building energy efficiency analysis, and equipment operation optimization.

The generation of intelligent buildings is based on the development of social building technology and the needs of building functions. It emphasizes user experience and has endogenous development momentum. Intelligent building refers to the organic combination of intelligent computer technology, communication technology, control technology, multimedia technology and modern architectural art by using system integration method to master the four basic elements of buildings, namely structure, system, service and management. With the optimized design, through the automatic monitoring of equipment, the management of information resources, and the optimized combination of users' information services and their building environment, an elegant, comfortable, convenient, and highly safe environment can be provided in the future.

In the process of intelligent buildings' development, building management system has become an important part of intelligent buildings with their unique advantages. Building equipment management system is a building management system that uses computer information technology to implement comprehensive management of building equipment, ensure comfort and safety in buildings, and reduce energy consumption of building equipment. With the increasing amount of equipment in buildings, the requirements for realizing the functions of the equipment become more and more complex, and the control requirements become higher and higher. The equipment includes central air conditioning equipment, water supply and drainage equipment, elevators, lighting equipment, electrical equipment, etc. It is inevitable for modern building management to use building equipment management system to realize the integrated control and management of the building equipment.

The development of science and technology also plays an important role in the decision-making of energy conservation in the process of architectural design. As one of the pillar industries, the construction industry is accompanied by high material consumption, high energy consumption and high carbon dioxide emissions in its rapid development. There are many factors affecting building energy consumption, such as local meteorological conditions, equipment system selection and operation, enclosure orientation and personnel activities. With the continuous development of technology, energy consumption simulation software has become one of the basic tools to assist building equipment systems in energy consumption calculation, design optimization, operation management, energy saving diagnosis and other aspects. Through simulation, building design can be effectively optimized to achieve the purpose of low-carbon emissions and energy conservation. For example, the outdoor wind field distribution in different seasons can be obtained through the simulation of the wind environment, which can be used as the basis for adjusting the building layout. The simulation results of outdoor wind field can also be used as the input parameters of building natural ventilation simulation, and then used to evaluate the indoor natural ventilation effect; through the simulation of the thermal performance of the building envelope, the design of the envelope can be optimized, such as the ratio of windows to walls, wall insulation structure, window selection, etc. So it has a suitable thermal performance, and reduces the operating energy consumption and the carbon emissions of the residence; through residential natural lighting simulation, the indoor lighting conditions can be evaluated, and lighting design can be optimized to reduce thermal energy consumption and achieve low carbon emissions.

To sum up, this book attempts to provide a complete introduction of how to coordinate the relationship between architecture, environment and human. Architects control the thermal and humidity environment of buildings based on energy-saving concepts to ensure a comfortable, healthy and energy-saving living environment. At the same time, this book also tries to establish the energy conservation concept that the building affects the building environment and the building environment adapts to the climate.

Chapter 2

Building Outdoor Environment

2.1 Natural Climate Elements and Characteristics

2.1.1 Solar Radiation

2.1.1.1 Definition of Solar Radiation

Solar radiation (kW/m^2) is the radiant energy emitted by the sun, especially electromagnetic energy. Approximately half of the radiation falls within the visible shortwave portion of the electromagnetic spectrum. Most of the other half is in the near infrared portion, with some in the ultraviolet portion of the spectrum.

Solar radiation (energy) is transformed to heat energy only after it is absorbed and re-emitted by matter (Figure 2-1). Thus terrestrial radiation (heat energy) is the result of the earth absorbing solar radiation and then re-emitting it into the atmosphere. If all other variables are held constant, higher elevations are colder than lower elevations (Figure 2-2).

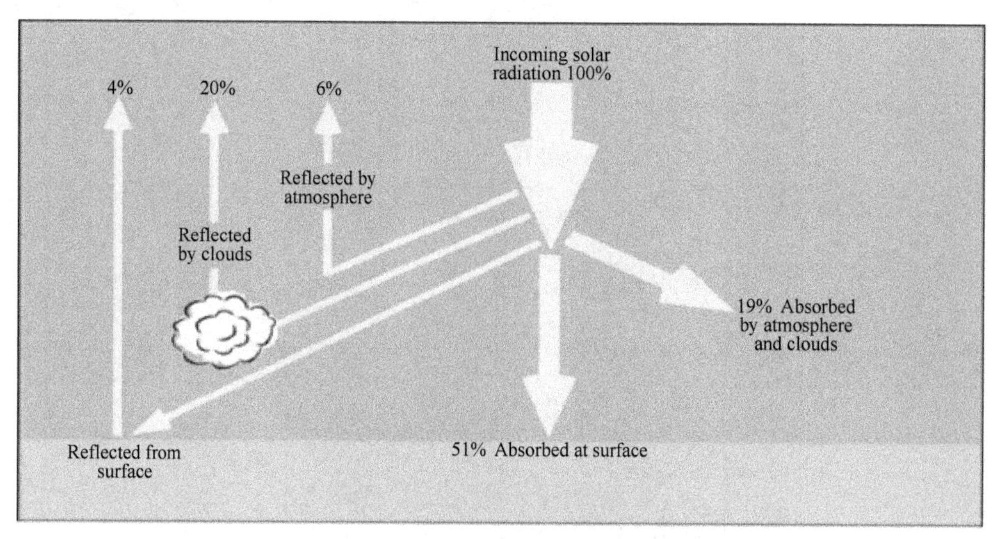

Figure 2-1 The absorption of solar radiation
(https://www.e-education.psu.edu/emsc100tsb/node/324)

Plants are able to absorb solar energy, and via a chemical process known as photosynthesis, store energy in plant tissue. Firewood is actually potential solar energy in storage. When the wood is burned, it creates heat energy (Figure 2-3).

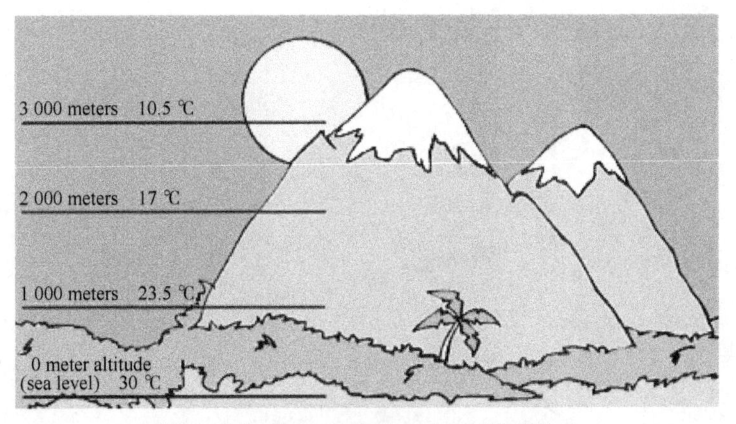

Figure 2-2 Temperature varies with altitude

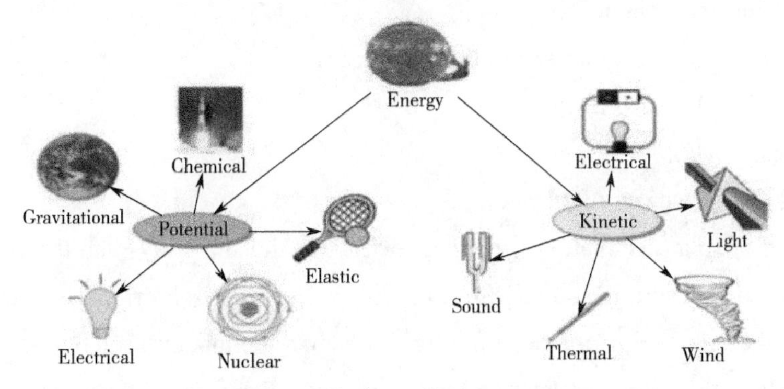

Figure 2-3 Use of solar energy

(www.blendspace.com)

2.1.1.2 Significance of Solar Radiation as a Design Basis

Solar radiation is original source of energy of the earth and important heat gain for buildings (penetrating windows or heating facade surfaces), increasing cooling load and overheating risks. In addition, it is also the important renewable energy for buildings, e.g. photovoltaic and solar collectors (Figure 2-4).

Figure 2-4 Solar radiation for buildings

1. Solar Principles

Solar gain (Figure 2-5) is beneficial in the heating season, but it is undesirable in the cooling season. The thermal capacity of air is low, so a small increase of thermal energy gives rise to a large increase in air temperature and consequent discomfort. Solar gain is often a significant part of the total internal heat gain of a space, and attention has to be paid to reducing it during the cooling season. There are already well-established parameters of understanding solar gains:

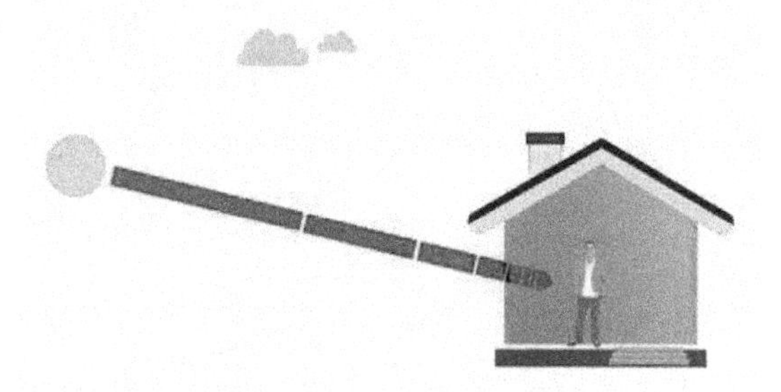

Figure 2-5 Solar gain

- Declination and latitude;
- Angle of the surface compared to the solar beam;
- Direct and diffuse light;
- Clearness index.

2. Solar Declination Angle

The angle of solar declination is the angle between the sun's rays and the earth's equator. The earth is rotating about the sun and has its own axis of rotation orbit. The earth is not perpendicular to the direction of the orbit. Instead, it has an inclination of about 23.45° . Because of this inclination, the earth's equator and the sun's rays always make a certain angle with each other (Figure 2-6).

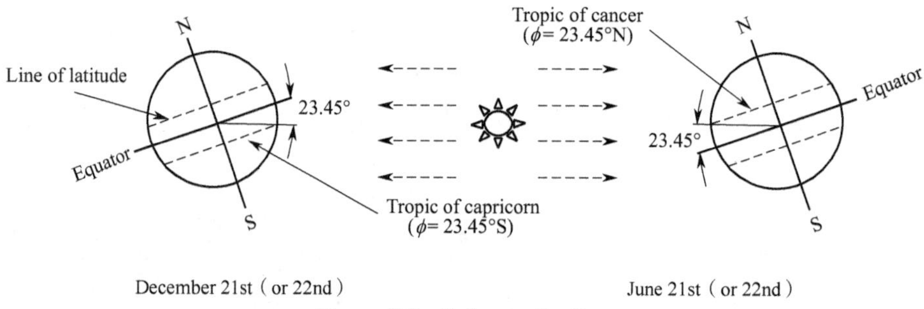

Figure 2-6 Solar declination

The angle of solar declination peaks on June 21st (or 22nd) . The day is referred to as the June solstice. On that day the earth is to the north of the sun, corresponding to a declination angle of 23.45° . In the northern hemisphere, this day is called the summer solstice, while in the southern hemisphere, it is called the winter solstice.

It has the lowest declination angle of 23.45° on December 21st (or 22nd). This December solstice is known in the northern hemisphere as the winter solstice, while in the southern hemisphere as the summer solstice.

As shown in Figure 2-7, declination angle ϕ = 23.45sin [360 × (284 + n)/365], where n= day number (January 1st = 1).

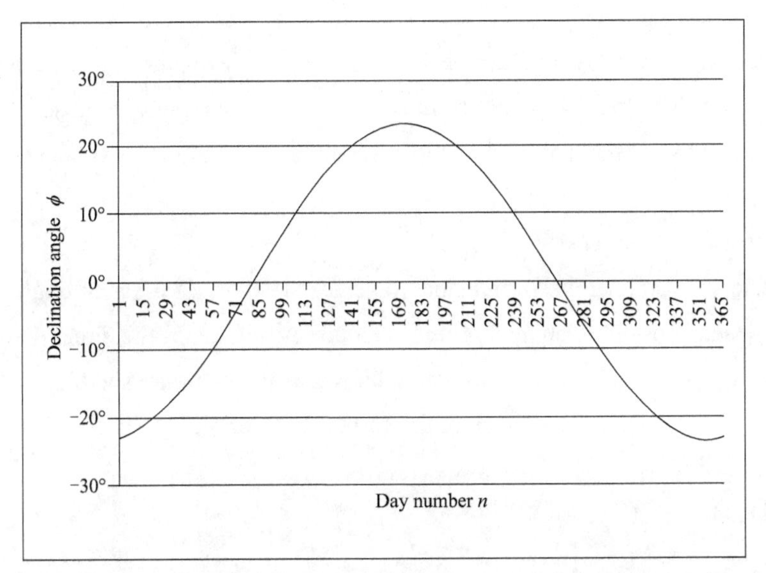

Figure 2-7 The relationship between declination angle and day number

3. Solar Zenith Angle

The angle of solar zenith is the angle between the direction of incidence of the sun's rays and the vertical direction of the zenith, varying from 0° to 180° , and its formula for calculation is related to the angle of solar declination, which is the latitude of the target surface, and the solar hour angle. For a horizontal plane, the angle of incidence and the zenith angle are the same (Figure 2-8).

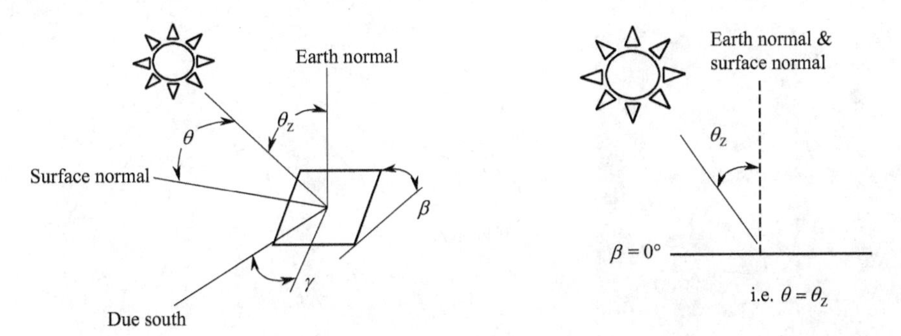

Figure 2-8 Solar zenith angle

4. Solar Hour Angle

Observing the sun from the earth, solar hour angle is an expression of time, expressed in angular measurement, usually degrees, from solar noon. If we take the earth as an example, at the same time on the earth, the sun corresponds to the same angle of time for people who live in the same longitude and different latitudes. Solar noon (informally high noon) is the moment when the sun touches the observer's meridian, reaching its highest position above the horizon on that day. At solar noon the hour angle is 0° , with the local time before solar noon expressed as negative degrees, and the local time after solar noon expressed as positive degrees. There are twenty-four hours in one day, and during that time the earth rotates 360° . So the earth spins 15° every hour, so that after every hour from solar noon the hour angle increases by 15° .For example, at 10:30 A.M. local apparent time the hour angle is −22.5° (15° per hour times 1.5 hours before noon).

5. Meridians

The meridian is an imaginary circle that traverses the earth's surface through the North and South Pole, which is used to denote longitude (Figure 2-9). Each of the circumferences passing through the two poles is divided into two parts, known as the meridian and the antimeridian.

Correct the solar beam to the plane of the surface of interest...

$$\cos\theta = \sin d \sin\phi \cos\beta - \sin d \cos\phi \sin\beta \cos\gamma + \cos d \cos\phi \cos\beta \cos\omega + \dots + \cos d \sin\phi \sin\beta \cos\gamma \cos\omega + \cos d \sin\beta \sin\gamma \sin\omega$$

From which, when $\beta = 0$, the zenith angle is defined...

$$\cos\theta_Z = \sin d \sin\phi + \cos d \cos\phi \cos\omega$$

where θ is surface-solar incidence; d is earth declination angle; ϕ is declination angle; w is hour angle.

The following is the simplified version.

If $\beta = 0$, i.e. horizontal surface, solar radiation outside the earth's atmosphere varies very slightly throughout the year, again due to the elliptic earth orbit.

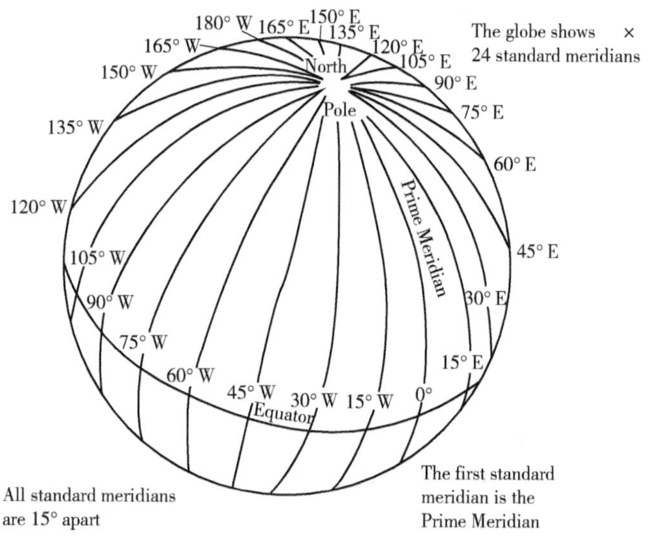

Figure 2-9 Meridians

$$I_e=I_0\times[1+0.033\cos(360n/365)]$$

where I_0 is the solar constant and it equals 1 367 W/m² — the energy from the sun outside the earth's atmosphere at the mean earth-sun distance.

6. Direct and Diffuse Solar Radiation

Suppose on clear days:

Global solar radiation= direct + diffuse components (Figure 2-10)

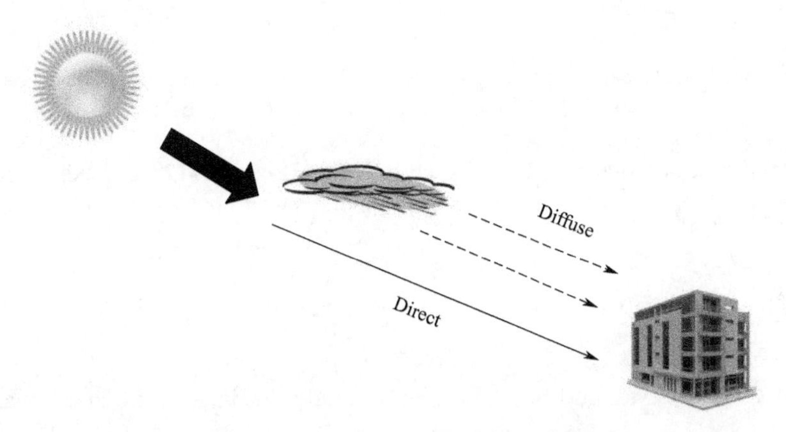

Figure 2-10 Direct and diffuse solar radiation

1) Direct component

E.g. Hottel's model — applicable at altitudes up to 2.5 km; good visibility up to 23 km.

$$I_{dir,0}=I_e\times[a_0+a_1\exp(-k/\cos\theta_Z)]$$

where $I_{dir,0}$ is the direct terrestrial solar radiation normal to the sun.

2) Diffuse component

Liu and Jordan's model can be used for the horizontal diffuse component.

$$I_{dif,h}=I_e\cos\theta_Z[0.271-0.294a_0+a_1\exp(-k/\cos\theta_Z)]$$

Model constants are obtained as follows:

$$a_0=r_0\times[0.423\,7-0.008\,21\times(6-A)^2]$$

$$a_1=r_1\times[0.505\,5+0.005\,95\times(6.5-A)^2]$$

$$k=r_k\times[0.271\,1+0.018\,58\times(2.5-A)^2]$$

The constants r_0, r_1 and r_k depend on latitude and time of year and A is the site altitude in km.

Under the more usual conditions of mixed cloud, the measurement of solar data must be resorted to the measuring instruments shown in Figure 2-11. Horizontal total components can be measured by using a pyranometer. The direct component of sunlight and the diffuse component of skylight falling together on a horizontal surface make up global horizontal irradiance (GHI). Direct radiation is best measured by use of a pyrheliometer, which measures radiation at normal incidence. Diffuse radiation can either be measured by shading a pyranometer from the direct radiation so that the thermopile is only receiving the diffuse radiation.

Pyranometer for both direct & diffuse components   Pyranometer for diffuse component only

Figure 2-11 Measuring instruments of solar data

Measured total, or global, horizontal solar radiation (Figure 2-12), I_h, leads to the definition of a clearness index, k_c.

$$k_c = I_h / I_{e,h}$$

where $I_{e,h}$ is the extraterrestrial solar radiation corrected to the horizontal plane and

$$I_{e,h} = I_e \cos \theta_Z$$

Typical values of k_c range from 0.25 (a very cloudy month, such as an average December in London) to 0.75 (a very sunny month, such as an average June in Phoenix).

Total solar radiation incident on a surface:

$$I_s = I_{dir,s} + I_{dif,s} + I_{gr,s}$$

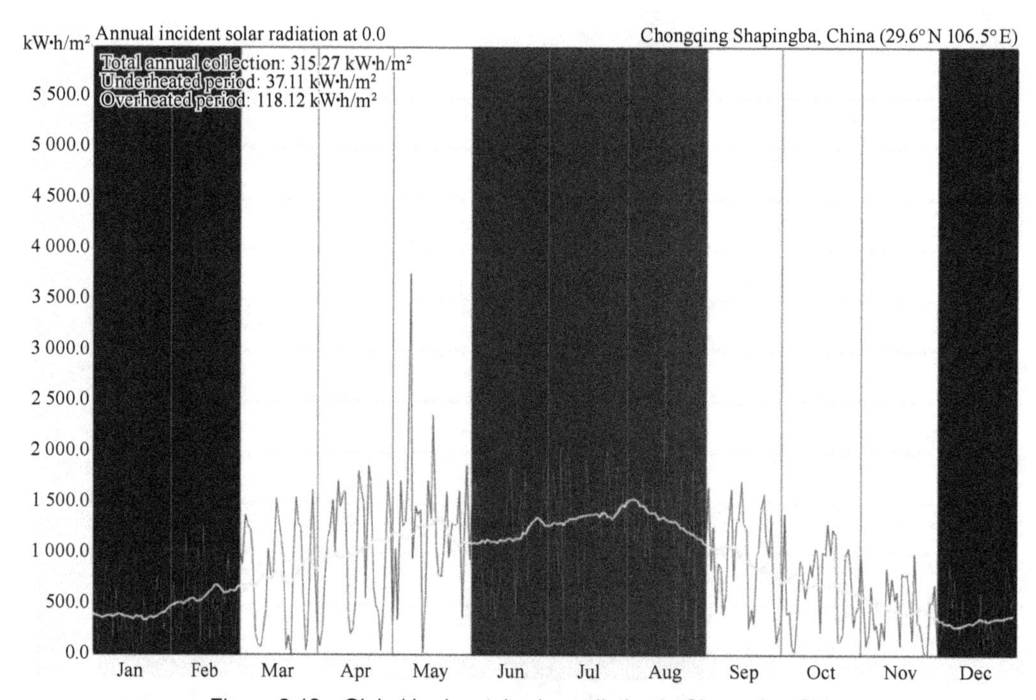

Figure 2-12 Global horizontal solar radiation in Chongqing,China

2.1.1.3 Sol-air Temperature

Solar energy absorbed at the outside surfaces of walls and roofs is partly transmitted to the interior of the building. This absorbed radiation has the same effect as a rise in the outside temperature and the calculation of energy gain is facilitated by the concept of "sol-air temperature" (Table 2-1). The sol-air temperature represents the equivalent outdoor air temperature that gives the same rate of heat flow to a surface as would the combination of incident solar radiation and convection/radiation with the environment.

Table 2-1 Air and corresponding sol-air temperatures (January 21st, London)

Time	Air temp. /℃	Sol-air temperature for stated orientation and surface color/℃																	
		Horizontal		North		North-east		East		South-east		South		South-west		West		North-west	
		Dark	Light	Dark	Light	Dark	Light	Dark	Light	Dark	Light	Dark	Light	Dark	Light	Dark	Light	Dark	Light
00:30	-3.2	-6.2	-6.2	-4.2	-4.2	-4.2	-4.2	-4.3	-4.3	-4.2	-4.2	-4.2	-4.2	-4.2	-4.2	-4.2	-4.2	-4.2	-4.2
01:30	-3.9	-7.2	-7.2	-5.0	-5.0	-5.0	-5.0	-4.1	-4.1	-5.0	-5.0	-5.0	-5.0	-5.0	-5.0	-5.0	-5.0	-5.0	-5.0
02:30	-4.1	-7.4	-7.4	-5.2	-5.2	-5.2	-5.2	-5.2	-5.2	-5.2	-5.2	-5.2	-5.2	-5.2	-5.2	-5.2	-5.2	-5.2	-5.2
03:30	-4.7	-8.0	-8.0	-5.8	-5.8	-5.8	-5.8	-5.8	-5.8	-5.8	-5.8	-5.8	-5.8	-5.8	-5.8	-5.8	-5.8	-5.8	-5.8
04:30	-4.4	-7.4	-7.4	-5.4	-5.4	-5.4	-5.4	-5.5	-5.5	-5.4	-5.4	-5.4	-5.4	-5.4	-5.4	-5.4	-5.4	-5.4	-5.4
05:30	-4.4	-7.7	-7.7	-5.5	-5.5	-5.5	-5.5	-4.6	-4.6	-5.5	-5.5	-5.5	-5.5	-5.5	-5.5	-5.5	-5.5	-5.5	-5.5
06:30	-4.7	-8.0	-8.0	-5.8	-5.8	-5.8	-5.8	-5.8	-5.8	-5.8	-5.8	-5.8	-5.8	-5.8	-5.8	-5.8	-5.8	-5.8	-5.8
07:30	-4.7	-8.0	-8.0	-5.8	-5.8	-5.8	-5.8	-5.8	-5.8	-5.8	-5.8	-5.8	-5.8	-5.8	-5.8	-5.8	-5.8	-5.8	-5.8
08:30	-4.1	-6.0	-6.6	-4.0	-4.5	-4.0	-4.6	-4.0	-4.5	-4.0	-4.5	-4.0	-4.5	-4.0	-4.6	-4.0	-4.5	-4.0	-4.5
09:30	-2.8	-0.7	-3.1	-1.7	-2.7	-1.7	-2.7	13.3	-5.7	23.7	11.4	19.2	8.9	2.5	-0.4	-1.7	-2.7	-1.7	-2.7
10:30	-1.0	4.9	0.8	0.9	-0.5	0.9	-0.5	15.4	7.6	37.2	19.7	37.5	19.9	16.1	8.0	0.9	-0.5	0.9	-0.5
11:30	0.5	8.3	3.4	2.9	1.3	2.9	1.3	7.9	4.1	36.1	19.8	44.6	24.5	28.3	15.5	2.9	1.3	2.9	1.3

2.1.2 Air Temperature

2.1.2.1 Definition of Air Temperature

Air temperature is a physical quantity expressing hot and cold (Figure 2-13). It is measured with a thermometer calibrated in one or more temperature scales. The most commonly used scales are the Celsius scale (formerly called centigrade) (denoted as ℃), Fahrenheit scale (denoted as ℉), and Kelvin scale (denoted as K, SI unit). Absolute zero is denoted as 0 K on the Kelvin scale, −273.15 ℃ on the Celsius scale, and −459.67 ℉ on the Fahrenheit scale. For an ideal gas, temperature is proportional to the average kinetic energy of the random motions of the constituent microscopic particles.

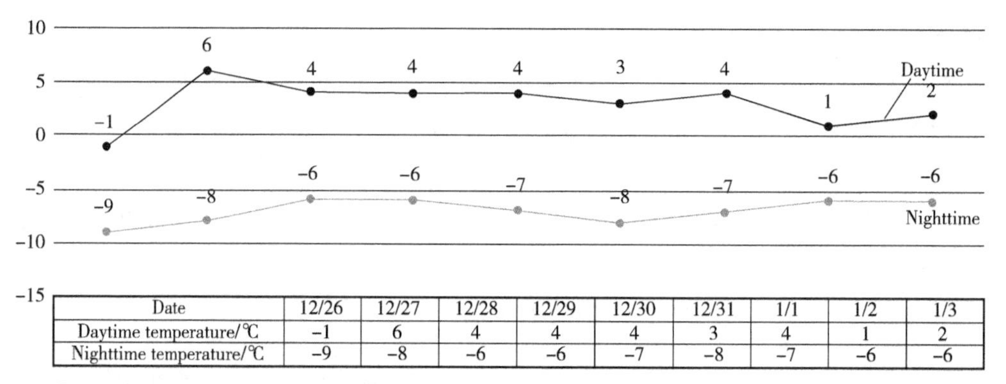

Date	12/26	12/27	12/28	12/29	12/30	12/31	1/1	1/2	1/3
Daytime temperature/℃	−1	6	4	4	4	3	4	1	2
Nighttime temperature/℃	-9	−8	−6	−6	−7	−8	−7	−6	−6

Figure 2-13 Temperature trend in Beijing from December 26th, 2021 to January 3rd, 2022

2.1.2.2 Factors Affecting Air Temperature

1. Natural Factors

• Degree of insolation: Influenced by the earth's rotation and revolution, the air temperature is necessarily affected by day or night, winter or summer.

• Latitude: In general, the closer people get to the poles, the lower the air temperature will be.

• Clouds: Clouds have an effect of counter radiation on sunlight. When there is a lot of clouds, the cloud layer is equivalent to the greenhouse cover, the air temperature is relatively low during the day and relatively high at night, the inverse radiation of the atmosphere is strong, so the air temperature difference is small, and vice versa is large.

• Type of surface: The drier the place is, the easier it is to raise the temperature, on the contrary, the wetter the place is, the less likely it is to change the temperature. For example, the coastal cities are by the sea, according to physics it is known that the specific heat capacity of water is greater than that of sand or earth, and the sun emits a certain amount of heat in the same amount of time, so the heat absorbed on the earth is also the same. The air temperature that will rise (or fall) is inversely proportional to the specific heat capacity, and the specific heat capacity of water is large, so it is said that the temperature drops slowly, this is why the coastal cities are warmer at night than inland cities.

• Altitude: The higher you go, the faster the temperature changes. For every 100 meters of altitude rise, the temperature drops by 0.6 ℃ . This is because as the altitude rises, the air pressure becomes lower and the air becomes thinner, absorbing less long wave reflections from the ground and lowering the temperature.

• Angle of illumination and surface area: According to the law of radiation, the closer the angle is to 90° for the same beam with the same energy, the less energy it radiates to the surface of the object.

2. Human Factors

• Properties of the urban subsurface: The large number of artificial structures in the city such as paved floors and various building walls changes the thermal properties of the subsurface. The low water content of the urban surface allows more heat to enter the air in the form of sensible heat, leading to warming of the air. At the same time, the urban surface has a higher absorption rate of sunlight than the natural surface and can absorb more solar radiation, but the side effect is that air receives more heat and the temperature rises.

• Urban air pollution: Industrial production, and the activities of large numbers of people in cities produce large quantities of nitrogen oxides, carbon dioxide, and dust, which can take up large amounts of energy from thermal radiation in the environment, producing the greenhouse effect and leading to further warming of the atmosphere.

• Artificial heat sources: Factories, motor vehicles, residential life, etc., burn all kinds of fuels and consume a lot of energy. At the same time, countless furnaces are burning and all of them are emitting heat.

• Reduction of natural substratum in the city: With the massive increase in buildings, squares, roads, etc. in the city and the corresponding decrease in natural factors such as green areas and water bodies, more heat is released and less heat is absorbed, weakening the ability to mitigate the heat island effect.

2.1.2.3 Role of Temperature in Guiding Various Stages of Design

(1) System selection during design stage, e.g. natural ventilation vs. air-conditioning.

(2) System sizing during design stage, e.g. total load of chillers or boilers.

(3) Change of heating & cooling loads during operation stage, e.g. mixed-mode strategy.

(4) Efficiency of natural ventilation during operation stage, e.g. buoyancy driven.

(5) Occupants' use of systems during operation stage, e.g. opening windows.

2.1.3 Atmospheric Humidity

2.1.3.1 Definition of Atmospheric Humidity

Atmospheric humidity is critical to organisms as along with temperature, wind speed and the radiation balance; it determines the potential rate of water loss from a surface. Atmospheric humidity is the amount or volume of water vapor that is present in the atmosphere. The main source of moisture in the air is the surface of the oceans and seas, where water evaporates constantly. Other sources of atmospheric moisture are lakes, glaciers and rivers, as well as the evapotranspiration processes of soil, plants and animals.

2.1.3.2 Types of Atmospheric Humidity

• Absolute humidity: It refers to the actual mass of water vapor in a given volume of air. It may be expressed as the number of grams of water vapor in a cubic meter of moist air or the mass of water vapor per unit volume of air.

• Specific humidity: It is defined as the moisture content of moist air as determined by the ratio of the mass of water vapor to the mass of moist air in which the mass of water vapor is contained.

• Relative humidity: A common parameter for expressing the water vapor content of air is relative humidity. This value is the percentage of water vapor present in the air compared to the saturated condition at a given temperature and pressure.

2.1.3.3 Significance of Atmospheric Humidity as a Design Basis

Outdoor air is a main source of moisture indoors through infiltration, ventilation and fresh air supply. The properties of air can be shown on a single graph known as a psychrometric chart. Figure 2-14 illustrates such a chart. The dry-bulb temperature of an air sample increases from left to right; the moisture content of the air sample increases from bottom to top and is shown in the chart as the humidity ratio. Too low or too high humidity level indoors will cause thermal discomfort of occupants: Low indoor humidity level will cause static electricity in rooms such as computer rooms; high indoor humidity level will cause damp and mould issues. In air-conditioning and ventilation systems, outdoor air can be used to control indoor humidity level.

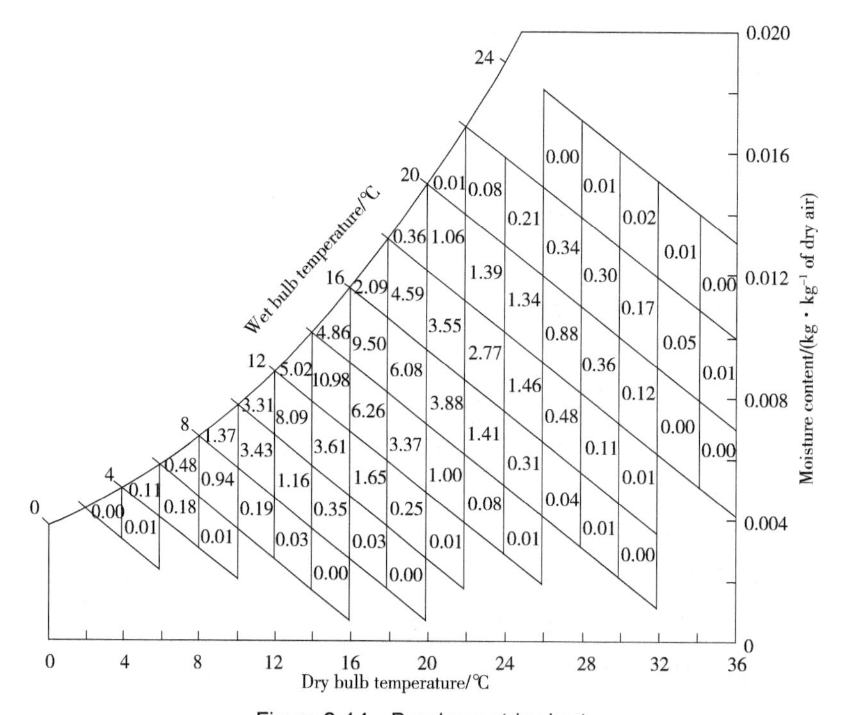

Figure 2-14 Psychrometric chart

2.1.4 Daylight Climate

2.1.4.1 Definition of Daylight Climate

Daylight is the combination of all direct and indirect sunlight during the daytime. This includes direct sunlight and diffuse sky radiation, and both of these are reflected by the earth and terrestrial objects, such as landforms and buildings.

• Diffuse illuminance (also called skylight): Solar radiation reaches the earth as a result of scattering in the atmosphere.

• Direct illuminance (also called sunlight): Solar radiation reaches the earth's surface as parallel rays, directly from the sun's disc, after selective attenuation by the atmosphere.

• Global illuminance: It contains both diffuse and direct illuminance (Table 2-2).

Table 2-2 Daylight intensity in different conditions

Illuminance/lx	Example
120 000	Brightest sunlight
111 000	Bright sunlight
20 000	Shade illuminated by entire clear blue sky, midday
1 000-2 000	Typical overcast day, midday
400	Sunrise or sunset on a clear day (ambient illumination)
<200	Extreme of darkest storm clouds, midday
40	Fully overcast, sunset/sunrise
<1	Extreme of darkest storm clouds, sunset/rise

2.1.4.2 Significance of Daylight Climate as a Design Basis

Throughout history, daylight has been a crucial factor in the design of buildings (CIBSE LG10: Daylighting and Window Design). Good daylight contributes to healthy building design, which in turn has implications for occupants' productivity. Quantitative daylight illuminance data are needed for daylighting design calculations including the sizing of windows, choice of glazing materials and the design of window shading systems. Also, daylight and electric lighting systems must be designed to operate interactively. The following table shows the UK global illuminance data from CIBSE Guide A (Table 2-3).

Table 2-3 Percentage of year for which stated global illuminance is exceeded in the UK

Global illum./ klx	Belfast	B'ham	Cardiff	Edin.	Glasgow	Leeds	London	Manch.	N'castle	Nor-wich	N'ham	Ply-mouth	S'ton	Swindon
1.0	99.6	99.5	99.5	99.6	99.6	99.5	99.5	99.5	99.5	99.5	99.5	99.5	99.5	99.5
3.0	92.2	93.2	93.6	91.6	91.6	92.7	93.4	92.9	92.5	92.9	93.0	94.0	93.6	93.4
5.0	86.6	88.5	89.0	85.5	85.5	87.6	88.8	87.9	87.2	88.1	88.1	89.4	89.1	88.7
7.0	80.2	82.3	83.0	78.9	78.9	81.3	83.0	81.8	80.9	82.1	82.0	83.8	83.4	82.7
9.0	74.3	76.8	77.6	73.2	73.1	75.5	77.6	76.0	75.0	76.4	76.3	78.3	78.0	77.3
11.0	70.4	73.1	73.8	69.2	69.1	71.8	73.8	72.3	71.2	72.7	72.6	74.6	74.3	73.5
13.0	66.5	69.3	70.2	65.2	65.2	68.0	70.2	68.4	67.4	68.9	68.8	71.0	70.7	69.8
15.0	62.6	65.4	66.2	61.3	61.3	64.0	66.3	64.5	63.5	65.1	65.0	67.1	66.8	65.9
17.0	58.7	61.4	62.3	57.3	57.2	60.1	62.3	60.6	59.5	61.2	61.0	63.1	62.9	62.0
19.0	54.8	57.6	58.4	53.4	53.4	56.3	58.5	56.8	55.7	57.3	57.1	59.2	59.0	58.1
21.0	51.0	53.8	54.6	49.8	49.7	52.5	54.6	53.0	51.9	53.6	53.4	55.3	55.2	54.3
23.0	47.3	50.0	50.8	46.1	46.0	48.7	50.9	49.3	48.2	49.7	49.6	51.5	51.4	50.5
25.0	44.4	46.6	47.4	43.5	43.5	45.6	47.5	46.0	45.1	46.5	46.3	48.1	48.0	47.1
27.0	42.1	44.2	44.9	41.2	41.1	43.2	45.0	43.6	42.8	44.0	43.9	45.5	45.4	44.6
29.0	39.8	41.8	42.3	39.0	38.9	40.9	42.4	41.2	40.4	41.6	41.5	43.0	42.9	42.2
31.0	37.5	39.4	40.0	36.7	36.7	38.6	40.1	38.9	38.2	39.3	39.2	40.5	40.4	39.8

The following table shows the UK diffuse illuminance data from CIBSE Guide A (Table 2-4).

Table 2-4 Percentage of year for which stated diffuse illuminance is exceeded in the UK

Diffuse illum./ klx	Belfast	B'ham	Cardiff	Edin.	Glasgow	Leeds	London	Manch.	N'castle	Nor-wich	N'ham	Ply-mouth	S'ton	Swin-don
0.5	97.7	97.5	97.4	97.9	97.9	97.5	97.4	97.5	97.6	97.5	97.5	97.4	97.4	97.4
1.5	93.5	93.8	94.1	93.1	93.2	93.5	93.9	93.7	93.5	93.6	93.7	94.3	94.0	93.9
2.5	90.3	91.1	91.4	89.8	89.8	90.7	91.2	90.9	90.5	90.9	91.0	91.7	91.4	91.2
3.5	87.0	88.4	88.8	86.2	86.2	87.7	88.7	88.0	87.5	88.1	88.2	89.2	88.9	88.6
4.5	83.1	84.8	85.6	82.3	82.2	83.9	85.4	84.3	83.6	84.5	84.5	86.1	85.8	85.3
5.5	79.3	81.3	81.9	78.1	78.1	80.3	81.8	80.7	79.9	81.0	80.9	82.5	82.2	81.7
6.5	76.3	78.7	79.4	75.3	75.2	77.4	79.3	77.8	76.9	78.3	78.2	80.0	79.7	79.1
7.5	73.5	75.9	76.8	72.3	72.3	74.7	76.7	75.2	74.3	75.6	75.5	77.7	77.3	76.4
8.5	70.6	73.3	74.1	69.4	69.4	72.0	74.1	72.5	71.4	73.0	72.9	74.9	74.6	73.8
9.5	67.7	70.5	71.5	66.4	66.3	69.1	71.5	69.6	68.6	70.1	70.0	72.3	72.0	71.2
10.5	64.7	67.6	68.5	63.4	63.4	66.2	68.5	66.7	65.5	67.3	67.1	69.5	69.1	68.1

Continued

Diffuse illum./ klx	Belfast	B'ham	Cardiff	Edin.	Glasgow	Leeds	London	Manch.	N'castle	Nor-wich	N'ham	Ply-mouth	S'ton	Swin-don
11.5	61.7	64.6	65.5	60.3	60.3	63.1	65.5	63.6	62.6	64.3	64.1	66.4	66.1	65.2
12.5	58.7	61.5	62.6	57.4	57.3	60.2	62.6	60.7	59.6	61.2	61.1	63.5	63.2	62.2
13.5	55.8	58.6	59.5	54.5	54.5	57.2	59.6	57.8	56.6	58.4	58.2	60.5	60.2	59.2
14.5	52.9	55.7	56.5	51.8	51.7	54.3	56.6	54.8	53.8	55.4	55.2	57.4	57.2	56.3
15.5	50.1	52.7	53.6	49.0	48.9	51.5	53.7	52.0	50.9	52.5	52.3	54.5	54.3	53.3

Values for other sites can be estimated from Tables 2-3 and 2-4 using linear interpolation based on latitude. However, the effects of urban pollution mean that the data for central London should not be used for latitude interpolation. The above cumulative frequency analysis must be based on defined working day lengths, i.e. 9:00 to 17:30 in winter and 8:00 to 16:30 for the period between April and October to allow for British Summer Time (Figure 2-15).

In China, we usually have the diffuse or overcast sky condition.

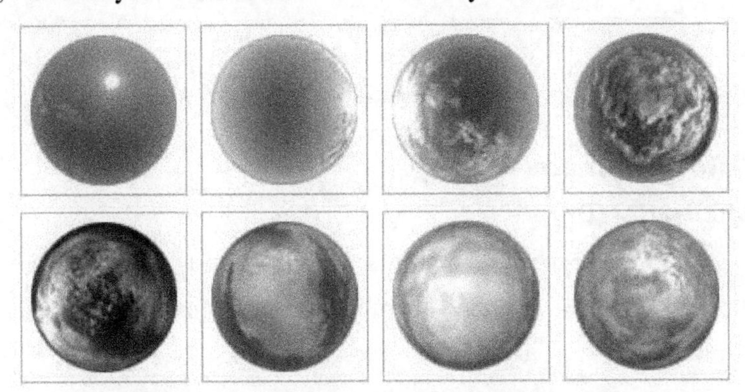

Figure 2-15 Examples of different sky distributions

(http://wiki.naturalfrequency.com/wiki/Sky Illuminance(access on 29 Oct. 2011))

2.1.4.3 Commission Internationale de l'Eclairage

CIE stands for Commission Internationale de I'Eclairage — International Commission on Illumination (Figure 2-16).

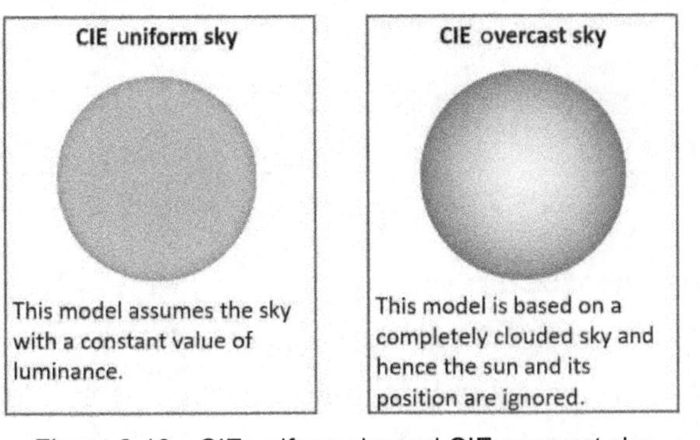

Figure 2-16 CIE uniform sky and CIE overcast sky

CIE standard overcast sky is non-uniform sky. Because of the temperate climate, the amount of direct sunlight in the UK is variable and unpredictable, so the overcast sky is generally used in the UK for daylight calculations.

The National Physics Laboratory of the UK has measured the horizontal illuminance, excluding that of direct sunlight, at varying times of the day over the year,in varying parts of the country and over a long period of time (Figure 2-17).

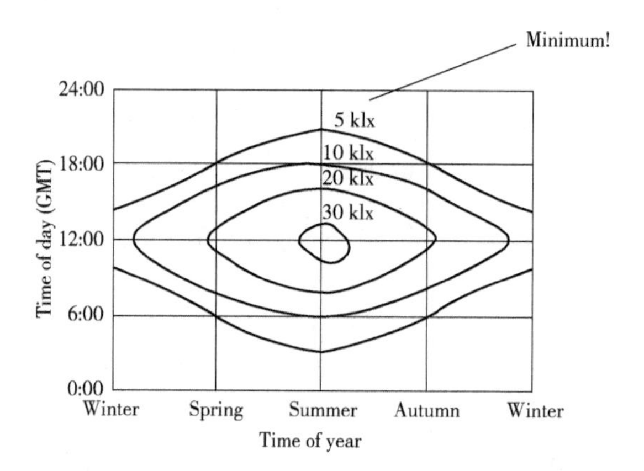

Figure 2-17 Daylight illuminance variation by time of year and time of day (excluding direct sunlight)

2.1.4.4 Related Parameters

1. Daylight Factor (DF)

Daylight factor is an architectural term that describes the amount of natural light within a room or structure (Figure 2-18). It is the ratio of the light level inside a structure to the light level outside the structure.

$$DF=\frac{\text{Horizontal illuminance}}{\text{Simultaneous horizontal illuminance outdoors}}\times 100\%$$

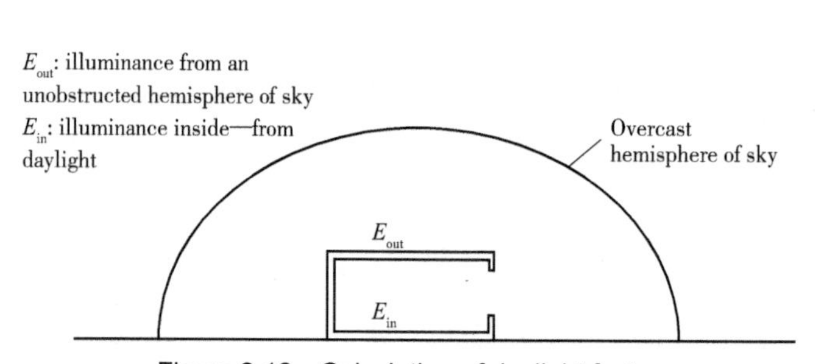

Figure 2-18 Calculation of daylight factor

2. Average Daylight Factor

The average DF will provide the designer with enough information to make decisions on the relationship between natural and electric/artificial lighting (Table 2-5).

$$\text{Average DF}=\frac{TW\theta \cdot \text{CF}}{A\left(1-R^2\right)}\times 100\%$$

where T is diffuse transmittance of glazing material (decimal); W is area of glazing (m^2); θ is angle from the center of the window subtended by visible sky (degrees); CF is correction factor (dirt and frame); A is total area of all indoor surfaces (m^2); R is average area weighted reflectance of all indoor surfaces (decimal).

Table 2-5 Lighting code

Range of Av. DF	Category	Comments
< 2%	Poor daylighting	Artificial (electric) lighting will be needed almost permanently
2%-5%	Well daylighting	Localized or local lighting could be advantageous with daylight as the general lighting
>5%	High level of daylighting	Artificial (electric) lighting would be mainly used in nighttime and darkness periods or when specific task requires a high level of illuminance

3. Limiting Depth (LD) of Daylight

Even if the average DF meets the recommended level, side-lit rooms may be too deep (window wall to rear wall) to realize daylight illuminance and often appear very dull in the deeper parts of the room (Figure 2-19).

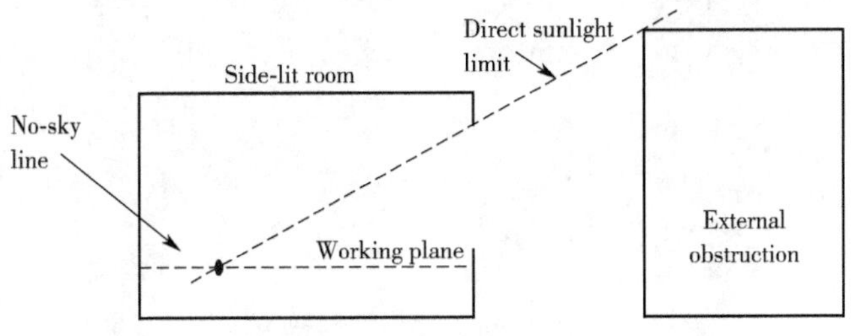

Figure 2-19 Limiting depth

$$\text{Limiting depth (LD)}=\frac{2wh}{(h+w)(1-R_b)}$$

where w is width of room parallel to window (m); h is height of window head above floor level (m); R_b is average weighted reflectance of half the room surface area away from the window (decimal).

2.1.4.5 Control of Daylighting

The purpose of architectural shading is to prevent direct sunlight, which has three benefits: it can prevent direct sunlight through the glass to make the room overheat, avoid overheating of the building envelope to cause thermal radiation to the indoor environment and prevent the strong glare caused by direct sunlight (Figure 2-20).

(1) External shading, i.e. overhang, louver, shutter, fin and awning.

(2) Internal shading, i.e. blind and curtain.

(3) PV shading.

(4) External windows with shading.

Figure 2-20 Control of daylighting by shadings

(https://data.london.gov.uk/dataset/london-s-urban-heat-island)

(5) Window/wall ratio (WWR).

Window/wall ratio has an influence on annual energy consumption for heating and air conditioning in residential buildings (Figure 2-21).

> 60%: Overheating in summer unless special measures are taken — solar glare problems.

25%-60%: Good daylighting with moderate heat gains (in summer) and losses (in winter).

< 25%: Glazing too small to contribute to much daylighting.

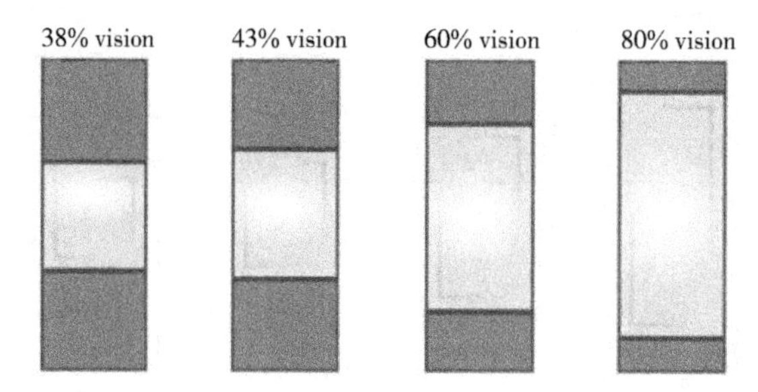

Figure 2-21 Influence of different window/wall ratios

(6) Window/floor ratio (WFR).

< 5%: Windows too small to contribute to daylighting.

5%-15%: Good daylighting with moderate heat gains (in summer) and losses (in winter).

>15%: Overheating in summer, large heat losses and gains unless special measures are taken.

2.1.5 Wind Direction and Wind Speed

2.1.5.1 Definition of Wind

Wind is the flow of gases on a large scale. Winds are commonly classified by their spatial scale, their speed, the types of forces that cause them, the regions in which they occur, and their effect. It occurs on a range of scales, from thunderstorm flows lasting tens of minutes, to local breezes generated by heating of land surfaces and lasting a few hours, to global winds resulting from the difference in absorption of solar energy between the climate zones on the earth. The two main causes of large-scale atmospheric circulation are the differential heating between the equator and the poles, and the rotation of the planet (Coriolis effect). Local winds often occur in coastal areas and those with special terrains, mountains and valleys.

2.1.5.2 Definition of Wind Speed

Wind speed, or wind flow velocity, is a fundamental atmospheric quantity caused by air moving from high to low pressure, usually due to changes in temperature. Generally in the unit of kn (knot), m/s (meters per second) or mph (miles per hour): 1 international knot = 1.15 miles per hour = 0.514 44 meters per second. Wind speed is affected mainly by four factors, i.e. pressure gradient, Rossby waves and jet streams, and local weather conditions. It is now commonly measured with an anemometer.

2.1.5.3 Definition of Wind Direction

Wind direction is reported by the direction from which it originates. For example, a northerly wind blows from the north to the south (Table 2-6). It is usually reported in cardinal directions or in azimuth degrees. Wind direction is measured in degrees clockwise from true north. A wind vane is a modern instrument used to measure wind direction.

Table 2-6 Wind direction

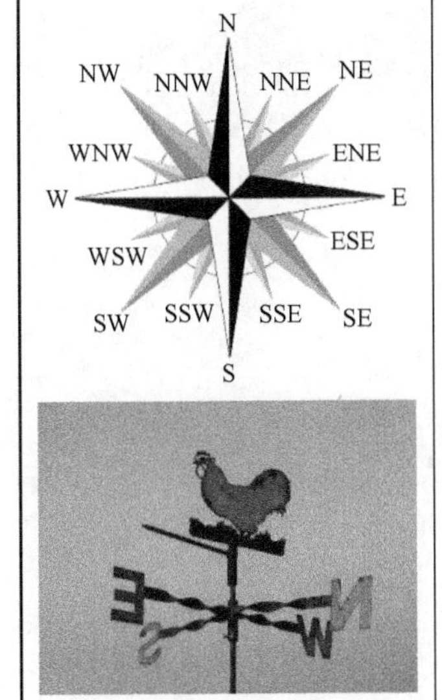

Degrees represented by each pointer of the compass

Direction	Degree	Direction	Degree
North	0°	South	180°
North-northeast	22.5°	South-southwest	202.5°
Northeast	45°	Southwest	225°
East-northeast	67.5°	West-southwest	247.5°
East	90°	West	270°
East-southeast	112.5°	West-northwest	292.5°
Southeast	135°	Northwest	315°
South-southeast	157.5°	North-northwest	337.5°

2.1.5.4 Impact of Wind on Buildings

Wind angles affect building's natural ventilation and energy consumption (Figure 2-22). Wind direction in the outdoor environment will influence the heat loss from the building during winter, whereas in summer the change in wind direction and velocity in the outdoor environment will influence the building's ventilation and the flow of indoor air.

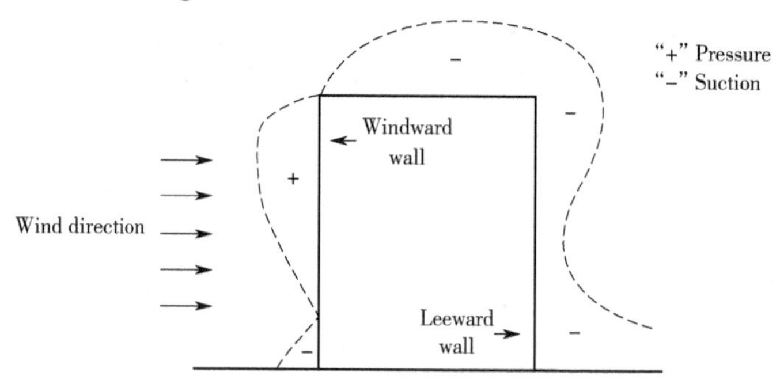

Figure 2-22 Wind driven natural ventilation & infiltration

(https://greenhome.osu.edu/natural-ventilation)

2.1.5.5 Wind Data

Wind data are measured with an anemometer and a wind vane mounted where possible 10 m above ground level or, in some cases, above the roof of a building. Most sites are in exposed open situations such as airports; very few data are available from city centers. The hourly values used to derive the statistics quoted in the tables are hourly mean speeds and median direction, derived either from anemograph charts or from a dedicated logging system that records data at 1-minute intervals.

2.1.5.6 Wind Rose

A wind rose diagram is a tool which graphically displays wind speed and wind direction at a particular location over a period of time (Figure 2-23). The diagram normally comprises of 8, 16 or 32 radiating spokes, which represent wind directions in terms of the cardinal wind directions and their intermediate directions.

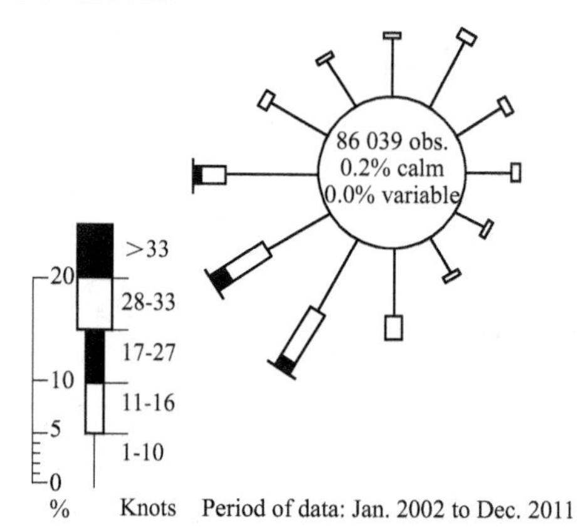

Figure 2-23 Wind rose

Influence of height and environment on mean wind speed is as follows:

$$v = v_m K_s z^a$$

where v is mean wind speed at height "z" ; v_m is mean wind speed at height of 10 m in open country; K_s is modifier for terrain; z is building height; a is coefficient between wind speed and height above ground.

Table 2-7 Wind speed/terrain/height parameters (BS 5925:1991)

Terrain	K_s	a
Open country	0.68	0.17
Country with scattered breaks in terrain	0.52	0.20
Urban	0.35	0.25
City	0.21	0.33

2.1.5.7 Wind Energy

Wind power is a clean, free and easily accessible renewable energy source (Figure 2-24). Every day, all over the world, wind turbines capture the winds energy and convert it into electricity. Wind energy production is playing an increasingly important role in how we feed our world cleanly and sustainably.

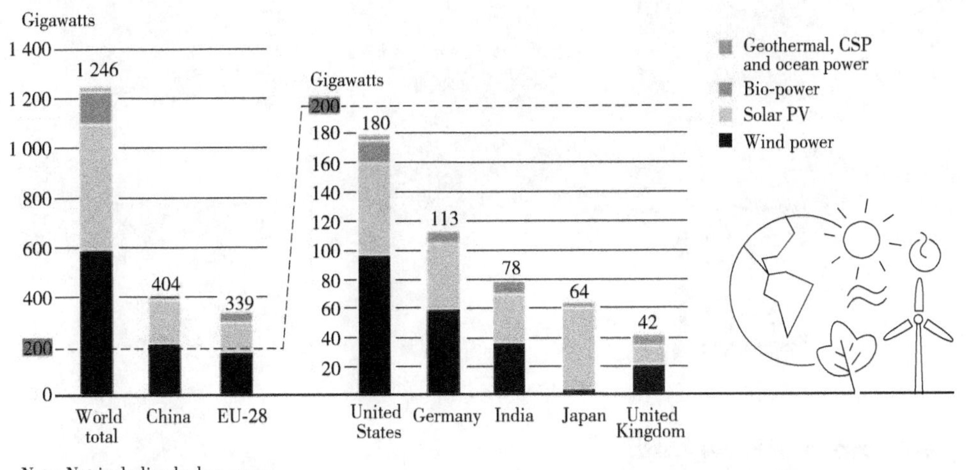

Figure 2-24 Wind power capacities in world, EU-28 and Top 6 countries in 2018

2.1.6 Condensation and Precipitation

2.1.6.1 Definition of Condensation & Precipitation

Water vapor in the air condenses when the air temperature falls below the dew point. Cooling in the air can occur through contact with cold surfaces, mixing with cold air and diffusion by blowing up. The first two result in dew and fog, while the latter results in precipitation.

Dew and fog: Air near the surface is cooled by contact with the cold surface where condensation occurs. Fog is produced when air is not in direct contact with a cold surface that is cooled below the dew point.

Precipitation: Precipitation includes rain, snow and hail (Figure 2-25). It is liquid or solid water that evaporates or condenses from the earth's surface and returns to the ground. Precipitation intensity is the total amount of precipitation in millimeters over a 24-hour period. It is influenced by factors such as temperature, topography, atmospheric circulation and distribution over land and sea, and is classified into different classes. Precipitation is the result of adiabatic cooling of air masses.

Figure 2-25 Precipitation

2.1.6.2 Precipitation Data

Rainfall is the depth of the water layer that accumulates on the surface of the water without evaporation, infiltration or loss of rainwater falling from the sky to the ground, generally in millimeters, and it can visually represent the amount of rainfall (Figure 2-26, Figure 2-27).

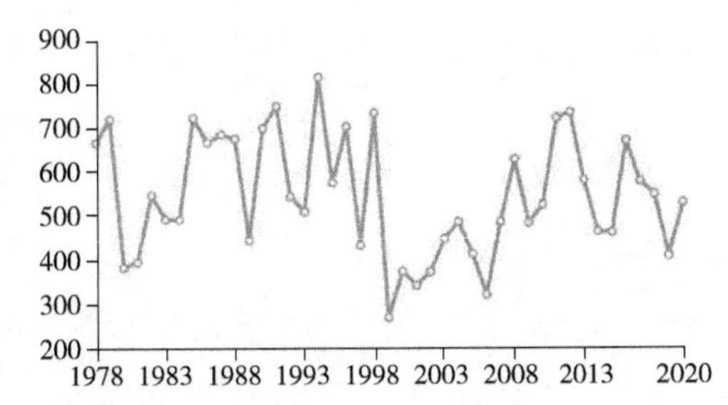

Figure 2-26 Average annual rainfall in Beijing from 1978 to 2020 (in mm)

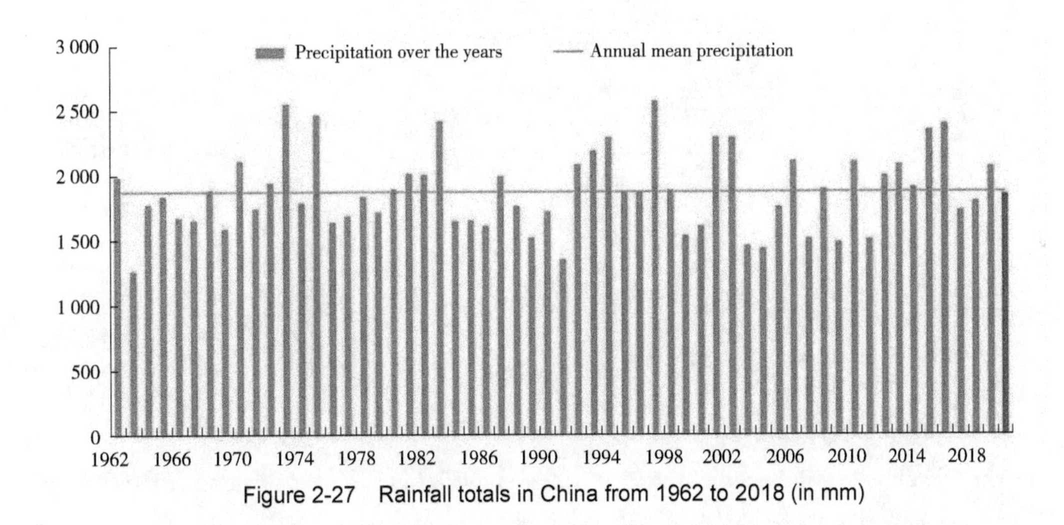

Figure 2-27 Rainfall totals in China from 1962 to 2018 (in mm)

2.2 Climate Modification and Monitoring

2.2.1 Climate Change

It is now generally accepted that the weather data used for building performance calculations should account for future as well as present day climate. Probabilistic approach has been used to calculate changes in climate variables.

Uncertainty in climate projections is due to:

- Natural variability;
- Incomplete understanding of the climate system;
- Imperfect representation in models.

2.2.1.1 Aspects of Future Climate Change

1. Future Temperature

The UK climate will become warmer. Temperature rise will be the greatest in the southeast, the lowest in the northwest, and greater in summer and autumn than in winter and spring (Figure 2-28).

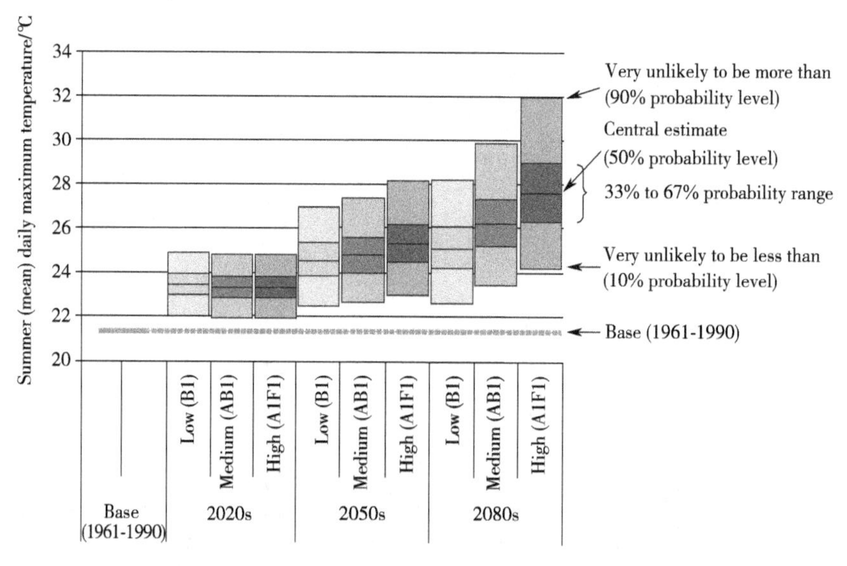

Figure 2-28 Probabilistic climate profile (ProCliP)

2. Future Precipitation

The UK climate will become wetter in winter and drier in summer: in summer the estimated changes across the UK range from −40% to +1%; in winter the estimated changes across the UK range from −2% to +33%.

3. Sea Level

Projected changes in absolute sea level by 2080s lie between 12 cm and 76 cm under the medium emissions scenario. Taking account of vertical land movements gives higher projections relative to the land in the south of the UK and lower increases in the north.

4. Relative Humidity

In 2080s for the medium emissions scenario, the relative humidity level will decrease in summer in parts of southern England. The variational range of relative humidity in winter is smaller than that in other seasons, and the degree of change is only a few percent.

5. Solar Radiation

In 2080s for the medium emissions scenario, summer mean cloud amount will decrease in parts of southern England, giving an increase in downward shortwave radiation. For some parts of northern Scotland, the cloud amount increases, hence reducing radiation levels.

6. Wind Speed

There are small reductions in summer wind speeds for most of the UK in 2050s. In winter the changes are not significant.

2.2.1.2 Effect on Buildings and Building Services (Table 2-8)

Table 2-8 Major impacts of climate change on built environment (CIBSE Guide A)

Climate change	Impact on built environment	Consequential impacts
Rising summer temperature	Overheating of buildings	· Reduced thermal comfort · Loss of staff hours due to high internal building temperature · Increased temperature mortality (heat)
	Energy demand for cooling	· Increase in cooling loads, reduction in chiller coefficient of performance
Rising winter temperature	Less demand for heating	· Reduction in heating loads · Reduced temperature mortality (cold)
Decrease in summer rainfall	Limitation of water supply	· Pressure on water consumption of building service systems
	Ground movement	· Increased structural damage
More intense rainfall and rising sea level		· Increased flooding, leading to property damage and casualties

2.2.2 Urban Heat Island

2.2.2.1 Definition of Urban Heat Island

Urban heat island (UHI) is a phenomenon whereby urban areas have higher temperatures than the rural surroundings.

UHI intensity is the temperature difference between an urban site and a rural or semi-rural site, which is mainly caused by:

(1) The form making up urban landscapes;

(2) The fabric making up urban landscapes.

UHI intensities have been shown to be related to the population and size of a city. A better indicator for estimating the urban heat island intensity is how urban fractions, building size and density and sky-view factors of a city vary.

2.2.2.2 Causes of Urban Heat Island Formation

(1) The geometry of street canyons and increased building density help trap incoming radiation, whilst reducing the rate of longwave radiant cooling at night; the orientation and density of structures within cities also reduce wind speeds, resulting in a decrease in the amount of convective heat transport away from the urban environment; street canyons have lower sky-view factors (i.e. the proportion of sky visible in a 180° field of view) due to their high aspect ratios. They store more heat, enabling them to maintain higher surface temperatures for longer periods of time.

(2) The fabric of the built environment has a greater heat capacity than its rural counter-

part, absorbing incoming solar radiation during the day and releasing it at night as sensible heat flux. Buildings, pavements and roads usually have darker surfaces with lower albedo and higher heat capacity compared to rural landscapes, thereby increasing their ability to absorb heat.

2.2.2.3 Negative Effects of Urban Heat Island

1. Increased Energy Consumption

Increased temperatures during summer in cities amplify energy demand for air conditioning. Studies reveal that electricity demand for air conditioning or cooling increases in the range of 1.5 to 2 percent for every 1 ℉ (0.6 ℃) increase in air temperatures (range of 68 to 77 ℉ (20 to 25 ℃)), implying that the community requires about 5 to 10 percent more electricity to cater for the urban heat effect.

This means the increased demands for cooling or air-conditioning during summer contribute to higher energy bills. Also, during exacerbated periods of urban heat islands, the resulting demand for air conditioning can overload systems, which can lead to power outages and blackouts.

2. Elevated Greenhouse Gas Emissions and Air Pollution

As explained earlier, urban heat island raises electricity demand during summer. As a result, power plants have to supply the needed extra energy, and since they rely on fossil fuel for energy production, there is an increase in greenhouse gas emissions and air pollution. The main greenhouse gases and pollutants include carbon monoxide (CO), carbon dioxide (CO_2), sulfur dioxide (SO_2), nitrogen oxides (NO_x), particulate matter and mercury (Hg).

Increased greenhouse gases lead to global warming and climate change, while pollutants have a negative impact on human health as well as air quality. Sometimes the urban heat island can also lead to the formation of ground-level ozone and acid rain. Research shows that high urban heat island correlates with increased levels and accumulation of air pollutants at night, affecting the next day's air quality.

3. Danger to Aquatic Systems

High temperatures within the urban areas mean elevated temperatures for pavements and rooftops. Accordingly, these heated surfaces can heat stormwater runoff. The heated stormwater is the runoff that flows into storm drainage systems and raises water temperatures as it is discharged into ponds, streams, rivers, lakes and oceans, resulting in the thermal pollution. Consequently, increased water temperatures affect the aquatic system, particularly the reproduction and metabolism of aquatic species and may even be lethal to aquatic organisms.

4. Discomfort and Danger to Human Health

Higher air pollution reduces nighttime cooling, and increased temperatures as outcomes of urban heat island can adversely affect human health. Human health is negatively impacted because of increased general discomfort, exhaustion, heat-related mortality, respiratory problems, headaches, heat stroke and heat cramps.

Because urban heat islands can also worsen the impacts of heatwaves, arise abnormal weather periods, which can seriously affect the health of sensitive and vulnerable populations such as older adults, children, and those with weather-responsive health conditions.

Exacerbated heat events or sudden temperature increases can result in higher mortality rates (Figure 2-29). Many deaths were registered owing to excessive exposure to heat.

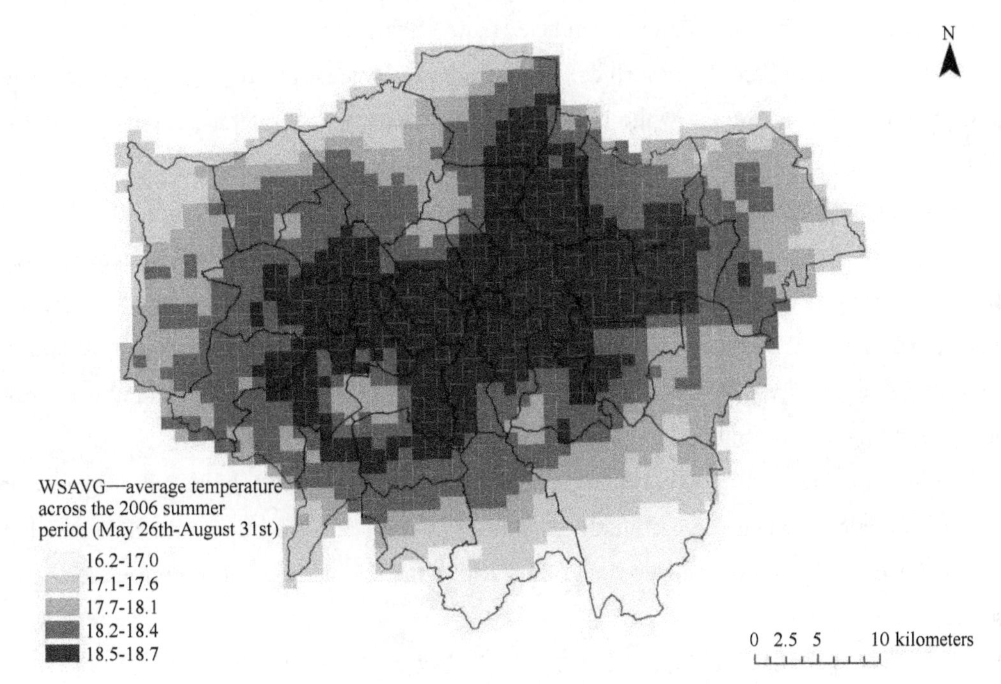

Figure 2-29 Urban heat island effect in London

5. Secondary Impacts on Weather and Climate

Besides the high-temperature increases, urban heat island can bring forth secondary effects on the local weather and climate. These include changes in local wind patterns, fog and cloud formation, rainfall rates, and humidity. The unusual heat caused by UHI contributes to a more intense upward wind movement that can stimulate thunderstorm and precipitation activity.

Furthermore, urban heat island creates a local low-pressure area where cool air from its adjacent areas converges that induces the formation of clouds and rain. This increases the total rainfall rates within cities. These changes may impact growing seasons within cities, especially by prolonging the growth of plants and crops.

6. Impacts on Animals

Most species need optimum temperatures to colonize, utilize and thrive in their ecosystems. When there is the existence of high temperatures due to urban heat island, harsh and cruel ecological surrounding is created which limits the essential activities of the organisms such as metabolism, breeding and reproduction. Adverse heat can also significantly reduce the availability of food, shelter, and water.

2.2.2.4 Solutions to Urban Heat Island

1. Use of Light-colored Concrete and White Roofs

Light-colored concrete and white roofs have been found to be effective in reflecting as much as 50% light and reducing ambient air temperature. These strategies have been shown to offer great solutions in reducing the urban heat island effect.

The black and dull colors absorb copious amounts of solar heat, leading to warmer surfaces. Light-colored concrete and white roofs may just as well reduce overall air-conditioning requirements.

2. Green Roofs and Vegetation Cover

Green roofs present an excellent method for mitigating urban heat island impacts. Green roofing is the practice of planting vegetation on top of a roof, in the same way that they are planted in a garden. The roof plants are excellent insulators during the summer months and decrease the overall urban heat island effect. In addition, plants cool the surrounding environments, reducing the demands of air conditioning.

Furthermore, air quality is improved as the plants absorb carbon dioxide and produce fresh air. Other practices that can be used include open space planting, street trees and curbside planting. All these practices produce a cooling effect within the urban areas and decrease the costs of temperature reduction.

3. Planting Trees in Cities

Tree planting practices inside and around cities are an amazing way to reflect solar radiation while diminishing the urban heat island effect. The trees provide shade, absorb carbon dioxide, release oxygen and fresh air, as well as provide a cooling effect. Deciduous trees are best suited for urban areas because they provide a cooling effect in summer, and they don't block the heat during the winter.

4. Green Parking Lots

Green parking uses green infrastructure strategies to limit urban heat island effect. Specifically, it cushions against the rise in pavement temperatures that can significantly prevent thermal pollution from stormwater runoff. Once this is in place, the hazard to aquatic systems is reduced.

5. Implementation and Sensitization of Heat Reduction Policies and Rules

The state implementation of environmental policies such as the Clean Air Act, low carbon fuel standards, uses of renewable energy, and clean car rule standards can impressively regulate the anthropogenic inducers of urban heat island effect.

With fewer emissions, the level of greenhouse gases in the atmosphere can be reduced, thus decreasing the effects of climate change and global warming. Education and outreach can also be done to ensure communities are aware of the economic and social benefits of planting trees and eco-roofing.

2.2.3 Weather Stations

2.2.3.1 Definition of Weather Stations

A weather station is a facility, either on land or at sea, equipped with instruments and equipment for the measurement of atmospheric conditions to provide information for weather forecasting and for the study of weather and climate (Figure 2-30, Figure 2-31). Measurements are made of temperature, atmospheric pressure, humidity, wind speed, wind direction and the amount of precipitation.

Located either 10 m above an open ground or 3 m above the building roof.

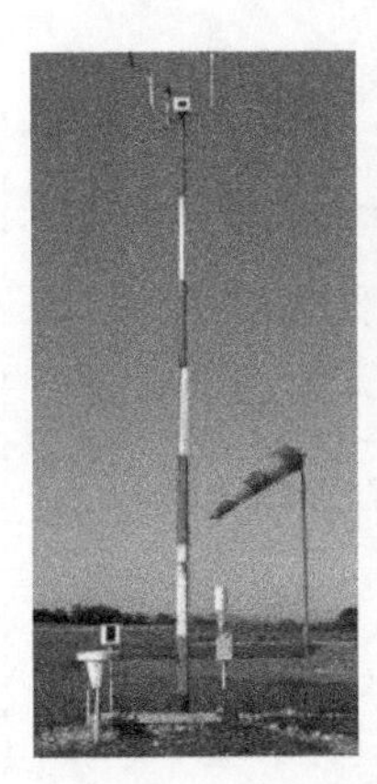

Figure 2-30 Weather stations

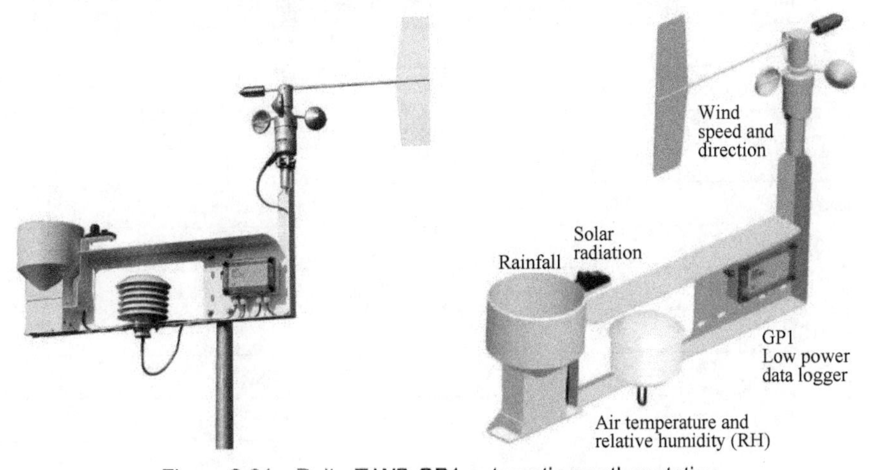

Figure 2-31 Delta-T WS-GP1 automatic weather station

2.2.3.2 Weather Station Specification

Weather station specification is listed in Table 2-9.

Table 2-9 Weather station specification

Air temperature RHT3nl-CA (combined RH sensor)		
Accuracy	± 0.3 ℃	−20 to 70 ℃
Solar radiation D-PYRPA-CA		
Accuracy	± 5 ℃	0 to 1.1 kW · m^{-2} 300 to 1 100 nm
Cosine response	± 1% at 45° ± 4% at 75°	At zenith angle
Data recording and power		
Logging frequency	1 s to 24 h	Logging status indicated by flashing LFD
Comms	To PC or laptop	RS232, USE or modem
Battery life	9 V alkaline	190 days typical, reading every 5 minutes
Physical		
Environmental	IP65 sealing	
Temperature	−20 to 60℃	If icing minimal
Cross-arm	White-painted stainless steel	Fits horizontal, or vertical pole (42-52 mm diameter)
Weight	14.3 kg	Complete with 2 m tripod and ground stakes
	Specification	Range/Note
Wind speed D-034B-CA (combined wind sensor)		
Range	0.4 to 75 m · s^{-1}	(0-167 mph)
Accuracy	± 0.1 m · s^{-1}	Up to 10.1 m · s^{-1}
	± 1.1% of reading	Over 10.1 m · s^{-1}
Starting threshold	0.4 m · s^{-1}	-30 to 70 ℃ (if icing minimal)
Wind direction D-034B-CA (combined wind sensor)		
Accuracy	± 4° 0.5° (resolution)	Mechanical: 0 to 360° Electrical: 0 to 356°
Starting threshold	0.4 m · s^{-1}	-30 to 70 ℃ (if icing minimal)
Rainfall RG2+WS-CA		
Sensitivity	0.2 mm/tip	160 mm funnel diameter
Humidity RHT3nl-CA (combined air temp sensor)		
Accuracy	± 2% RH	5% to 95% RH
	± 2.5% RH	<5% RH, >95% RH

Chapter 3

Building Thermal Environment

3.1 Physical Factors Affecting Indoor Thermal Environment

3.1.1 Building Fabric and Thermal Mass

3.1.1.1 Definition and Classification of Building Fabric

Building fabric is an essential component of any building since it both protects the occupants of the building and plays a major role in the regulation of the indoor environment (Figure 3-1). The fabric controls the flow of energy between the interior and exterior of the building, and it can be generally categorized into two parts.

- Transparent: External window, roof light.
- Opaque: External wall, roof, ground floor, external door, ceiling.

Figure 3-1 Building fabric

3.1.1.2 Roles of Building Fabric

Building fabric serves to (Figure 3-2):

- Protect the building occupants from the weather, such as wind, rain, solar radiation and snow;
- Regulate the indoor environment in terms of temperature, humidity, moisture and so on;
- Provide privacy for building occupants;
- Prevent the transmission of noise;
- Provide security for building occupants and assets;
- Provide safety, for example preventing the spread of fire or smoke;
- Provide views into and out of the building;
- Provide access between the inside and the outside of the building.

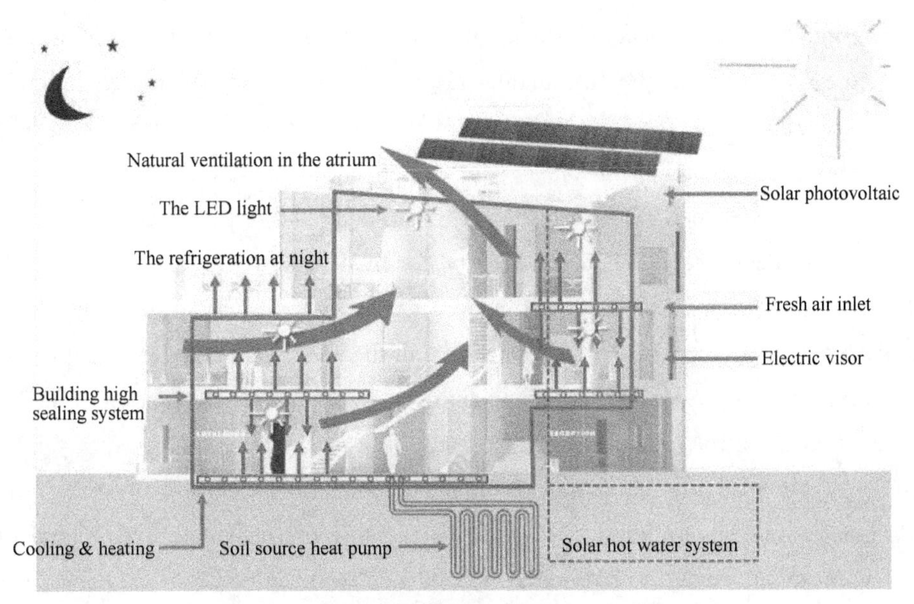

Figure 3-2 Protective effects of building fabric

3.1.1.3 Heat Gains/Losses in Buildings

In a typical building, the building enclosure is subjected to a number of heat gains and/or losses (Figure 3-3). During typical daytime hours, solar gain through windows and skylights can be significant.

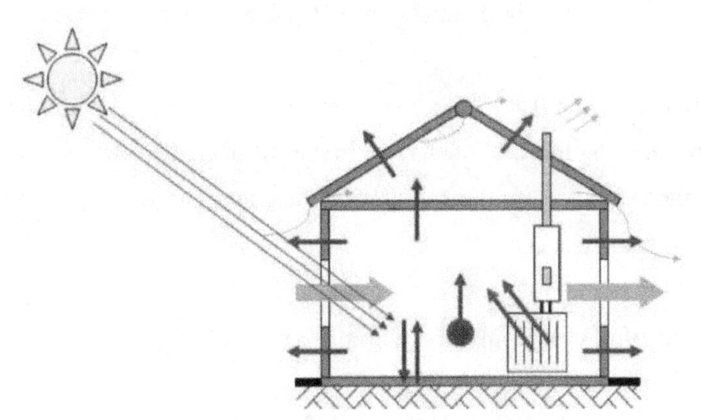

Figure 3-3 Heat gains/losses in buildings

(https://www.sciencedirect.com/science/article/abs/pii/S037877881300683X)

When it is cooler outside the building than inside, heat is lost through windows, walls, roofs, and floors. Losses also result from infiltration or air leakage through the building envelope.

1. Heat Gains

(1) Solar: Mainly due to the heat entering the room from solar radiation. In summer, solar radiation transfers heat to the room through glass windows, causing increased heat in the room.

(2) People: Heat is conducted between human body and environment. When heat is dissipated by the body through the surrounding air by means of convection, heat is conducted through layers of air or the surface of the garment close to the body's skin. Heat conduction also occurs when the human body is in contact with a solid surface of the wall. When there is a temperature difference between the surface of the human body and the surface of the surrounding environment, the heat exchange between them occurs by means of radiation.

(3) Equipment: For example, when indoor lighting is turned on for extended periods of time, this results in an increase in the temperature of the equipment and heat dissipation to the outside, with some of the electrical energy being converted to heat. In addition, miscellaneous appliances (ovens, refrigerators, irons, radios, TVs) and other heat sources (hot fluid passes through pipes, etc.) can cause increased heat in the building.

(4) Conduction through building fabric ($T_o>T_{in}$): The transfer of heat in the material is carried out by conduction from the hot side to the cold side. When the outdoor temperature is higher than the indoor temperature, the heat from outside is transferred into the room through the building envelope such as walls and windows.

(5) Ventilation ($T_o>T_{in}$): When the building is ventilated, the outdoor air enters the room with its own temperature and mixes with the indoor air during the flow, and heat is exchanged with the indoor surfaces according to the temperature difference between the indoor and the outdoor. Ventilation can raise the temperature when the indoor temperature is lower than the outdoor temperature.

(6) Infiltration ($T_o>T_{in}$): Air infiltration takes place primarily through inadequate and imperfect sealing between building elements, the frames of openings and their host construction as well as between the different components of individual door and window assemblies. It can also result from ill-fitting and unsealed door frames, window frames, outlets, walls, floors and roofs, as well as through exhaust appliance such as roof vents, bathroom fans, range hoods and dryer vents. Reduced air infiltration combined with proper ventilation can not only increase energy efficiency but also improve the quality of indoor air.

2. Heat Losses

(1) Radiation: The biggest reason of heat loss is radiation. It accounts for around two thirds of all energy lost through windows. This is because glass is an exceptional conductor of heat. Warmth generated within the house is absorbed by the window's inner pane. It is then transmitted to the outer pane's cooler surface and eventually to the outside, where it is lost irretrievably.

(2) Conduction through building fabric ($T_o<T_{in}$): Heat transport through a material takes place by conduction from the warm side to the cold side. When the outdoor temperature is lower than the indoor temperature, the indoor heat is dissipated into the outdoor area through the

building envelope such as walls and windows.

(3) Ventilation ($T_o<T_{in}$): The heat capacity of air is very low, so when the building is not ventilated, the indoor temperature will be close to the temperature level of the inner surface of building envelope. Ventilation can cool down the temperature when the indoor temperature is higher than the outdoor temperature.

(4) Infiltration ($T_o<T_{in}$): The term refers to the uncontrolled flow of air into a space through adventitious or unintentional gaps and fissures in the building envelope. However, it can also refer to leakage through individual components or building assemblies such as doors, windows and skylights. The corresponding loss of air from a closed space is referred to as "exfiltration" . The rate at which air seeps depends on the porosity of the building envelope and the magnitude of the natural driving forces of the wind and temperature.

3.1.1.4 Building Heating & Cooling Load Determination

If heat gains are greater than the envelope and ventilation losses, the building or space has a net cooling load (the building is too hot).

If heat losses are greater than the internal gains, the building or space has a net heating load (the building is too cold).

Cooling load is defined as, "the total heat required to be removed from the space in order to bring it to the desired temperature by air conditioning and refrigeration equipment" .

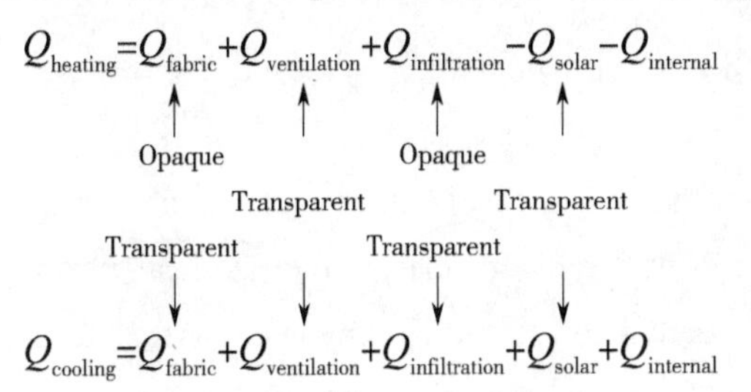

The purpose of calculating cooling load is to choose the equipment in the cooling system cycle correctly and economically. If the cooling system elements are selected correctly, the system will work efficiently and for many years in a way that is expected.

3.1.1.5 Definition of Thermal Mass

Thermal mass is a property of a building's mass that allows it to store heat, provide "inertia" against temperature fluctuations, which may help to reduce temperature fluctuations throughout the day; thereby reducing the heating and cooling demands of the building itself. Thermal mass materials achieve this effect by absorbing heat during periods of high solar insolation, and releasing heat when the surrounding air begins to cool(Figure 3-4). When incorporated into passive solar heating and cooling technologies, thermal mass can play a large role in reducing a building's energy use.

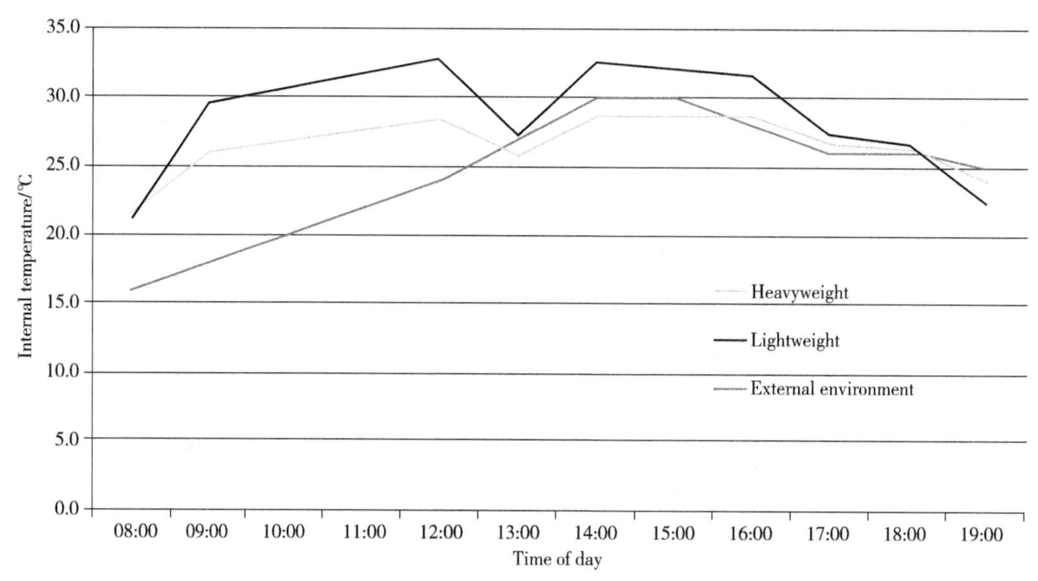

Figure 3-4 Effect of heavyweight and lightweight constructions on the internal temperature of a naturally ventilated school classroom

(https://en.wikipedia.org/wiki/Thermal_mass)

3.1.1.6 Selection of Materials for Thermal Mass

Ideal materials for thermal mass are supposed to have a combination of three basic characteristics.

• A high specific heat capacity (kJ/(kg · K)): Under such condition, the heat squeezed into every kilogram is maximized.

• A high density (kg/m^3): The heavier the material is, the more heat it can store by volume.

• A moderate thermal conductivity: With this characteristic, the rate heat flows in and out of the material is roughly in step with the daily heating and cooling cycle of the building.

Heavyweight construction materials like masonry and concrete have these features. It combines a large storage capacity with a moderate thermal conductivity. In other words, heat travels between the material's surface and its interior at a rate that roughly corresponds to the building's daily cycle of heating and cooling. Although some materials, such as wood, have a high specific heat capacity, their thermal conductivity is relatively low, which limits the rate at which heat can be absorbed during the day and rejected during the night. Steel can store a lot of heat, but conducts it too rapidly to be practically useful; thus comparatively little is used in buildings. However, a modest amount of thermal mass may still be provided if concrete floors are used in steel frame construction, although these are usually limited to a depth of only 100 mm and are usually covered by a false ceiling, limiting their ability to absorb and release heat.

3.1.1.7 Advantages and Disadvantages of Thermal Mass

1. Advantages

• Fewer spikes in heating and cooling requirements are observed since thermal mass slows down the response time and moderates the fluctuations in indoor temperature.

• For climates with a large daily temperature swing, a thermally massive building consumes less energy than a similar low-mass building because of the reduced heat transfer through the massive features, regardless of the level of insulation in the low-mass building.

• Thermal mass may shift energy demand to off-peak periods when utility tariffs are lower. Because power plants are designed to deliver power at peak loads, shifting the peak load may reduce the number of power plants needed.

2. Disadvantages

• In summer, thermal mass is only beneficial if nighttime ventilation (or some other means of cooling) can be used to remove the heat absorbed by the building fabric during the day. The provision of adequate ventilation can be challenging in some environments, particularly urban locations.

• In winter, older heavyweight buildings with comparatively low levels of insulation and poor airtightness will require a longer preheat period to warm up the fabric, resulting in more energy being used than in a similar lightweight building. However, for newer buildings the greatly improved standard of fabric performance means this is no longer the problem it once was, as the fabric retains most more of its warmth during periods when the heating is off. In practice, the ability of thermal mass to enhance summertime performance in many building types is of much greater significance.

3.1.1.8 Definition of Phase Change Materials

A phase change material is a substance that absorbs and releases thermal energy when it changes phase (known as latent heat). As a material melts, it transitions from solid to liquid phase. During the phase transition, many materials are able to absorb a significant amount of heat energy. The opposite is true when the material freezes and solidifies: the material will give out the heat that it has absorbed when it melts. Different materials will melt and solidify at different temperatures and are able to absorb different amount of heat energy.

3.1.1.9 Application of Phase Change Materials

Due to climate change, China's summer is becoming hotter and hotter, more houses with no air-conditioners will face overheating risks; under this circumstance, it is preferable to use phase change materials as a passive method to store heat when the room is hot and release the stored heat when the room is cold.

Phase change materials are widely used in the field of architecture in the use of solar energy, recovering industrial waste heat and regulating the temperature of buildings to increase thermal mass. Incorporating phase change materials into the building improves the room's thermal performance, saves energy and reduces CO_2 emissions, etc. Phase change materials in building materials are beneficial during extremely hot and cold climates, store solar gains when hot and release stored heat when cold. They help maintain the internal conditions of buildings to a comfortable level, thus reducing energy consumption.

3.1.1.10 Study on PCM Use

In this study, DesignBuilder has been adopted to predict the performance of the building under various simulation scenarios. In DesignBuilder, PCMs can be modeled by using some general thermal properties, e.g. thickness, conductivity and density, with some specific properties, i.e. the material temperature-enthalpy curve.

3.1.2 Transfer of Heat and Moisture

The transfer of heat and moisture is a natural phenomenon that is widespread in nature. As long as there is a temperature difference between objects or between different parts of the same object, the transfer of heat and moisture will occur and continue until the temperature is the same. The only condition for the transfer of heat and moisture to occur is the existence of a temperature difference, independent of the state of the objects and whether they are in contact with each other. The result of the transfer is that the difference in temperature disappears, i.e. different objects between which heat transfer occurs or different parts of the object reach the same temperature.

There are three mechanisms of heat transfer: conduction, convection and radiation (Figure 3-5).

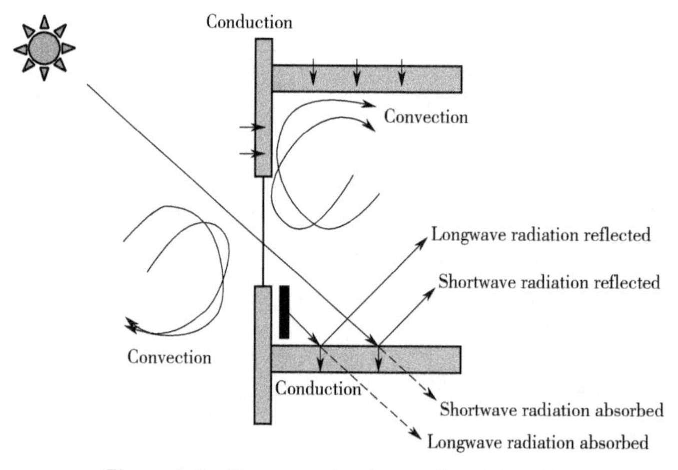

Figure 3-5 Three mechanisms of heat transfer

3.1.2.1 Definition of Radiation & Radiation Heat Transfer

Radiation heat transfer is the transfer of heat by electromagnetic radiation, such as insolation, without the need for matter to be present in the space between bodies, which originates due to the temperature of a body. Thermal radiation is described as energy transport through electromagnetic waves. In contrast to conduction and convection, radiative heat transfer is not bound with matter, and it can even take place in vacuum. Any body with a temperature higher than absolute zero radiates energy as electromagnetic radiation. The spectral distribution of the energy radiated from any body depends on the temperature of its surface. The higher the temperature of the body, the smaller the wavelength of the radiation that makes up the bulk of the total radiation emitted.

3.1.2.2 Radiation Heat Transfer Rate

$$Q_r = h_{reff} A_s (T_s - T_c)$$

where h_{reff} is radiation heat transfer coefficient (only for longwave radiation between surfaces) (W/(m² · K)); A_s is surface area (m²); T_s is mean surface temperature (warm) (Figure 3-6); T_c is mean surface temperature (cold). h_{reff} is a query value according to the mean surface temperature.

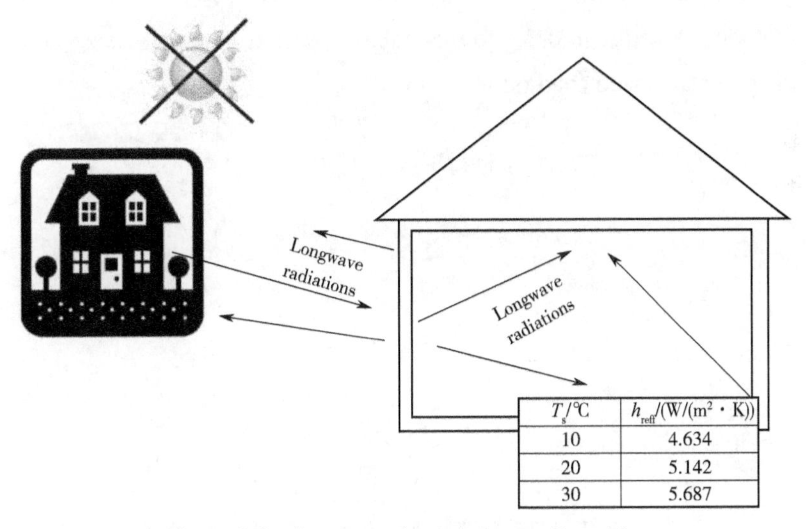

T_s/℃	h_{reff}/(W/(m² · K))
10	4.634
20	5.142
30	5.687

Figure 3-6 Correspondence between T_s and h_{reff}

3.1.2.3 *U*-value (Thermal Transmittance)

Thermal transmittance, also known as *U*-value, is the rate of transfer of heat through a structure (which can be a single material or a composite), divided by the difference in temperature across that structure (Figure 3-7). The unit of measurement is W/(m² · K).

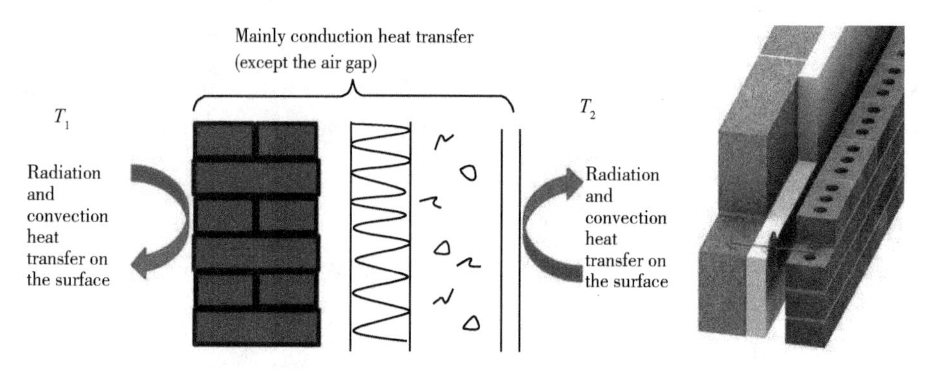

Figure 3-7 *U*-value for a cavity wall

The *U*-value depends primarily on the thermal conductivity of the solid (heat transfer by thermal conduction), but also on the coefficient of heat transfer between fluid and solid or solid and liquid (heat transfer by thermal convection). In addition, heat transfer by thermal radiation also occurs. In practice, however, the *U*-value for different components is usually not determined on the basis of the thermal conductivity or the heat transfer coefficient, but is determined experimentally.

The thermal transmittance of a building envelope is the principal factor in the determination of the steady state heat loss and heat gain, which will ultimately determine the capacity of heating and cooling systems, in order to maintain the desired internal design conditions, with corresponding external design conditions (Figure 3-8).

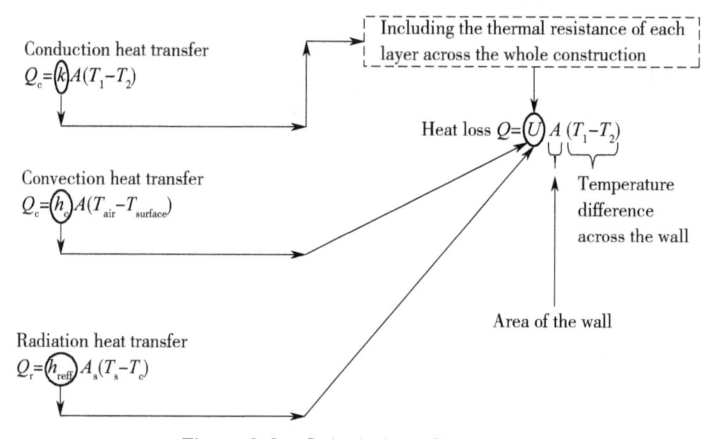

Figure 3-8 Calculation of heat loss

Most rooms are made up from surfaces and elements with different *U*-values, e.g. walls, windows, roofs.

3.1.2.4 Calculation of *U*-value (Thermal Resistance)

The formula we have seen needs slight modification to allow for these different *U*-values:

$$Q=\Sigma UA\cdot\Delta T$$

where Q is the rate of heat loss from the room (W); ΣUA is the sum of all the products of *U*-value times area for every external element (W/(m^2 · K)).

$$U = \frac{1}{R_{so} + R_1 + R_a + R_2 + R_{si}}$$

where R_{so} and R_{si} are the external and internal surface resistances, respectively; R_i is the thermal resistance of each individual component, i=1, 2,...; R_a is the thermal resistance of any air gap.

3.1.2.5 Calculation of *R*-value (Thermal Resistance)

R-value (thermal insulance factor) is a measure of the thermal resistance. The larger the value of *R*, the greater the insulation efficiency. Thermal insulance has the units (m^2 · K)/W in SI units or (ft^2 · ℉ · h)/Btu in British units, so it is important to note that thermal insulance has the following properties: thermal resistance is the unit area thermal resistance of a material; the *R*-value depends on the insulation type, thickness, and density. Surface area and temperature difference are needed to solve for the heat transferred.

$$R = \frac{L}{k}$$

where *R* is the ability of a material to prevent heat from passing through it per unit area at a specified temperature; *k* is material thermal conductivity (W/(m · K)); *L* is material thickness (m).

Parameters such as *R*-value (resistance) are used in the construction industry, which is expressed by the material thickness normalized to the thermal conductance, and under uniform conditions it is the ratio of the difference in temperature across an insulation and the density of heat flow through it: $R(x) = \Delta T/q$. The larger the value of *R* is, the more heat transfer is prevented by a material. It can be seen that the strength depends on the thickness of the material.

Some typical thermal conductivity values:

Air 0.025 W/(m · K)

Sandstone 2.4 W/(m · K)

Copper 401 W/(m · K)

3.1.2.6 Typical Heat Loss from a Building

In general, heat loss in buildings occurs via walls, roofs, windows, doors, slabs, thermal bridges, and the air exchange between indoor and outdoor areas (Figure 3-9).

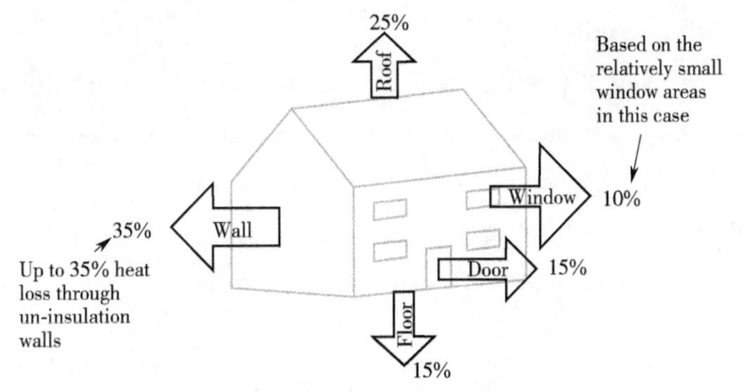

Figure 3-9 Heat loss in buildings

3.1.2.7 Significance of *U*-value

U-value is particularly important for the envelope of buildings. In building engineering, the fluid is air. In this case, windows, masonry, plaster or other insulating materials serve as heat transmitting solids. These components should prevent the heat transfer between the interior of the house and the environment as much as possible. With regard to thermal insulation, the aim is always to use materials with the lowest possible *U*-values in order to achieve the greatest possible insulation effect.

For example, a building's heating need can be minimized according to the *U*-value (Table 3-1).

Table 3-1 Limiting fabric parameters (*U*-values) (UK Building Regulations Part L)

Component	*U*-value/(W/(m^2 · K))
Roof	0.25
Wall	0.35
Floor	0.25
Window	2.2
Vehicle access/large door	1.5
High usage entrance door	3.5
Roof ventilator	3.5
Air permeability	10 m^3/h at 50 Pa

The lower the *U*-value, the better the building's thermal performance.

3.1.2.8 Definition of Convection & Convection Heat Transfer

When heat is transported by a fluid, like air or water, this is called convection. Thus above a radiator, hot air ascends and heats the upper part of the room. The extent of convective heat transfer depends on a number of things, like the position of the surface (horizontal or vertical), but mainly on the speed of the passing air. Outdoors the speed is determined by wind speed and direction, which are very variable. When the air is driven by an outside force such as the wind, this is called "forced convection".

When there is no wind, convection will occur by temperature or density differences. This is called "free convection". Hot air ascending above a radiator is an example of free convection. Room air is heated by the radiator and ascends because the density of the hot air next to the radiator is lower than the density of the cooler air in the rest of the room. This results in the warmed air rising and being displaced by the cooler air. Downdraught is also an example of free convection.

The heat transfer through forced convection in general is bigger than that through free convection, because of higher air speed. So the heat transfer from a surface to the outdoor air will generally be bigger than the heat transfer from the indoor surface to the indoor air.

3.1.2.9 Convection Heat Transfer Rate

$$Q_c = h_c A\left(T_{air} - T_{surface}\right)$$

where h_c is convection heat transfer coefficient (W/(m² · K)).

External h_c is dominated by the wind velocity (Figure 3-10).

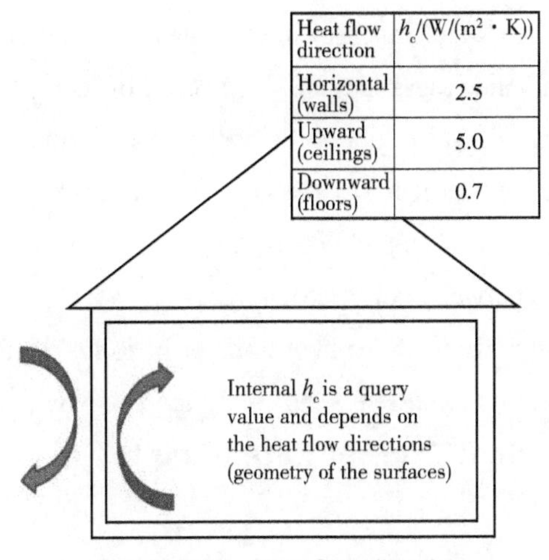

Figure 3-10 Heat flow direction

3.1.2.10 Forced Convection vs. Natural Convection

According to heat transfer mechanism, convection can be divided into two types: forced convection and natural convection.

Natural convection depends on differences in fluid density produced by temperature differences in the liquid. In electronics that use natural convection based cooling, air is usually the natural surrounding liquid. The air near the electronics absorbs heat from its surface. The warmer the air is, the higher it rises, because of density differences. Cooler air displaces warmer air, thus developing natural convection. Natural convection-based cooling is aided by atmospheric air circulation and local weather conditions.

When forced convection is used to cool the electronics, the surrounding liquid is in motion. Forced convective heat transfer allows large amounts of thermal energy to be efficiently transferred to the fluid. The moving fluid may be air motion created by employing cooling fans in the circuit. Devices may also be liquid-cooled using a cold base plate.

In forced convection heat transfer, fluid motion occurs with an external driver. Cooling fans, for example, force the fluid that is air to the materials in order to cool them. So, there is a forced convection heat transfer phenomenon here.

In natural convection, there is no external factor for fluid flow. The fluid is free to transmit thermal energy. An example of the phenomenon of natural convection is the transfer of heat between the hot object and the cold medium.

3.1.2.11 Definition of Conduction & Conduction Heat Transfer

Heat conduction (also known as thermal conductivity) is the phenomenon of energy transfer through microscopic vibrations, displacements and mutual collisions of molecules, atoms and electrons within an object when there is a temperature difference between different objects or within the same object. The mechanism of internal heat conduction varies in different phases of matter. Heat conduction inside a gas is mainly the result of its internal molecules doing irregular thermal motion as a result of mutual collisions; in a non-conductive solid, vibration near the equilibrium position of its lattice structure transfers energy to neighboring molecules to achieve heat conduction; and heat conduction in a metallic solid is accomplished by virtue of the motion of free electrons between lattice structures.

Heat conduction is the main mode of heat transfer in solids. In fluids such as gases or liquids, the process of heat conduction often occurs simultaneously with heat convection.

3.1.2.12 Conduction Heat Transfer Rate (Figure 3-11)

$$Q_c = \frac{kA(T_1 - T_2)}{\delta x}$$

where k is thermal conductivity (W/(m · K)); A is surface area (m^2); δx is thickness of the wall (m).

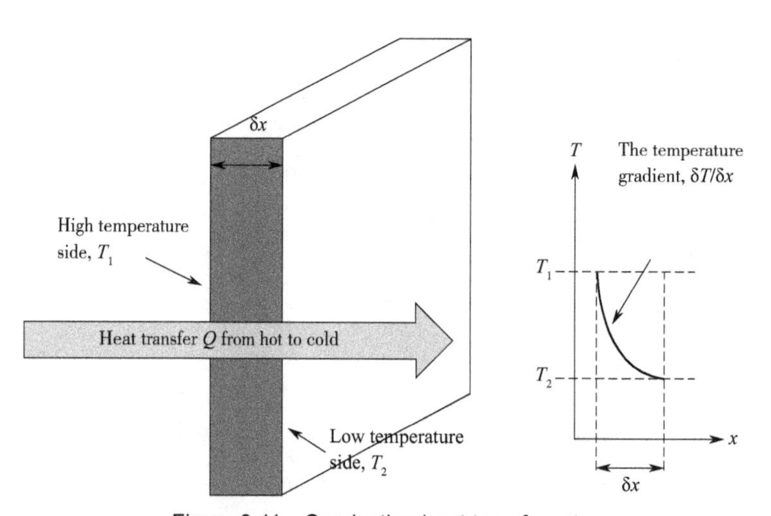

Figure 3-11 Conduction heat transfer rate

3.1.2.13 Factors That Affect the Thermal Conductivity of Materials

The thermal conductivity of a material depends on the following factors.

1. Presence of Free Electrons

Free electrons readily move into the material from which it can transfer heat at a faster rate. Therefore, as the number of free electrons present in the material increases, so does the thermal conductivity of the material. As a result, metals have higher thermal conductivity due to higher number of free electrons present in metals.

2. Presence of Impurities

In its pure form, the material has a higher thermal conductivity. As impurities or alloy elements increase, the thermal conductivity of the material decreases. For example, pure copper has a higher thermal conductivity than brass.

3. Density

Due to the increased rate of molecular collisions, the thermal conductivity of the material increases as its density increases.

4. Temperature Range

As the temperature of solid materials increases, the vibration of the molecules also increases, and consequently the motion of the free electrons decreases, and thus the thermal conductivity decreases. For liquids, as temperature increases, the density of the liquid decreases. As a result, the thermal conductivity of the liquid decreases.

5. Material Structure

The crystalline structure has a higher rate of thermal conductivity while the amorphous material has a lower value of thermal conductivity.

6. Presence of Humidity

The thermal conductivity of material increases with an increase in moisture content in the material.

Typical thermal conductivity of building materials is listed in Table 3-2.

Table 3-2 Typical thermal conductivity of building materials

Structural and finishing materials (Always check manufacturer's details—variation will occur depending on product and nature of materials)	Thermal conductivity/(W/(m · K))
Acoustic plasterboard	0.25
Aerated concrete slab (500 kg/m^3)	0.16
Aluminium	237
Asphalt (1 700 kg/m^3)	0.50
Bitumen-impregnated fibreboard	0.05
Brickwork (outer leaf 1 700 kg/m^3)	0.84
Brickwork (inner leaf 1 700 kg/m^3)	0.62

Continued

Structural and finishing materials (Always check manufacturer's details—variation will occur depending on product and nature of materials)	Thermal conductivity/(W/(m · K))
Dense aggregate concrete block (1 800 kg/m³, exposed)	1.21
Dense aggregate concrete block (1 800 kg/m³, protected)	1.13
Calcium silicate board (600 kg/m³)	0.17
Concrete general	1.28
Cast concrete (heavyweight, 2 300 kg/m³)	1.63
Cast concrete (dense 2 100 kg/m³ typical floor)	1.40
Cast concrete (dense 2 000 kg/m³ typical floor)	1.13
Cast concrete (medium, 1 400 kg/m³)	0.51
Cast concrete (lightweight, 1 200 kg/m³)	0.38
Cast concrete (lightweight, 600 kg/m³)	0.19
Concrete slab (aerated, 500 kg/m³)	0.16

3.1.3 Thermal Performance of Semi-transparent Structures

3.1.3.1 Double Skin Facade (DSF)

A double skin facade is a building system in which two floors or facades are arranged in such a way that air circulates through the air gap between them (Figure 3-12). This space (from 20 cm to several meters) acts as a barrier against extremes of temperature, wind, and noise, and improves the thermal efficiency of the building during high and low temperatures.

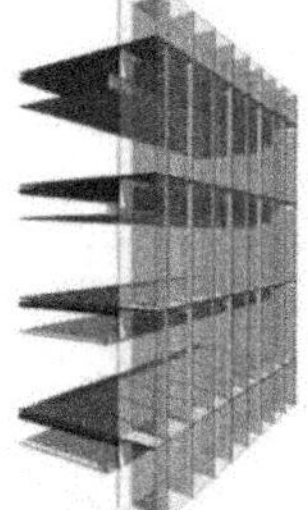

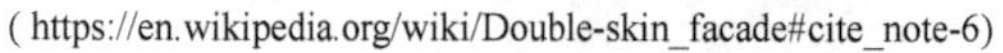

(https://en.wikipedia.org/wiki/Double-skin_facade#cite_note-6) (http://smartbuildings.unh.edu/project-a/)

Figure 3-12 Double skin facade

Furthermore, double skin facades are adaptable to cooler and hotter weather. This versatility is what makes them so interesting: through slight modifications, such as the opening or closing of inlet or outlet fins or the activation of air circulators, the behavior of the facades is altered.

In warmer seasons the system exhausts warm air up from the inside and draws in cooler breezes from the bottom. In warm climates, the cavity can be vented out of the building to mitigate solar gain and reduce the cooling load. Excess heat is drained by a process known as the stack effect, where differences in the density of air create circular motion that causes hotter air to escape. As the temperature of the air in the cavity increases, it is pushed outward, bringing a gentle breeze to the surroundings while isolating against the heat gain. In winter the system draws in cold air and heats it, thus ventilating the space and keeping it warm (Figure 3-13).

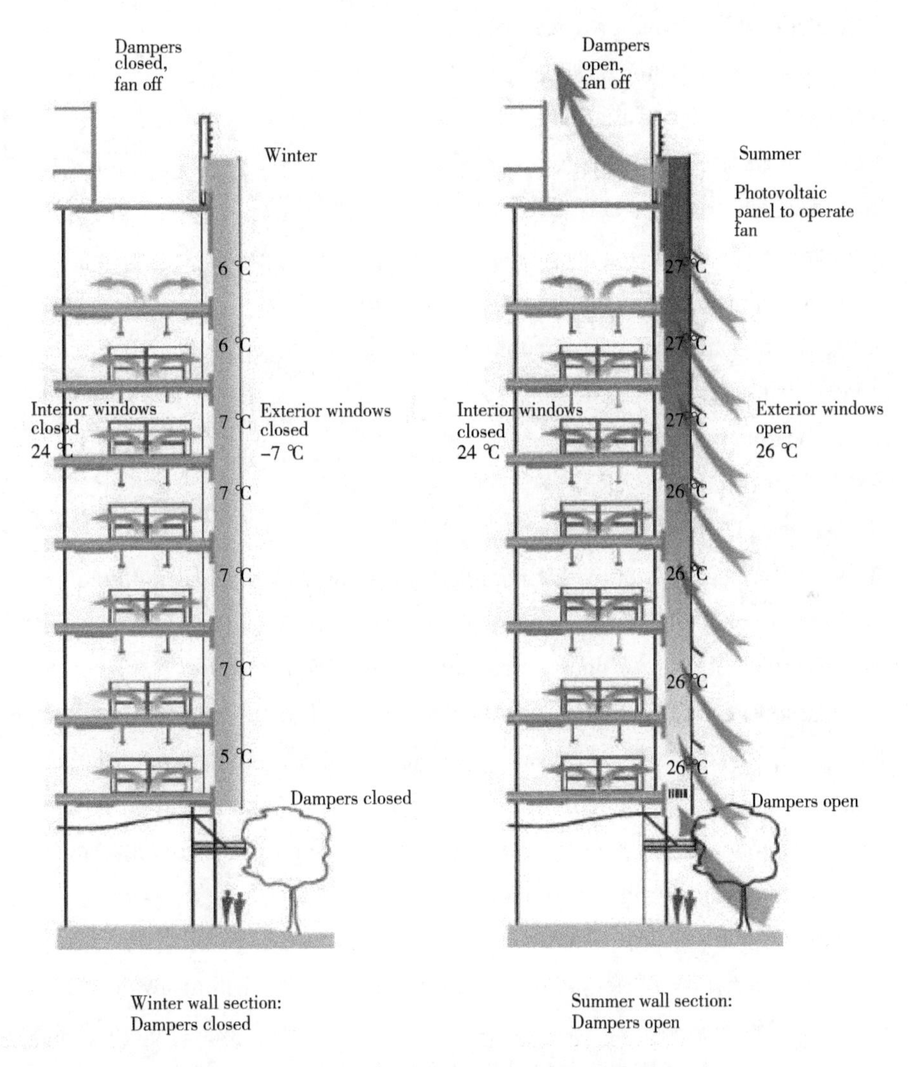

Figure 3-13 Working principle of double skin facade

(https://www.pinterest.co.uk/pin/450289662716935801/)

Overall, double skin facades depend heavily on external conditions (solar radiation, external temperature, etc.) that directly influence internal comfort and life quality of users. Therefore, careful design is essential for each case, requiring detailed knowledge of solar orientation, context, local radiation, temperature conditions, building occupancy, and much more.

3.1.3.2 Advantages of DSF

• Lower construction cost compared to solutions that can be provided by the use of high-tech windows.

• Acoustic insulation.

• Thermal insulation.

• Better protection of the shading or lighting devices.

• Reduction of the wind pressure effects.

• Transparency (architectural design).

• Natural ventilation availability.

• Less heat loss and solar gain.

3.1.3.3 Disadvantages of DSF

• Higher construction cost compared to a conventional facade.

• Reduction of rentable office space.

• Additional maintenance and operation costs.

• Increased weight of the structure.

• Overheating problem.

3.1.3.4 Classifications of DSF

1. External Circulation Type

The outer curtain wall is made of single-layer glass with air inlet in the lower part and air exhaust in the upper part. The inner curtain wall uses hollow glass and heat insulation profiles, and has openable windows or doors. It requires no special mechanical equipment and relies entirely on natural ventilation to exhaust solar radiant heat to the outdoors through the exhaust ports on the pipes. This saves energy as well as mechanical operation and maintenance costs. In summer, the upper and lower vents are opened for natural exhaust cooling. In winter, close the vents to allow solar radiant heat to accumulate in the cavity, and heat loss can be reduced by heat entering the room through windows and doors.

2. Internal Circulation Type

The outer curtain wall adopts hollow glass and heat insulation profiles to form a closed state. The inner curtain wall adopts single-layer glass or single-layer aluminum alloy doors and windows to form an openable state. Using mechanical ventilation, the air enters the channel from the air outlet on the floor or underground and flows into the roof through the upper exhaust outlet.

Since the incoming air is indoor air, the air temperature inside the channel is basically the same as the indoor temperature, so it can save energy for heating and cooling, which is more beneficial to the heating area. Since the internal ventilation requires mechanical equipment and photoelectric control of blinds or shading systems, there are higher technical requirements and costs.

3. Comprehensive Internal and External Circulation Type

It is not just stuck to the single circulation method of internal or external circulation, but has more flexible adaptability to external weather and climate conditions. For example, it has a better balance of summer and winter, and reduces the dependence on other systems (such as fresh air system, cooling system), which is conducive to improving the comprehensive energy-saving effect.

The width of the passage is 100-300 mm for ventilation only, 500-900 mm for maintenance and cleaning, and over 900 mm for rest, viewing and walking, with grille.

3.2 Human Response to Thermal Environment

3.2.1 Human Thermal Sensation

3.2.1.1 Definition of Human Thermal Sensation

Thermal sensation is defined in ASHRAE Standard 55-2010 as a person's subjective description of "cold" and "hot", i.e. the condition of mind which expresses satisfaction with the thermal environment. Although people can evaluate the "coldness" and "warmth" of their surroundings, what they actually feel is the temperature of the nerve endings located under the skin surface, not the real temperature of the environment (Figure 3-14).

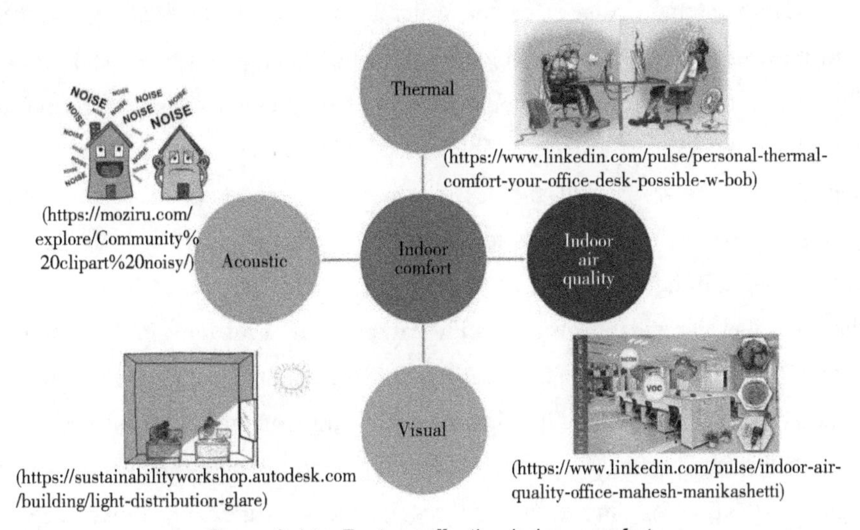

Figure 3-14 Factors affecting indoor comfort

3.2.1.2 Calculation of Occupant Satisfaction/Productivity & Indoor Comfort

$$l = 1.24 + 0.39S_{\mathrm{w}} + 0.36S_{\mathrm{aq}} + 0.16S_{\mathrm{am}} + 0.13S_{\mathrm{n}} + 0.12S_{\mathrm{h}} + 0.05S_{\mathrm{l}}$$

where l is occupants' satisfaction with indoor thermal comfort; S_w is satisfaction code for warmth; S_{aq} is satisfaction code for air quality; S_{am} is satisfaction code for air movement; S_n is satisfaction code for noise; S_h is satisfaction code for humidity; S_l is satisfaction code for lighting.

3.2.1.3 Human Thermoregulatory System

The human body, as a complex organic whole, is constantly exchanging substances and energy with the external environment, which is the fundamental factor on which human life depends. The body maintains a relatively stable core temperature as a basic condition for metabolism and normal life activities. The maintenance of normal body temperature is the result of a dynamic balance between heat production and heat dissipation in the human body, a process that is achieved by the thermoregulatory system within the scope of thermal comfort. The human thermoregulatory system is divided into two categories, physiological regulation and behavioral regulation, according to their different regulatory mechanisms.

Physiological regulation refers to the mode of maintaining a stable body temperature by increasing or decreasing the blood flow in the skin, sweating, chilling and other physiological regulatory responses under the control of the hypothalamus regulatory center. The hypothalamus is the central system of thermoregulation and is part of the brain. It plays a major role in a number of autonomic functions such as food intake, water balance, and thermoregulation. Physiological regulation is carried out by the body's own regulatory system, the biological control system. This system integrates the signals received by the thermoreceptors distributed throughout the body to the thermoregulatory center, which then drives the body's blood vessels, sweat glands and muscles to produce physiological thermoregulatory responses. And all signals are fed back to the central system synchronously to eventually produce cold and heat sensation and behavioral regulation requirements to establish body heat balance and stabilize body temperature. When the temperature is high, the heat dissipation center sends out instructions to secret sweat glands and expand blood vessels to enhance heat dissipation, thus enabling the body to achieve thermal equilibrium.

Behavioral thermoregulation is the modification of the heat transfer coefficient between the body and its surroundings through extracorporeal regulation, such as dressing or purposeful use of external energy to reduce the physiological heat strain on the body, thus keeping the body temperature within the normal range. Behavioral regulation of body temperature is based on physiological regulation to compensate for the deficiencies of the body's own regulation. The most primitive and simple behavioral regulation is postural change and place migration. Clothing is an important tool for human behavioral thermoregulation, which greatly enhances the ability of humans to adapt to nature, such as adding or removing clothing in different environmental conditions and creating artificial climates for thermoregulation purposes.

3.2.1.4 Core Body Temperature

The core body temperature is approximately 37 ℃. When it exceeds 37 ℃, body is overheated, while when it is lower than 34 ℃, body is undercooled.

Main health issues caused by overheating and undercooling:

- Heat stoke;
- Heat exhaustion;
- Frostbite;
- Hypothermia.

Main physiological mechanisms in our body thermoregulatory system:

- Sweating—release fluid onto the skin's surface where it can evaporate to cause cooling;
- Vasodilatation—blood vessels close to the skin expand, allowing blood from the core to the skin, so the skin temperature increases;
- Vasoconstriction—a reverse mechanism to vasodilatation;
- Shivering—enable the muscle system to produce heat.

3.2.2 PMV-PPD Model

3.2.2.1 Fanger's PMV

The full name of PMV is predicted mean vote, that is, the predicted average number of votes. The PMV value is an evaluation index representing the human thermal response (cold and heat sensation) proposed by Professor Povl Ole Fanger (1934-2006) from Technical University of Denmark (Figure 3-15).

PMV is a measure of the body's thermal response to heat, and represents the average of most people's sensations of heat and cold in the same environment. It is an index that seeks to predict the average value of votes of a group of people on a seven point scale of thermal sensation. Thermal equilibrium is achieved when an individual's internal heat output is the same as its heat loss. PMV is influenced by both environmental factors, including air temperature, relative humidity, air flow rate and mean radiation temperature, and human factors, including human metabolic rate and clothing thermal resistance. Of the environmental factors, air temperature has the most significant effect, with PMV values increasing dramatically when temperatures rise.

Figure 3-15 Povl Ole Fanger

(www.ie.dtu.dk)

3.2.2.2 Fanger's Experiment

The PMV thermal comfort model is based on theories of thermoregulation and heat balance. According to these theories, the body maintains a balance between metabolic heat production and body heat loss through physiological processes such as sweating, cold shivering and regulation of blood flow to the skin. Maintaining thermal equilibrium is the primary condition for achieving thermoneutrality, and Fanger believes that the body's thermoregulation is so efficient that it can achieve thermal equilibrium within a wide range of environmental parameters, even in uncomfortable environments.

In order to predict thermoneutrality, Fanger investigated the physiological processes that occur when the body is close to thermal neutrality and concluded that the factors that influence thermal equilibrium at this time are sweating rate and average skin temperature, and the thermal neutrality depends on the level of human movement. He derived a linear relationship between exercise level and sweat rate based on data from McNall et al. and from a poll of 183 subjects who felt thermoneutral at a given exercise level. Fanger obtained a linear relationship between skin temperature and level of exercise, under the thermoneutral conditions studied by McNall et al. and without direct polling of the subjects for thermal comfort, which has led some to question the formulation of the PMV thermal comfort model.

The comfort equation only predicts what happens when people feel thermally neutral, but in practice it is also important to know what happens when people do not feel thermally neutral. Fanger combined the experiments of Nevins and McNall et al. with his own research to extend the comfort equation based on data consisting of a total of 1 396 subjects in Denmark and the USA carried out in thermal comfort chambers. This equation links the environmental condition with the seven-point thermal sensation scale (Figure 3-16).

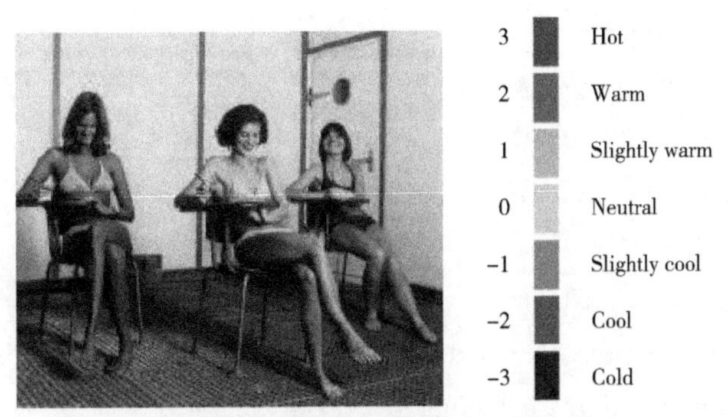

Figure 3-16 Indicators of thermal comfort

(http://blogs.mentor.com)

Since the introduction of the PMV thermal comfort model, a large number of thermal comfort studies have been carried out, both experimental and field studies. Some of the results of the experimental studies agree well with the predictions of the PMV thermal comfort model, while others deviate significantly from the predicted values. In general, the actual thermal sensations in field studies are more likely to deviate from the predictions of the PMV thermal comfort model. However, as the most comprehensive indicator for evaluating the thermal environment to date, it is still used extensively in the evaluation of comfort in buildings, etc.

3.2.2.3 Calculation of PMV Equation

Professor Fanger pointed out that the human body has to meet three conditions in order to achieve a state of thermal comfort:

- The body has to be in energy balance;
- The skin surface temperature has to meet certain requirements;
- There should be a certain rate of perspiration in order to achieve a state of comfort, except in a state of minimal activity, such as at rest.

$$\begin{aligned}\mathrm{PMV}=&\left[0.303\exp(-0.036M)+0.028\right]\Big\{(M-W)-3.05\times10^{-3}\left[5\,733-6.99(M-W)-\right.\\&\left.p_{\mathrm{a}}\right]-0.42\left[(M-W)-58.15\right]-1.7\times10^{-5}M(5\,867-p_{\mathrm{a}})-0.001\,4M(34-t_{\mathrm{a}})-\\&3.96\times10^{-8}f_{\mathrm{cl}}\left[(t_{\mathrm{cl}}+273)^4-(\bar{t}_{\mathrm{r}}+273)^4\right]-f_{\mathrm{cl}}h_{\mathrm{c}}(t_{\mathrm{cl}}-t_{\mathrm{a}})\Big\}\end{aligned}$$

$$t_{\mathrm{cl}}=35.7-0.028(M-W)-I_{\mathrm{cl}}\left\{3.96\times10^{-8}f_{\mathrm{cl}}\left[(t_{\mathrm{cl}}+273)^4-(\bar{t}_{\mathrm{r}}+273)^4\right]+f_{\mathrm{cl}}h_{\mathrm{c}}(t_{\mathrm{cl}}-t_{\mathrm{a}})\right\}$$

$$h_{\mathrm{c}}=\begin{cases}2.38|t_{\mathrm{cl}}-t_{\mathrm{a}}|^{0.25} & \text{for}\quad 2.38|t_{\mathrm{cl}}-t_{\mathrm{a}}|^{0.25}>12.1\sqrt{v_{\mathrm{ar}}}\\12.1\sqrt{v_{\mathrm{ar}}} & \text{for}\quad 2.38|t_{\mathrm{cl}}-t_{\mathrm{a}}|^{0.25}\leqslant12.1\sqrt{v_{\mathrm{ar}}}\end{cases}$$

$$f_{\mathrm{cl}}=\begin{cases}1.00+1.290l_{\mathrm{cl}} & \text{for}\quad l_{\mathrm{cl}}\leqslant 0.078\ \mathrm{m}^2\cdot\mathrm{kW}\\1.05+0.645l_{\mathrm{cl}} & \text{for}\quad l_{\mathrm{cl}}>0.078\ \mathrm{m}^2\cdot\mathrm{kW}\end{cases}$$

where M is metabolic rate (W/m^2); W is effective mechanical power (W/m^2); I_{cl} is clothing insulation (m^2·K/W); f_{cl} is clothing surface area factor; t_{a} is air temperature (℃); $\bar{t}_{\mathrm{r}}$ is mean radiant temperature (℃); v_{ar} is relative air velocity (m/s); p_a is partial pressure of vapor (Pa); h_c is convective heat transfer coefficient (W/(m^2·K)); t_{cl} is clothing surface temperature (℃).

As can be seen from the comfort equation, the factors affecting thermal comfort can be summarized in the following two ways.

1. Human Factors

• Function of clothing, including I_{clo}—clothing thermal resistance, f_{cl}—clothing area factor, i.e. the ratio of the surface area of the body wearing clothing to the exposed body;

• Function of human activity, including M—metabolic rate (W/m^2).

2. Environmental Variables

Environmental variables include: t_a—air temperature (℃); $\bar{t}_r$—mean radiant temperature (℃); v_{ar}—relative air velocity (m/s); p_a—water vapor partial pressure (mmHg).

3.2.2.4 Predicted Percentage of Dissatisfication

The predicted percentage of dissatisfaction (PPD) is to describe the ratio of potential complainers in buildings, i.e. a prediction of the percentage of a group of people who are uncomfortable with a given thermal environment on a mean thermal sensory scale as a percentage of the total number of people. The PPD value is a function of the predicted mean thermal sensory scale based on the number of votes cast by many people on the thermal sensory scale of the environment. It is plotted as a curve to reflect the percentage of people who feel dissatisfied. People's thermal perception of the environment varies, and the curve reflects that 5% of people are dissatisfied even when PMV = 0 under comfortable conditions (Figure 3-17).

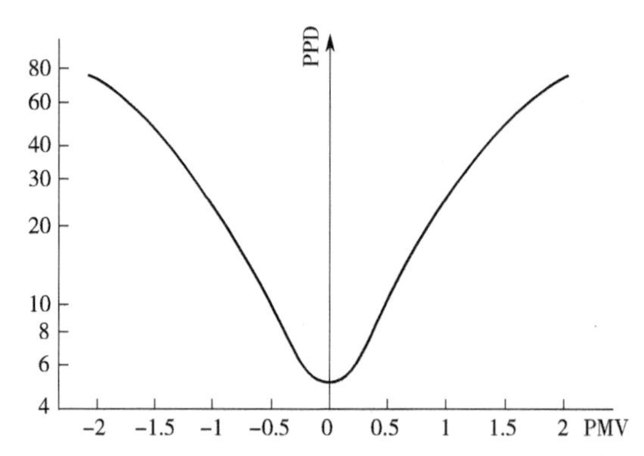

Figure 3-17 Analytical determination and interpretation of thermal comfort using calculation of the PMV and PPD indices and local thermal comfort criteria (ISO 7730)

3.2.2.5 Thermal Environment Category

The desired thermal environment for a space may be selected from three categories, A, B and C according to Table 3-3. All the criteria should be satisfied simultaneously for each category.

Table 3-3 Building design criteria from ISO 7730

Category	PPD/%	PMV
A	< 6	$-0.2 < \text{PMV} < +0.2$
B	< 10	$-0.5 < \text{PMV} < +0.5$
C	< 15	$-0.7 < \text{PMV} < +0.7$

- Category A: High level expectation (used for spaces occupied by very sensitive persons)
- Category B: General expectation (for new buildings and renovations)
- Category C: Moderate expectation (for existing buildings)

3.2.2.6 Influencing Factors of Thermal comfort

1. Natural Factors

1) Mean radiant temperature

All objects with a temperature above absolute zero emit heat radiation. When people are indoors, there is a radiative heat exchange between the surfaces of the objects in the room and the human body. The average radiative temperature is the average of the temperatures of the surfaces in the room that have an influence on the radiative heat exchange with the human body (Figure 3-18).

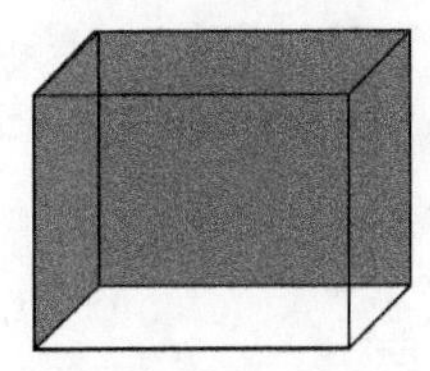

Figure 3-18 Average effect of temperature from surrounding surfaces

The average radiation temperature depends on the surrounding surface temperature. In the actual production and living environments, air temperature and average radiation temperature are not always uniform or equal, people often encounter a situation that a part of the body is cold and hot, such as indoor temperature up and down obviously asymmetric, the human side of the radiation heat source, etc. The amount of radiant heat lost or received by the human body is the sum of all the radiant fluxes exchanged with the surrounding sources, and is called the radiant heat flux.

- Calculation of mean radiant temperature (MRT) is as follows:

$$\bar{t}_{\mathrm{r}} = \frac{A_1 t_1 + A_2 t_2 + A_3 t_3}{A_1 + A_2 + A_3}$$

where A_i is area of surface (m^2); t_i is temperature of surface (K).

2) Relative humidity

The relative humidity of air is the ratio of the actual moisture content of air to the water vapor content of saturated air at a given temperature and pressure. Air humidity directly affects the evaporation of heat from the human skin and the diffusion of water through the skin, which in turn affects the body's metabolic balance, and thus affects its temperature and thermal sensation. At low humidity, extreme dryness of the skin can lead to damage, roughness and maladjustment; at high humidity, the skin becomes more hydrated and shuts down sweat glands, reducing perspiration and affecting the human sense of touch on fabrics.

3) Air velocity

Air velocity affects the convection and evaporation heat transfer to the human body, thus affecting thermal comfort. Air velocity affects two sensory elements of the thermal environment that are closely related to human comfort—skin surface wetness and skin surface temperature. When the air flow rate increases, it not only increases the heat transfer coefficient between the skin and the environment, but also accelerates the evaporation of sweat, which rapidly reduces the temperature and wetness of the skin surface.Therefore in summer when sweating, blowing on a fan will make you feel cooler.

4) Air temperature

The air temperature in a room is determined by the heat balance consisting of heat gain and heat loss in the room, the temperature of the inner surfaces of the envelope and ventilation, which also directly determines the heat balance between the human body and the surrounding environment. Air temperature and average radiation temperature influence the dry heat exchange of the human body through convection and radiation. Under conditions of constant water vapor pressure and air velocity, the body's response to an increase in ambient temperature is mainly in the form of an increase in skin temperature and an increase in the rate of perspiration. Changes in ambient temperature alter the subjective thermal sensation. Air temperature is thus the main factor influencing thermal comfort, which directly affects the body's apparent heat exchange through convection and radiation.

2. Human Factors

1) Metabolic rate

The body can produce heat through metabolism to maintain a certain body temperature and therefore the metabolic rate is one of the most important factors influencing the body's thermal comfort. The metabolism of the human body is influenced by many factors such as age, gender, health status, activity level and the surrounding thermal environment.

2) Clothing insulation

The thermal resistance of clothing is an important factor affecting the body's thermal sensation, not only in terms of insulation and protection from the cold, but also in terms of radiation and convection heat exchange between the body and its surroundings.

3.2.2.7 Operative Temperature

Operative temperature is the uniform temperature of an enclosure in which an occupant exchanges the same amount of heat by radiation plus convection as in the existing non-uniform environment. (defined in ISO 7726)

Operative temperature depends on a variety of natural and artificial environmental conditions. These factors include the weather, humidity, wind, insolation, altitude, and location of equipment use. Artificial conditions include enclosure, load, equipment in close proximity, whether equipment is used indoors or outdoors, additional cooling options, and many other factors.

$$\theta_{op} = \frac{\theta_{ai}\sqrt{10v + \theta_r}}{1+\sqrt{10v}}$$

where θ_r is mean radiant temperature (℃); θ_{ai} is indoor air temperature (℃); θ_{op} is operative temperature (℃); v is indoor air velocity (m/s).

When v<0.2 m/s,

$$\theta_{op} = \frac{\theta_{ai} + \theta_r}{2}$$

3.2.2.8 Problem of PMV

As for the circumstances and conditions of application, the PMV equation is applicable to the evaluation of human thermal comfort in steady-state thermal environments, but not in dynamic, or, in other words, transitional thermal environments. This is because the experimental data for the introduction of PMV by Professor Fanger were obtained in artificial climate chambers where the indoor parameters were strictly controlled by an air-conditioning system. Therefore, the PMV experimental regression equation is only applicable to thermal environments where the indoor parameters are stable and uniformly distributed around the human body.

In general PMV indicators are predicted with high accuracy for air-conditioned buildings close to the thermal comfort environment (Figure 3-19), while different degrees of deviation exist for buildings with natural ventilation (Figure 3-20).

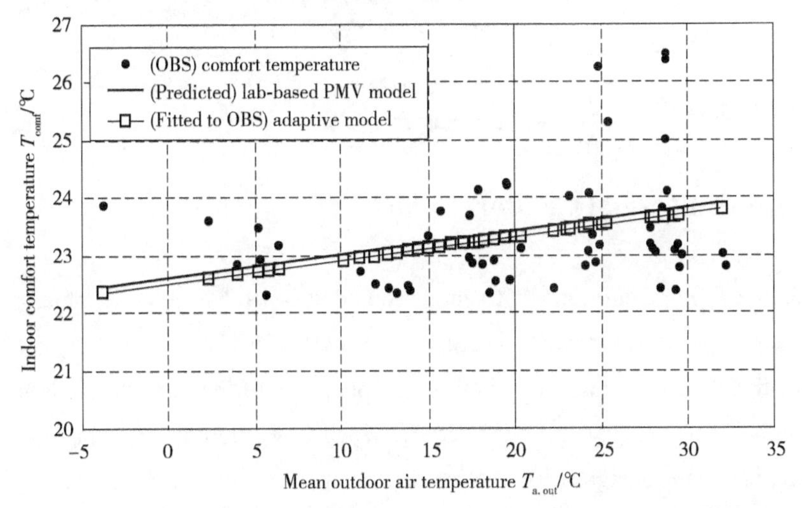

Figure 3-19 Indoor comfort temperature of air-conditioned buildings

(http://dx.doi.org/10.1016/S0378-7788(02)00005-1)

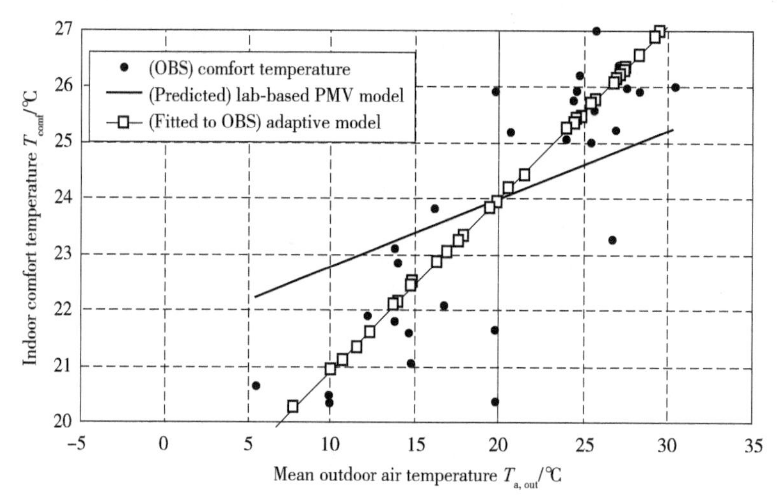

Figure 3-20 Indoor comfort temperature of naturally ventilated buildings

(http://dx.doi.org/10.1016/S0378-7788(02)00005-1)

3.2.2.9 Three Modes of Adaptation

Adaptation is the physical or behavioral characteristic of an organism that helps an organism to survive better in the surrounding environment (Figure 3-21).

• Behavioral adjustment adaptation: including all adaptive actions a person will take when he/she feels uncomfortable.

• Physiological adaptation: including all physiological responses from people to adapt to the thermal environment by either heritage or acclimatization.

• Psychological adaptation: altering occupants'perception to the indoor thermal environment based on habituation and expectation.

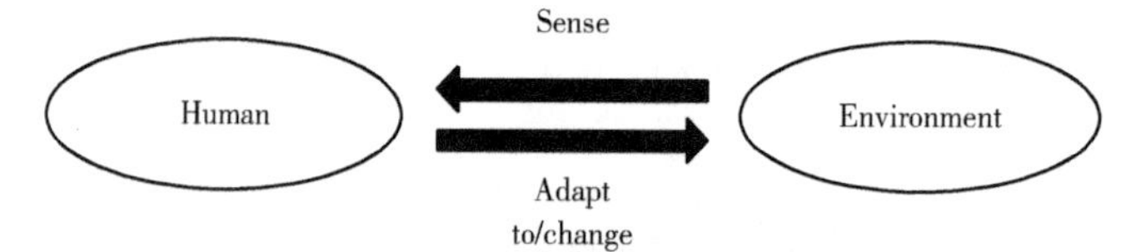

Figure 3-21 Human and environment

3.2.3 Adaptive Thermal Comfort Model

The adaptive thermal comfort (ATC) model was developed based on field measured data in real buildings, not by laboratory work where no adaptive actions can be taken (Fanger's PMV).

The adaptive thermal comfort model links people's comfort temperature indoors with temperature outdoors in naturally ventilated buildings when they are in free-running mode, which is a model relating indoor design temperatures or ranges of acceptable temperatures to outdoor weather or climatological parameters.

The adaptive thermal comfort model has been adopted by major building design standards, such as CIBSE Guide A, BS EN 15251, ASHRAE 55, except ISO 7730.

The original papers on the adaptive model were published by Humphreys and Nicol in the 1970 s when they were working for the UK Building Research Establishment. Their statistical analyses of comfort questionnaire data from building occupants are what we now recognize as "adaptive models" . They described a strong relationship of "comfort temperature" (a. k. a. "neutrality") inside a building, on the mean temperatures prevailing inside the building at the time of the survey. For naturally ventilated buildings and buildings operating in "free-running" mode the indoor comfort temperature was also noted to strongly correlate with the mean monthly temperature outdoors at the time of the survey. This concept suggests that people are able to adapt to the wider range of thermal conditions than what was generally considered before. For instance, human can tolerate higher temperature even feel more comfortable when they are under hotter environment. In addition, ASHRAE 55-2010 adapts mean monthly outdoor air temperature to evaluate the indoor comfortable operative temperature.

However, ASHRAE 55 is not always as the same as before. In the latest ASHARE 55-2013, it replaces the mean monthly temperature with prevailing mean air temperature. On the other hand, for the EN 15251, it adapts the running mean outdoor air temperature to predict indoor comfortable operative temperature.

$$\theta_{\mathrm{com}} = a\theta_{\mathrm{rm}} + b$$

where θ_{com} is comfort temperature; θ_{rm} is running mean outdoor air temperature on day n; a and b are two constants in the correlation.

3.2.4 Local Thermal Discomfort Models

Background: Both Fanger's PMV/PPD model and the adaptive model express warm and cold discomfort for the body as a whole. However, thermal dissatisfaction can also be caused by unwanted cooling or heating of local part of the human body. This is defined as local discomfort.

Officially defined local discomfort drivers in ISO 7730 are listed below.

(1) Draught: Thermal discomfort caused by draught usually happens when there is air movement when the body feels cold.

$$\mathrm{DR} = \left(34 - t_{\mathrm{a,l}}\right)\left(\overline{v}_{\mathrm{a,l}} - 0.05\right)^{0.62}\left(0.37\overline{v}_{\mathrm{a,l}} \cdot \mathrm{Tu} + 3.14\right)$$

where $t_{\mathrm{a,l}}$ is local air temperature, in degrees Celsius, 20 ℃ to 26 ℃ ; $\overline{v}_{\mathrm{a,l}}$ is local mean air velocity, in meters per second, <0.5 m/s; Tu is local turbulence intensity, in percent, 10% to 60% (if unknown, 40% may be used).

For $\overline{v}_{a,l} < 0.05$ m/s, use $\overline{v}_{a,l} = 0.05$ m/s;

For DR < 100%, use DR = 100%.

(2) Vertical air temperature difference ($\Delta t_{a,v}$): A high air temperature difference between head and ankles can cause discomfort (Figure 3-22).

$$PD = \frac{100}{1 + \exp\left(5.76 - 0.856\Delta t_{a,v}\right)}$$

where PD is percentage of dissatisfication (%); $\Delta t_{a,v} < 8$ ℃.

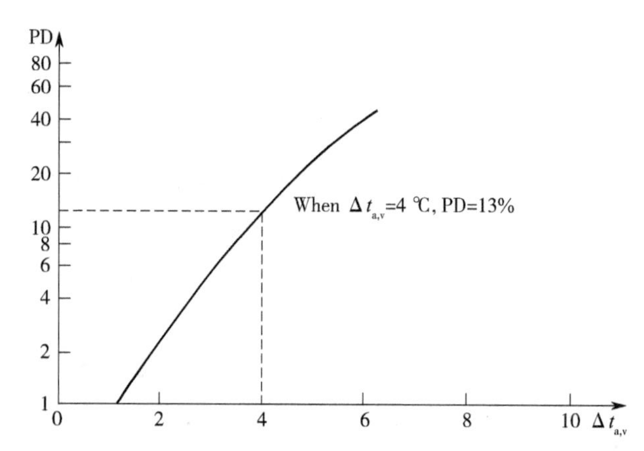

Figure 3-22 Vertical air temperature difference (ISO 7730)

(3) Too warm or too cool a floor: If the floor is too warm or too cool, the occupants will feel uncomfortable due to thermal sensation of their feet (light floor covering).

$$PD = 100 - 94\exp\left(-1\,387 + 0.118t_f - 0.002\,5t_f^2\right)$$

where t_f is floor temperature.

(4) Too high a radiant temperature asymmetry.

Radiant asymmetry (Δt_{pr}): The radiation heat gain/loss from different sides of the body is not similar (Figure 3-23).

Warm ceiling:

$$PD = \frac{100}{1 + \exp\left(2.84 - 0.174\Delta t_{pr}\right)} - 5.5 \qquad (\Delta t_{pr} < 23\ ℃)$$

Cool wall:

$$PD = \frac{100}{1 + \exp\left(6.61 - 0.345\Delta t_{pr}\right)} \qquad (\Delta t_{pr} < 15\ ℃)$$

Cool ceiling:

$$PD = \frac{100}{1 + \exp\left(9.93 - 0.50\Delta t_{pr}\right)} \qquad (\Delta t_{pr} < 15\ ℃)$$

Warm wall:

$$PD = \frac{100}{1 + \exp\left(3.72 - 0.052\Delta t_{pr}\right)} - 3.5 \qquad (\Delta t_{pr} < 35\ ℃)$$

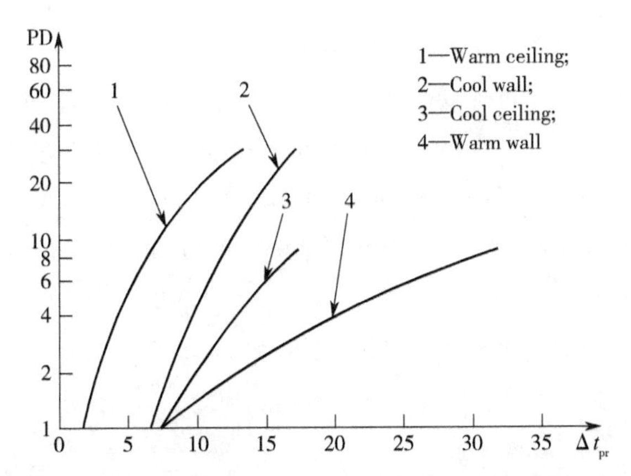

Figure 3-23 The radiation heat gain/loss from different sides of the body (ISO 7730)

3.3 Thermal Comfort Evaluation Methods

3.3.1 Important Standards

3.3.1.1 ISO 7730 (Theory)

This international standard covering the evaluation of moderate thermal environments was developed in parallel with the revised ASHRAE 55 and is one of a series of ISO documents specifying methods for the measurement and evaluation of the moderate and extreme thermal environments to which human beings are exposed (Figure 3-24).

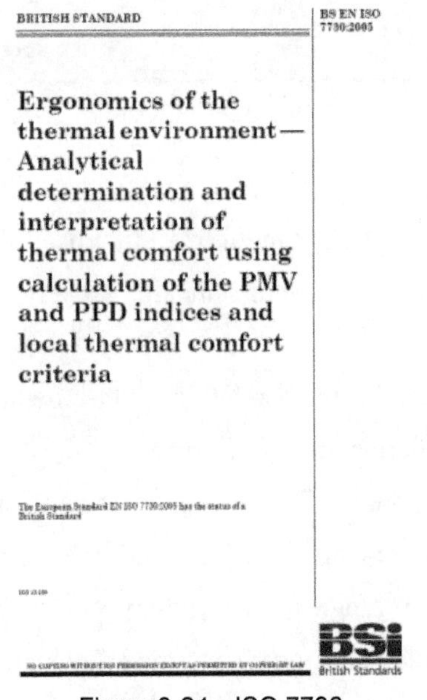
BRITISH STANDARD

BS EN ISO 7730:2005

Ergonomics of the thermal environment— Analytical determination and interpretation of thermal comfort using calculation of the PMV and PPD indices and local thermal comfort criteria

BSi

British Standards

Figure 3-24 ISO 7730

The PMV-PPD indicator is used to describe and evaluate the thermal environment in the ISO 7730 standard. This indicator takes into account six factors such as the level of human activity, the thermal resistance of clothing, air temperature, air humidity, average radiation temperature and air flow rate, and is based on a subjective sensory test to determine the level of warmth and cold sensation for the majority of people, conditional on satisfying the equations for human thermal balance.

3.3.1.2 ISO 7726 (How to Carry Out Objective Measurement)

This document is one of a series of international standards intended for using in the study of thermal environments (Figure 3-25).

BRITISH STANDARD

BS EN ISO 7726:2001

Ergonomics of the thermal environment— Instruments for measuring physical quantities

BSi

Figure 3-25 ISO 7726

This series of international standards deals in particular with:

• The finalization of definitions for the terms to be used in the methods of measurement, testing or interpretation, taking into account standards already in existence or in the process of being drafted;

• The laying down of specifications relating to the methods for measuring the physical quantities which characterize thermal environments;

• The selection of one or more methods for interpreting the parameters;

• The specification of recommended values or limits of exposure for the thermal environments coming within the comfort range and for extreme environments (both hot and cold);

• The specification of methods for measuring the efficiency of devices or processes for personal or collective protection from heat or cold.

3.3.1.3 ISO 10551 (How to Evaluate Thermal Comfort Subjectively)

The present international standard forms part of a series of standards on the assessment of thermal stress and strain in the work environment (Figure 3-26).

This series is concerned in particular with:

• Establishing specifications on methods for measuring and estimating the characteristic physical parameters of climatic environments, thermal properties of clothing and metabolic heat production;

• Establishing methods for assessing thermal stress in hot, cold and temperate environments.

This international standard proposes a set of specifications on direct expert assessment of thermal comfort/discomfort expressed by persons subjected to various degrees of thermal stress during periods spent in various climatic conditions at their workplace. The data provided by this assessment will most probably be used to supplement physical and physiological methods of assessing thermal loads. The methods belong to a psychological approach consisting of gathering, as appropriate, the on-site opinions of persons exposed to the conditions under consideration(diagnosis) and thus may complete data provided by predictive approaches described elsewhere in this series.

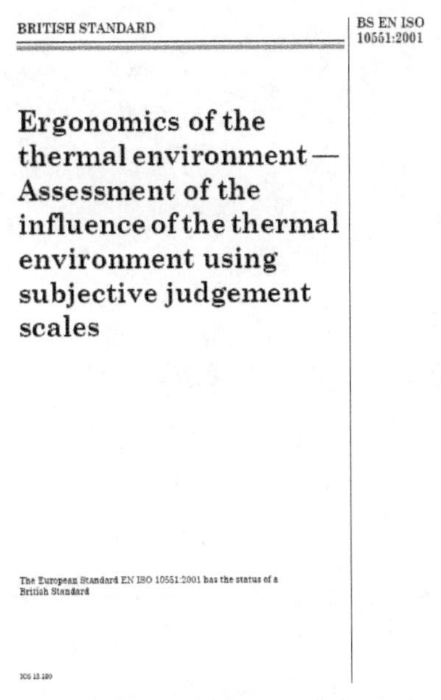
BRITISH STANDARD

BS EN ISO 10551:2001

Ergonomics of the thermal environment — Assessment of the influence of the thermal environment using subjective judgement scales

The European Standard EN ISO 10551:2001 has the status of a British Standard

ICS 13.180

Figure 3-26 ISO 10551

3.3.1.4 ASHRAE (USA Design Criteria)

For buildings, the aim is to create a reasonably comfortable built environment in which to work and live, while achieving energy saving and emission reduction. There are currently many codes and standards that go into great depth in this area. There are many factors that affect thermal comfort and many criteria for judging it, such as the PMV-PPD index, which we are most familiar with.

Metabolic rate and clothing thermal resistance are both factors that show the properties of the human body, and people with different properties have different perceptions of the same building environment, so metabolic rate and clothing thermal resistance determine the level of thermal comfort to a certain extent.

Other factors such as temperature, wind speed and humidity are all external conditions imposed by the built environment on thermal comfort. In ASHRAE 55-2017 (Figure 3-27) ,there are three main methods for determining thermal comfort:

• Graphic Comfort Zone Method;

• Analytical Comfort Zone Method;

• Elevated Air Speed Comfort Zone Method.

THE ASHRAE MEMBER'S SURVIVAL GUIDE
Design/Build

Mark Diamond, Esq.
Associate Member ASHRAE

E. Mitchell Swann, P.E.
Member ASHRAE

The information contained in this document represents the opinion of the author. It is not intended to, and does not, constitute legal advice, nor does it represent the opinion of ASHRAE or any of its bodies.

Figure 3-27 ASHRAE

3.3.2 Subjective Measurement

3.3.2.1 Scale of Perceptual Judgements on Personal Thermal State

The preference scale is used in the evaluation as it provides a judgment of "value" from subjects (Figure 3-28). If a subject rates a sensation as "slightly warm" for example, they do not indicate whether they want to be "slightly warm" or not. The preference score compares how the subject is to how they would like to be. "No change" denotes a form of acceptability, preference and satisfaction. Other scales may be useful depending on the objectives of the experiment. If a percentage of satisfaction is needed then a "forced" (the subject must choose) yes or no answer to "Are you satisfied?" would yield a direct measure. Ratings of enjoyment may be of interest.

They may be confounded by visual stimuli (for example, driving through the countryside on a sunny day) but solar radiation may elicit both pleasant and unpleasant responses and should therefore be taken into account. Acceptability ratings are useful for vehicle manufacturers. They require a sophisticated judgment based on what a subject would consider acceptable in this context. A combination of scales embedded in a questionnaire provides a useful measurement tool. The scales complement one another and provide a detailed profile of comfort. Subjective ratings of individual body parts provide insight into why subjects gave their "global" ratings.

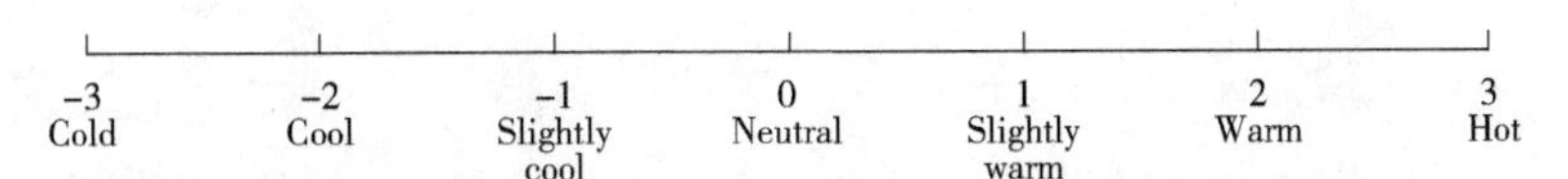

Figure 3-28 Scale of perceptual judgements on personal thermal state

3.3.2.2 Thermal Stress Assessment

Thermal stress can be defined as the heat load placed on the human body by a hot environment (Table 3-4). High temperatures can affect thermal comfort, productivity and physical and mental health, and can even lead to death. The risk of death from high temperatures increases significantly in the elderly and in people with heat-discharge deficits.

Table 3-4 Thermal stress assessment (ISO 10551)

No.	1	2	3	4	5
Type of judgement	Perceptivity	Affective evaluation	Thermal preference	Personal acceptability	Personal tolerance
Subject under judgement	Personal thermal state			Thermal ambience	
Wording	"How do you feel (at this precise moment)?" 7 or 9 degrees, from very (or extremely) COLD to very (or extremely) HOT	"Do you find it...?" 4 or 5 degrees, from COMFORTABLE to very (or extremely) UNCOMFORTABLE	"Please state how you would prefer to be now?" 7 (or 3) degrees, from (much) COLDER to (much) WARMER	"How do you judge this environment (local climate) on a personal level?" 2 degrees, GENERALLY ACCEPTABLE, GENERALLY UNACCEPTABLE	"Is it...?" 5 degrees, from perfectly TOLERABLE to INTOLERABLE

The following table summarizes the various judgements which are recommended for an assessment of comfort or stress based on subjective data.

3.3.3 Objective Measurement

3.3.3.1 Proper Measuring Equipment

The indoor of the residence, due to the combined effect of the envelope such as roof, floor, windows, doors and walls, as well as the artificial air conditioning equipment in the room, creates a microclimate indoors different from that outdoors, and each climate element can be measured by selecting proper measuring equipment (Figure 3-29).

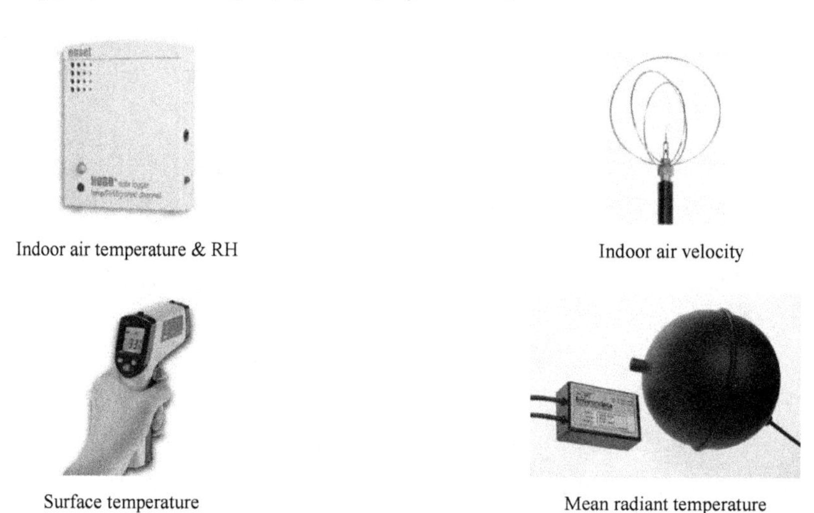

Indoor air temperature & RH　　Indoor air velocity

Surface temperature　　Mean radiant temperature

Figure 3-29 Measuring equipment

3.3.3.2 Considerations of Equipment and Measurement

- Main equipment specifications need to consider accuracy, response time, measurement range and resolution.
- For most cases, calibration is needed before the experiment.
- Three types of sensors for indoor air temperature: thermocouple, thermistor and PT100.
- Two types of sensors for indoor air velocity: omni-directional and uni-directional.
- Temperature and RH measurement should avoid direct sunshine and local heating/cooling resources.
- Weather station should be put at least 3 m higher than the roof level.

3.3.3.3 Specifications Relating to Measuring Methods

- Measuring heights and weighting (Table 3-5).

Table 3-5 Measuring heights for the physical quantities of an environment (ISO 7726)

Locations of the sensors	Weighting coefficients for measurements for calculation mean value				Recommended heights (for guidance only)	
	Homogeneous environment		Heterogeneous environment		Sitting	Standing
	Class C	Class C	Class C	Class S		
Head level			1	1	1.1 m	1.7 m
Abdomen level	1	1	1	2	0.6 m	1.1 m
Ankle level			1	1	0.1 m	0.1 m

Note: Class C—Comfort ; Class S—Thermal stress.

- Measurement of metabolic rate (Table 3-6).

Table 3-6 Metabolic rates (ISO 7726)

Activity	Metabolic rate	
	W/m²	met
Reclining	46	0.8
Seated, relaxed	58	1.0
Sedentary activity (office, dwelling, school, laboratory)	70	1.2
Standing, light activity (shopping, laboratory, light industry)	93	1.6
Standing, medium activity (shop assistant, domestic work, machine work)	116	2.0
Walking on level ground:		
2 km/h	110	1.9
3 km/h	140	2.4
4 km/h	165	2.8
5 km/h	200	3.4

- Measurement of clothing insulation (Table 3-7, Table 3-8).

Table 3-7 Thermal insulation for typical combinations of garments (ISO 7726)

Work clothing	I_{cl}		Daily wear clothing	I_{cl}	
	clo	m² · K/W		clo	m² · K/W
Underpants, boiler suit, socks, shoes	0.70	0.110	Panties, T-shirt, shorts, light socks, sandals	0.30	0.050
Underpants, shirt, boiler suit, socks, shoes	0.80	0.125	Underpants, shirt with short sleeves, light trousers, light socks, shoes	0.50	0.080
Underpants, shirt, trousers, smock, socks, shoes	0.90	0.140	Panties, petticoat, stockings, dress, shoes	0.70	0.105
Underwear with short sleeves and legs, shirt, trousers, jacket, socks, shoes	1.00	0.155	Underwear, shirt, trousers, socks, shoes	0.70	0.110
Underwear with long legs and sleeves, thermo-jacket, socks, shoes	1.20	0.185	Panties, shirt, trousers, jacket, socks, shoes	1.00	0.155

Continued

Work clothing	I_{cl}		Daily wear clothing	I_{cl}	
	clo	m²・K/W		clo	m²・K/W
Underwear with short sleeves and legs, shirt, trousers, jacket, heavy quilted outer jacket and overalls, socks, shoes, cap, gloves	1.40	0.220	Panties, stockings, blouse, long skirt, jacket, shoes	1.10	0.170
Underwear with short sleeves and legs, shirt, trousers, jacket, heavy quilted outer jacket and overalls, socks, shoes	2.00	0.310	Underwear with long sleeves and legs, shirt, trousers, V-neck sweater, jacket, socks, shoes	1.30	0.200
Underwear with long sleeves and legs, thermo-jacket and trousers, parka with heavy quilting, overalls with heave quilting, socks, shoes, cap, gloves	2.55	0.395	Underwear with short sleeves and legs, shirt, trousers, vest, jacket, coat, socks, shoes	1.50	0.230

Table 3-8 Thermal insulation for garments and changes of optimum operative temperature (ISO 7730)

Garment	I_{cl}		Change of optimum operative temperature/℃
	clo	m²・K/W	
Underwear			
Panties	0.03	0.005	0.2
Underpants with long legs	0.10	0.016	0.6
Singlet	0.04	0.006	0.3
T-shirt	0.09	0.014	0.6
Shirt with long sleeves	0.12	0.019	0.8
Panties and bra	0.03	0.005	0.2
Shirts/Blouses			
Short sleeves	0.15	0.023	0.9
Light-weight, long sleeves	0.20	0.031	1.3
Normal, long sleeves	0.25	0.039	1.6
Flannel shirt, long sleeves	0.30	0.047	1.9
Light-weight blouse, long sleeves	0.15	0.023	0.9
Trousers			
Shorts	0.06	0.009	0.4
Light-weight	0.20	0.031	1.3
Normal	0.25	0.039	1.6
Flannel	0.28	0.043	1.7
Dresses/Skirts			
Light skirts (summer)	0.15	0.023	0.9
Heavy skirt (winter)	0.25	0.039	1.6
Light dress, short sleeves	0.20	0.031	1.3
Winter dress, long sleeves	0.40	0.062	2.5
Boiler suit	0.55	0.085	3.4
Sweaters			
Sleeveless vest	0.12	0.019	0.8
Thin sweater	0.20	0.031	1.3
Sweater	0.28	0.043	1.7
Thick sweater	0.35	0.054	2.2

Continued

Garment	I_{cl}		Change of optimum operative temperature/℃
	clo	m²·K/W	
Jackets			
Light, summer jacket	0.25	0.039	1.6
Jacket	0.35	0.054	2.2
Smock	0.30	0.047	1.9
High-insulative, fiber-pelt			
Boiler suit	0.90	0.140	5.6
Trousers	0.35	0.054	2.2
Jacket	0.40	0.062	2.5
Vest	0.20	0.031	1.3
Outdoor clothing			
Coat	0.60	0.093	3.7
Down jacket	0.55	0.085	3.4
Parka	0.70	0.109	4.3
Fiber-pelt overalls	0.55	0.085	3.4
Sundries			
Socks	0.02	0.003	0.1
Thick, ankle socks	0.05	0.008	0.3
Thick, long socks	0.10	0.016	0.6
Nylon stockings	0.03	0.005	0.2
Shoes (thin soled)	0.02	0.003	0.1
Shoes (thick soled)	0.04	0.006	0.3
Boots	0.10	0.016	0.6
Gloves	0.05	0.008	0.3

Measurement of thermal insulation of chairs in direct contact with people (Table 3-9).

Table 3-9 Contribution from chairs (ISO 7730)

Type of chair	I_{cl}	
	clo	m²·K/W
Net/metal chair	0.00	0.000
Wooden stool	0.01	0.002
Standard office chair	0.10	0.016
Executive chair	0.15	0.023

3.3.3.4 Computer Tool Used for the PMV & PPD Calculation

PMV can be calculated from different combinations of metabolic rate, clothing thermal resistance, air temperature, mean radiation temperature, wind speed and air humidity.

The PMV index is derived from steady-state conditions, but good approximations can be obtained when there are small fluctuations in one or more parameters, provided that the time-weighted average of the first 1 h of the parameters is used.

The PMV index can be used to check whether a given thermal environment meets comfort guidelines and to establish environmental requirements for different levels of acceptability. By setting PMV = 0, an equation can be established that predicts the combination of activity, clothing and environmental parameters that provide generally moderate thermal sensations (Figure 3-30).

The accuracy of the tables is better than 0.1 PMV, provided the difference between air and mean radiant temperature is less than 5 ℃ . The tables apply for a relative air humidity of 50% (Figure 3-31).

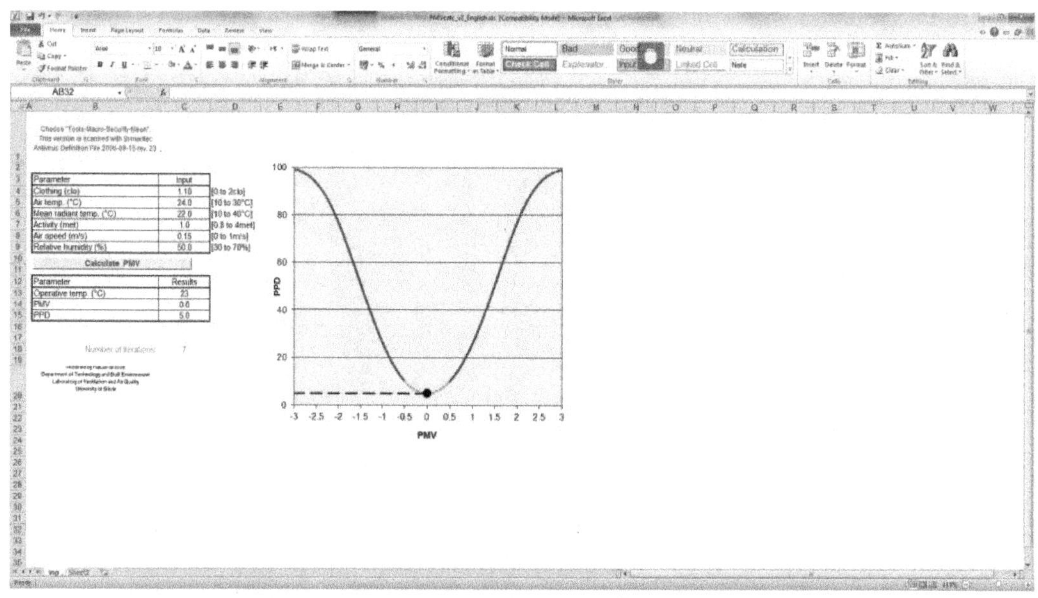

Figure 3-30 Computer tool used for the PMV & PPD calculation

(Downloadable Excel sheet, from www.lumasense.dk. (on blackboard, PMVcalc_V2_English. xls))

(t_a & t_{mr})

Table E.2 — Activity level: 58 W/m² (1 met) — (RH=50%)

Clothing		Operative temperature	Relative air velocity							
			m/s							
clo	m² · K/W	℃	<0.10	0.10	0.15	0.20	0.30	0.40	0.50	1.00
0	0	26	-1.62	-1.62	-1.96	-2.34				
		27	-1.00	-1.00	-1.36	-1.69				
		28	-0.39	-0.42	-0.76	-1.05				
		29	0.21	0.13	-0.15	-0.39				
		30	0.80	0.68	0.45	0.26				
		31	1.39	1.25	1.08	0.94				
		32	1.96	1.83	1.71	1.61				
		33	2.50	2.41	2.34	2.29				
0.25	0.039	24	-1.52	-1.52	-1.80	-2.06	-2.47			
		25	-1.05	-1.05	-1.33	-1.57	-1.94	2.24	-2.48	
		26	-0.58	-0.61	-0.87	-1.08	-1.41	-1.67	-1.89	-2.66
		27	-0.12	-0.17	-0.40	-0.58	-0.87	-1.10	-1.29	-1.97
		28	0.34	0.27	0.07	-0.09	-0.34	-0.53	-0.70	-1.28
		29	0.80	0.71	0.54	0.41	0.20	0.04	-0.10	-0.58
		30	1.25	1.15	1.02	0.91	0.74	0.61	0.50	0.11
		31	1.71	1.61	1.51	1.43	1.30	1.20	1.12	0.83

PMV

Figure 3-31 Determination of PMV/PPD

(ISO 7730: Ergonomics of the thermal environment—Analytical determination and interpretation of thermal comfort using calculation of the PMV and PPD indices and local thermal comfort criteria)

Chapter 4

Indoor Air Quality

4.1 Indoor Air Pollution

4.1.1 Sick Building Syndrome

4.1.1.1 Definition of Sick Building Syndrome

Sick building syndrome (SBS) describes a range of symptoms thought to be linked to spending time in a certain building, most often a workplace, but no specific cause can be found (defined by NHS). This is a syndrome caused by prolonged exposure to dust, water vapor, gaseous fungi, bacteria and other air pollutants that accumulate in buildings. Other causes of SBS include inadequate lighting, uncomfortable thermal conditions and excessive noise. It often occurs in people working in offices, or in buildings with a high concentration of people. It also occurs most often in new, enclosed buildings where windows are not opened to reduce heat dissipation. Generally, when patients leave these buildings, their syndromes disappear.

4.1.1.2 Symptoms of Sick Building Syndrome

Many people in sick buildings complain of feeling uncomfortable living in indoor environments, which is mainly manifested in the following aspects (Figure 4-1).

- Mucous-membrane irritation: eye irritation, throat irritation, cough.
- Neurotoxic effects: headaches, fatigue, lack of concentration.
- Respiratory symptoms: shortness of breath, cough, wheeze.
- Skin symptoms: rash, pruritus, dryness.
- Chemosensory changes: enhanced or abnormal odor perception.
- Visual disturbances.

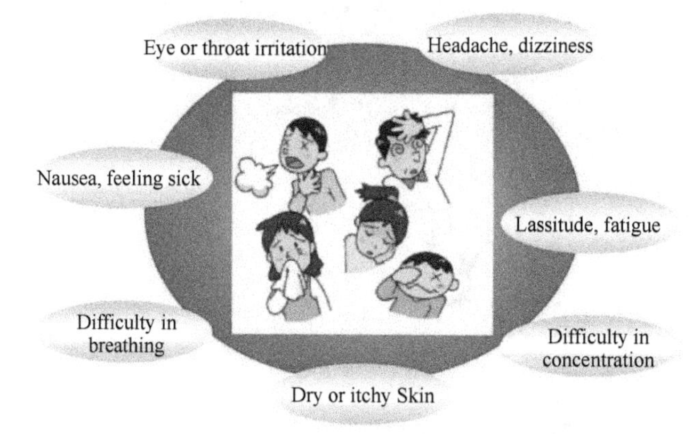

Figure 4-1 Symptoms of sick building syndrome

(http: //www.socalindustrialrealestateblog.com/how-green-building-combats-sick-building-syndrome/)

4.1.1.3 Causes of Sick Building Syndrome

1. Personal Factors

(1) Sedentary: Sitting still for a long time and slowing down blood circulation will lead to insufficient blood supply to the brain, injuring the mind and damaging the brain, producing mental aversion. Prolonged exposure to this state of stress and fatigue can therefore exacerbate pathological symptoms.

(2) Poor eating habits: For people who are excessively partial in their diet, the chances of suffering from poor building syndrome are greatly increased. These people do not consume enough nutrients to meet the normal activities of the body, resulting in a weakened body that is unable to resist external aggressive factors. If they are in a bad environment, they are vulnerable to the effects of the environment and may suffer from poor construction syndrome.

(3) Stress and repetitive work: If a person is under a high pressure and repetitive state for a long time, and often without timely adjustment, it will lead to anxiety, depression and other symptoms over time, or in severe cases, psychological disorders or mental illness. From a physiological point of view, if a person is always highly stressed, it will lead to endocrine dysfunction and a decrease in immunity, which will easily lead to sick building syndrome.

2. Physical/Environmental Factors

(1) Indoor air pollution: It is a major cause of SBS. With the extensive use of ventilation systems such as air conditioning, buildings are becoming more and more enclosed. Combined with the use of various chemical-based upholstery materials, various types of volatile organic compounds are entering the indoor spaces of buildings that do not have a high degree of circulation. These volatile gaseous organic compounds can cause chemical toxins that cause neurological reactions at high concentrations, and residents' health is easily compromised when they live in such an environment for a long time.

(2) Noise: Long-term exposure to high noise levels can cause people to suffer from hearing fatigue, fatigue, anxiety and irritability. Noise can disrupt the nervous system, accelerate the ageing of the heart and even lead directly to the development of certain diseases. In the industrial sector, strong noise can lead to acoustic fatigue in machines, equipment and certain industrial structures, the long-term effects of which will shorten their service life and even lead to production accidents.

(3) Lighting: People who stay indoors for long periods of time and rarely come into contact with light commonly suffer from symptoms such as physical weakness, slow reactions and reduced bone quality. In addition, due to the effect of light hours on hormone production levels, people are also prone to depression in winter, showing symptoms such as depression, fatigue, slow brain response and lethargy. And if the working environment is too bright, it can be

irritating to the eyes.

(4) Overheating: Indoor thermal pollution comes from two main sources. The first is caused by radiant energy from the outdoors entering the room, for example in hot areas with high summer temperatures, mainly from radiation from the inner surfaces of walls and roofs, and especially from solar heat entering through windows. The second is the size and direction of radiation from indoor objects, which has a great impact on the quality of the thermal environment. Workshops such as smelting and hot rolling have strong indoor sources of radiant heat, resulting in a high-temperature environment. When the human body is not sensitive to the heat radiation sensed by the environment, it can cause uncomfortable symptoms such as heat stroke or cold.

(5) Lack of greenery in the building: The green landscape has been proven to alleviate visual fatigue and mental stress to a certain extent. A large number of modern architectural designs have completely isolated people from nature. People are in a building space surrounded by all kinds of modern materials, without access to the natural green landscape, which can bring huge psychological problems to people who are in it for a long time, such as frequent boredom, lack of concentration, pessimism, etc. At the same time, the lack of activity space in office buildings leaves people who work long hours without the right amount of physical activity.

3. Design Factors

(1) Low floor/ceiling height: If the height of the floor is relatively low, combined with the space occupied by the ceiling, it will certainly look depressing. When the air circulation in the room is affected, it can also make the occupants of the room feel uncomfortable and affect their daily life and mood.

(2) Open plan: Due to the insufficient ventilation in many buildings, the impact of pollution sources is expanded, especially in office buildings where central air conditioning is installed, in order to save money, when the central air conditioning stops running, the ventilation system is not turned on, resulting in poor indoor air quality in office buildings. Fresh air rate is an important factor affecting the indoor air quality of office buildings. The ozone emitted by photocopiers in offices, the carbon dioxide exhaled by people, the inhalable particles from cigarette smoke, the microorganisms and bacteria adsorbed by people and carpets, and even the formaldehyde and other pollutants emitted by office furniture, all need to be diluted by ventilation. If people inhale too much, they will suffer from chest congestion and coughing.

(3) Artificial lighting: In our daily lives, we are exposed to computers, TVs and various light environments, and if we do not use artificial lighting scientifically, we can easily endanger our mental and physical health. Low illumination or dazzle can make people feel unhappy and easily fatigued, which can also affect work efficiency. In severe cases, dazzle can not only cause

loss or reduction in the ability to see objects, but can also cause discomfort such as tearing, pain, eyelid cramps and even optic neuritis. Poor lighting, especially low and uneven illumination, can increase the incidence of myopia and even cause accidents due to operational errors.

4. Organizational Factors

Poor maintenance and management: Healthy indoor air quality must be created jointly by architects, interior designers, building operators and users. The attitude of building owners and business operators is particularly important because it is difficult for employees to make specific improvements at the individual level. Healthy indoor environment depends on the owner's willingness to pay for environmental inspections. In fact, all buildings have problems with air quality and temperature, especially during the seasonal change of the year, which is a challenge for such a large, sealed building. Usually, environmental protection personnel can nip in the bud according to indoor gas quality survey, inspection around the office, search for possible pollution sources and assessment, and on-site sampling steps, so as to prevent the raging of building syndrome (Figure 4-2).

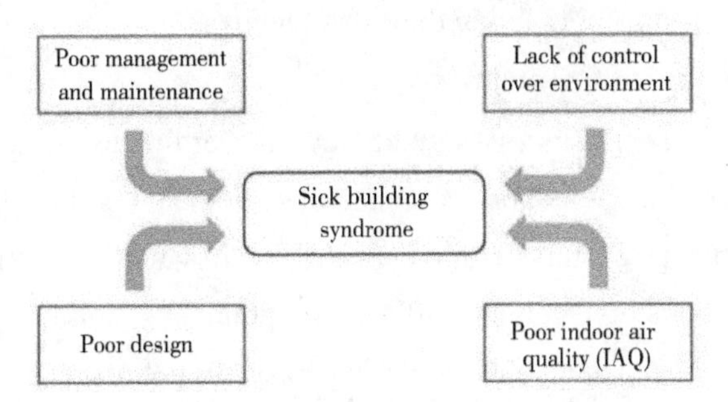

Figure 4-2 Causes of sick building syndrome

4.1.1.4 Precautionary Safety Measure

(1) From a human point of view, people are urged to take regular physical exercise, adjust their work rhythm, coupled with proper rest in the midst of their busy schedule, and it is essential to find ways to vent or transfer psychological stresses, such as traveling and talking. In addition, people can also take health food and complement with vitamins and minerals.

(2) From an environmental point of view, it is urgent to improve the air quality of the office environment by banning smoking, doing a good job of ventilation and greenery, opening windows as much as possible to allow fresh air to circulate and introducing houseplants that help reduce carbon dioxide and remove toxins from the atmosphere. Air conditioner equipped with fresh air filtering system provides operators with a nice working condition.

(3) From a design point of view, it is essential to reduce interior pollution through technical means, advocate adornment of green environmental protection, and pay attention to the applica-

tion of material of new-style environmental protection.

(4) From a regulatory point of view, there is a need for stricter regulatory control on new buildings. Regularly arrange for environmental experts to conduct indoor air quality surveys, patrol the office perimeter, identify possible sources of pollution and take an assessment, then timely take actions to address areas that do not meet standards.

4.1.2 Classification of Pollutants

4.1.2.1 Definition of Pollutants

A pollutant can be defined as a substance that enters the environment and changes the normal composition of the environment in a way that is directly or indirectly harmful to the growth, development and reproduction of living organisms. Pollutants act on all living things, including people. Environmental pollutants are substances that enter the environment as a result of human activities and cause changes in the normal composition and properties of the environment that are directly or indirectly harmful to living things and humans.

4.1.2.2 Classification of Pollutants

Pollutants can be classified in a variety of ways. According to the source of the pollutant, pollutants can be divided into pollutants from natural sources and pollutants from anthropogenic sources; some pollutants (e.g. sulphur dioxide) have both natural and anthropogenic sources. According to the environmental elements affected by pollutants, pollutants can be divided into atmospheric pollutants, water pollutants, soil pollutants, etc. Pollutants can be divided into gaseous pollutants, liquid pollutants and solid wastes according to their form. According to the nature of pollutants pollutants can be divided into chemical pollutants, physical pollutants and biological pollutants; chemical pollutants can be divided into inorganic pollutants and organic pollutants; physical pollutants can be divided into noise, microwave radiation, radioactive pollutants, etc. ; biological pollutants can be divided into pathogens, allergens pollutants, etc. According to the change of physical and chemical properties of pollutants in the environment, they can be divided into primary and secondary pollutants. In addition, to emphasize certain harmful effects of pollutants on the human body, they can also be divided into teratogens, mutagens and carcinogens, inhalable particles and malodorous substances.

For example, pollutants can be in solid, liquid or gaseous form.

• Solids pollutants such as dusts, fumes and smokes are relatively large particles that are suspended in air when they are turbulent. Dusts are solid particles such as mineral, metal, vegetable and flour, which are smaller than 100 mm diameter.

• Liquid pollutants can be mists, fogs and smokes that are produced by tars in cigarette smoke.

• Gaseous pollutants mainly include sulfur dioxide (SO_2), carbon monoxide (CO), and nitrogen oxides (NO_x) as well as ozone (O_3). The primary source of these gases is the combustion of fossil fuels in power plants, various industrial processes, and motor vehicles and equipment.

A brief overview of common pollutants in each category of pollution is given below.

4.1.2.3 Common Indoor Air Chemical Pollutants

1. Hazardous Combustion Product

The stoves used in daily household life are one of the sources of harmful gases. The gases, heat, visible smoke and so on produced by combustion are called combustion products, that is, all substances produced by combustion or thermolysis. Smoke is the main product of combustion, with water vapor and CO_2 being the main components. Water vapor and CO_2 are usually not significant for humans within certain specified limits, and the products that can cause damage to human health at low concentrations are mainly CO, NO_x and SO_x.

• CO: CO is a product of incomplete combustion. It is a colorless, odorless, strongly toxic combustible object that is difficult to dissolve in water. As carbon monoxide is toxic, it can replace oxygen from the blood's oxyhaemoglobin and combine with haemoglobin to form carbon monoxide haemoglobin, thus causing severe oxygen deprivation (Table 4-1).

Table 4-1 Effect from carbon monoxide(CO)

Concentration/ppm	Effect
50	The maximum allowable concentration for continuous exposure for healthy adults in any 8-hour period, according to OSHA
200	Slight headache, fatigue, dizziness, nausea after 2-3 hours
400	Frontal headaches within 1-2 hours, life threatening after 3 hours
800	Dizziness, nausea and convulsions within 45 minutes. Unconsciousness within 2 hours. Death within 2-3 hours
1 600	Headache, dizziness and nausea within 20 minutes. Death within 1 hour
3 200	Headache, dizziness and nausea within 5-10 minutes. Death within 25-30 minutes
6 400	Headache, dizziness and nausea within 1-2 minutes. Death within 10-15 minutes
12 800	Death within 1-3 minutes
http: //blog.julieacarda.com/2010_01_01_archive.html	
http: //www.onetemp.com.au/kimo-co110-co-sensor	

• CO_2: CO_2 is a colorless, odorless gas that is 1.5 times heavier than air under normal conditions. When CO_2 is low, it is not harmful to the human body, but when it exceeds a certain amount, it affects people's respiratory system due to the increase of carbolic acid concentration in the blood, which increases acidity and produces acidosis(Table 4-2).

Breathing is the largest source of indoor carbon dioxide in our lives. In addition to this, daily actions such as cooking, bathing and smoking produce carbon dioxide. Especially at night, when humans, animals and plants breathe frequently, the concentration of carbon dioxide in indoor air increases dramatically. For this reason, nighttime is often a high incidence of respiratory diseases, brain hemorrhage, and brain infarction.

Table 4-2 Effect from carbon dioxide(CO_2)

Concentration/ppm	Effect
350-450	Normal outdoor level
<600	Acceptable levels
600-1 000	Complaints of stiffness and odors
1 000	ASHRAE① and OSHA② standards
1 000-2 500	General drowsiness
2 500-5 000	Adverse health effects may be expected
5 000-10 000	Maximum allowed concentration within a 8-hour working period
30 000	Maximum allowed concentration within a 15-minute working period
30 000-40 000	Slightly intoxicating, breathing and pulse rate increase, nausea
50 000	Above plus headaches and sight impairment
100 000	Unconsciousness, further exposure death
Source: ① ASHRAE is the abbreviation of American Society of Heating Refrigerating and Air-conditioning Engineers; ② OSHA is the abbreviation of Occupational Safety and Health Administration	
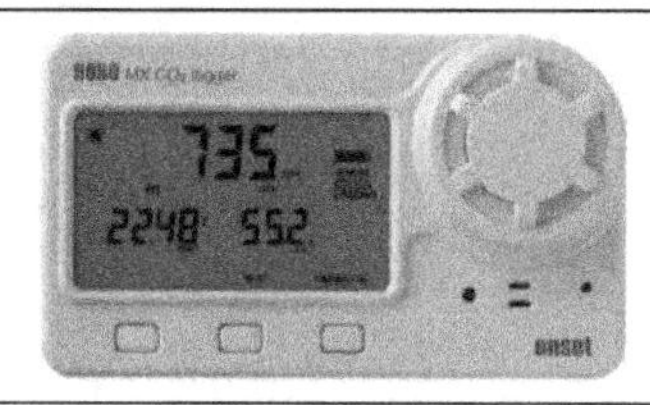	
http: //www.onsetcomp.com/products/loggers/mx1102	

• NO_x: NO_x Includes a wide range of compounds such as N_2O, NO, NO_2 and N_2O_5. Natural emission of NO_x, mainly from the decomposition of organic matter in soil and sea, belongs to the natural nitrogen cycle process. Most of the NO_x emissions from anthropogenic activities come from the burning of fossil fuels. When cooking, the level of nitrogen dioxide in the kitchen is three times higher than outdoors. This is because the combustion of gas and natural gas in the kitchen produces a high flame temperature, which generates nitrogen oxides (NO, NO_2). The higher combustion temperature, the higher concentration of nitrogen oxides.

Although the nitrogen oxides produced by cooking are not immediately fatal to the body, they can slowly invade the body and cause various illnesses. Nitrogen dioxide can make people more sensitive to allergens, such as pet dander; it can also damage skin tissue, which can become saggy, wrinkled, grey and rough over time; and it can lead to asthma or other respiratory diseases when people are constantly exposed to this environment. Nitrogen oxides can also directly reach the deep bronchi and alveoli of the respiratory tract, causing pulmonary oedema and, in severe cases, lung cancer.

• SO_x: The main sulphur oxides in life are SO_2 and SO_3, both of which are acidic gases. SO_2 comes mainly from atmospheric pollutants produced by burning coal and oil, and is easily soluble in water. It can be oxidized to SO_3 under certain conditions, after which it dissolves in rainwater and becomes acid rain. For people with long-term exposure to low concentrations of SO_2, SO_2 is inhaled into the trachea and bronchi and irritates the mucous membrane of the upper respiratory tract, causing respiratory diseases such as bronchitis, bronchial asthma and emphysema. If SO_2 is mixed with water and then in contact with the skin, frostbite may occur. When in contact with the eyes, it can cause red and swollen.

2. Volatile Organic Compounds

According to the World Health Organization, volatile organic compounds (VOCs) are various organic compounds with boiling points of 50 ℃ to 260 ℃ at normal temperature. Most volatile organic compounds have an unpleasantly distinctive odor and are poisonous, irritating, teratogenic and carcinogenic.

There are many sources of VOC pollution, and a common classification is that of the nature of the source of pollution. The main categories include the following.

• Organic solutions. Organic solutions are composed of organic matter as the medium of solvent. Common organic solutions are household cosmetics, shampoos and detergents, as well as adhesives, paints, and so on.

• Construction materials. Building materials are materials used in construction work that are prone to volatile odor, including paints, plastic meal boxes, foam insulation materials, man-made panels and other materials used indoors and outdoors in buildings.

• Interior decoration materials. Interior decoration materials refer to the interior paint of buildings or some other materials used in interior decoration that are prone to volatile odor, including wall paint, wallpaper, murals and other materials.

• Fibrous materials. Fibrous materials are materials made from natural or synthetic fibers, usually used for carpets, tapestries, chemical curtains and other items.

• Office supplies. Some office supplies are inherently volatile, such as ink. However, some supplies are not volatile, but in the working process they will emit a lot of heat, and a large number of harmful gases will be emitted along with the heat, such as photocopiers and printers.

3. Formaldehyde

Formaldehyde is a colorless gas with an irritating odor, indoor formaldehyde comes mainly from building materials, furniture, man-made panels, various adhesive paints and synthetic textiles, among others. The amount of formaldehyde emitted by burning fossil fuels is small, but smoking is an important source of formaldehyde emissions. It has been determined that each cigarette can emit about 2.4 mg of formaldehyde, and the concentration of formaldehyde in the smoke inhaled directly from cigarettes may exceed the warning concentration by more than 400 times. Formaldehyde is a strong irritant to the mucous membranes of the eyes, nose and throat, with the most common symptoms being eye irritation and headaches, and in severe cases, allergic dermatitis and asthma. Formaldehyde can inhibit cellular functions as it reacts with proteins to coagulate and denature proteins in cells.

4. Ammonia

Ammonia is a colorless and strongly irritating odorous gas that is lighter than air. Ammonia is an alkaline substance which has a corrosive and irritating effect on all skin tissues it comes into contact with, absorbing water from skin tissues, denaturing tissue proteins, saponifying tissue fats and destroying cell membrane structures. In addition to its corrosive effects, it can cause cardiac arrest and respiratory arrest at high concentrations. Ammonia is usually inhaled as a gas into the alveoli of the lungs, where it easily enters the bloodstream and binds to haemoglobin, destroying oxygen transport functions.

Ammonia is extremely soluble and therefore has an irritating and corrosive effect mainly on the upper respiratory tract of animals or humans, weakening the body's resistance to disease. It mainly comes from concrete admixtures used in building construction, such as concrete antifreezer, which contain large amounts of ammonia. The volatilization of ammonia in concrete walls will lead to an increase in the concentration of ammonia in the indoor air. Ammonia in indoor air can also come from interior decoration materials, for example, most of the additives and whitening agents used in furniture finishing use ammonia, which has become a necessary commodity in the building materials market.

4.1.2.4 Common Physical Contaminants

1. Particulate Matter

Solid or liquid particles suspended in the air that can be hazardous to biological and human health. There are more than 130 kinds of harmful substances, such as carbon black, asbestos, silica, iron, aluminium, cadmium and arsenic. In addition to the usual dust, more than 50 kinds are often detectable indoors. The particles are generally physical pollution; sometimes the particles are involved in chemical reactions or adsorb harmful chemicals, which can also cause chemical pollution.

Long-term exposure to airborne pollution particles increases the risk of lung cancer, and short-term increases in concentrations of these particles or other air pollutants also increase the risk of heart disease. European epidemiologists have found a clear association between lung cancer and air pollution particles in localized areas.

2. Fiber Materials

Fiber materials are also a type of indoor pollutant and are often derived from acoustic or insulation materials such as roofing, acoustic insulation and pipe linings. Common indoor polluting fibrous materials include asbestos, fiberglass and paper pulp.

Asbestos itself is not toxic, and its greatest danger comes from its dust. When this fine dust is inhaled into the body, it adheres to and deposits in the lungs, causing lung disease, and asbestos has been confirmed as a carcinogen by the International Agency for Research on Cancer. On the other hand, the extremely tiny asbestos dust that is dispersed into the air and inhaled into the human lungs can easily induce lung diseases such as lung cancer after a latent period of 20 to 40 years. This is the public health problem of asbestos, which is of varying degrees of concern in countries around the world.

If exposed to glass fiber for a long time, it may destroy the skin barrier of the human body, causing contact dermatitis, and at the same time, blisters may appear, which may lead to skin necrosis and ulcers after rupture; due to the small diameter of glass fiber, prolonged exposure may lead to the deposition of glass fiber in the mucous membrane of the eye, which may cause eye lesions for a long time, such as conjunctivitis, keratitis and other symptoms; furthermore, if the glass fibers enter the lungs through breathing and cannot be actively expelled or absorbed, they may also stimulate the lung tissue to develop fibrous inclusions and nodules. In addition, prolonged exposure to glass fibers may irritate the bronchi and other parts of the body, causing bronchitis and asthma.

Even the original fresh pulp contains a certain amount of lead and cadmium, which can easily breed microorganisms, and constitute a threat to human health.

3. Radon (Rn)

Rn is a naturally occurring colorless, odorless and tasteless inert gas which is radioactive. Radon, which is mainly derived from the natural decay of uranium 238, is one of the 19 major carcinogens listed by the World Health Organization (WHO) and is the second most important cause of lung cancer in humans after cigarettes. For most people, the majority of exposure to radon comes from the home. There are three impact factors for radon to enter the home: the rate of exchange between indoor and outdoor air, which depends on the construction of the house, the ventilation habits of the occupants and the degree of sealing of the windows.

When radon and its derivatives enter the human body through the respiratory tract, they often remain in the whole respiratory tract of the human body for a long time, which is one of the important causes of human respiratory system diseases. The danger of radon to human health is mainly manifested as follows: under the exposure of high radon concentration, the organism shows changes in blood cells. Radon has a high affinity to human fat, especially when radon is combined with the nervous system, it is more harmful. Another manifestation is the occurrence of tumors. As radon is a radioactive gas, when people inhale it into their bodies, it can induce lung cancer.

4.1.2.5 Indoor Air Microorganism Contamination

Indoor air microorganism contamination is caused by microorganisms such as bacteria, mould and virus. It is one of the indoor air quality indicators. The degree of contamination is related to the factors of surrounding environment, such as living density, indoor air temperature, humidity, dust content, lighting and ventilation. The main method to prevent microbial contamination is not only to remove the source of contamination and eliminate the growth of microorganisms, but to choose the right amount of ventilation and to keep the room dry.

The toxic effects of microbial pollution include three main aspects:

• Microorganisms multiply and compete with other organisms for nutrients or living space;

• Microorganisms infest other organisms through some means in state of being;

• Microorganisms produce and release large quantities of toxic products through their metabolic activities, thus damaging the ecological environment and killing other organisms.

Indoor microenvironments are diverse and of uneven size, with the multifarious activities of the people within them. Therefore, the spatial and temporal distribution of indoor airmicroorganisms is more complex than that of atmospheric microorganisms. Indoor air microorganisms have three features: pathogenicity, variability and controllability.

Microbial pollution is the source of major infectious diseases and thus indoor air microbial pollution is also the main cause of respiratory diseases. Air is an indispensable factor for human survival activities, but there are also a considerable number of microorganisms in the air, among which pathogenic microorganisms can spread through dust, droplets and affect human health. For example, influenza A virus and coronavirus are spread through the air.

4.1.3 Sources and Routes of Pollution

1. Pollutants from Indoor Things

• Building material including sealants, adhesives and paints: The interior decoration materials with toxic gas pollution materials accounting for 68%, containing more formaldehyde, benzene, xylene and other toxic organic substances, will evaporate harmful substances, threatening human health. When these harmful substances spread through the respiratory tract or skin into the human body, they will make people produce weakness, dizziness, nausea, vomiting, nasal congestion, and even mental trance and other symptoms.

• Cleaning materials, solvents and other consumer products: Various household cleaners, insecticides, perfumes, etc. often contain compounds derived from petroleum, which, when released into the environment, can have a huge impact on the environment. Detergents contain a variety of potential lung irritants such as bleach and ammonia, which may be inhaled into the respiratory tract by long-term use of these detergents, causing fibrosis of the lung tissue and affecting the body's immune system.

• Furnishings and fabrics: Fabrics have a high absorption capacity, not only for harmful gases such as formaldehyde and other free states released by pollutants, but also for pathogens, microorganisms and dust. Pure wool carpets, for example, are ideal for dust mites to breed and hide in.

• Furniture: Newly purchased furniture generally contains special odors such as formaldehyde, ammonia, benzene, TVOC and other volatile organic gases, which, when absorbed by the human body, can cause dizziness, nausea and other uncomfortable conditions.

• Devices such as photocopiers, printers, and document binders: They produce an electrostatic effect in their working state, which excites the oxygen in the air into ozone. Ozone has an irritating odor and can cause harm to the human body when it reaches a certain concentration. For example, it irritates the respiratory tract, causing respiratory stress symptoms such as coughing and bronchitis, and in severe cases it may even cause cancer. Apart from radiation, the biggest hazards of these devices are dust particles and gas pollution. When working, photocopiers emit a dust that is invisible to the naked eye. This dust contains a large amount of toner and iron powder, which can damage the lungs when inhaled in large quantities over a long period of time. Household appliances produce electromagnetic radiation, which can also cause dizziness, salivation, weakness and memory loss if the radiation intensity is high.

• Gas cookers, heaters, and other unflued fuel-burning appliances: Gas stoves and gas water heaters, which are mainly burned, can easily burn inadequately and produce carbon monoxide due to their varying quality. And as the combustion time continues, more harmful substances will be dispersed into the air, polluting the indoor air and causing serious health risks. For example, frying at high temperatures is a unique culinary custom in China, and cooking oil and food undergo cracking at high temperatures, producing large amounts of harmful fumes. When the temperature of the oil in the pan rises above 150 ℃, the oil undergoes complex chemical changes, generating substances such as low-grade aldehydes, ketones and carboxylic acids, forming fumes. When the temperature of the oil rises above 200 ℃, a mixture of peroxides and hydroperoxides are produced as a result of the catalytic effect of trace metal elements in the food, with a high content of hydroperoxides, a harmful substance that causes cells to senescence.

• Glues: Most of the adhesives also contain harmful substances such as formaldehyde, which will gradually evaporate over a long period of time, thus causing pollution to the indoor air and causing some harm to humans.

• House dust mites: Dust mites prefer dark, warm, humid places where sweat, dander and hair accumulate. Dust mite is one of the strongest allergens found so far, which is an important factor causing indoor allergic diseases. Dust mites float in the air and enter the nose, throat and lungs through breathing, attaching themselves to the nasal mucous and bronchi of the lungs, reacting with the body's immune system and thus triggering an alarm in the body's immune system. This eventually manifests itself in the form of a runny nose, sneezing and coughing. Most likely to cause allergic rhinitis, allergic asthma and allergic dermatitis and so on.

• Moulds and bacteria: Moulds are generally in the range of 2 to 10 μm in size and bacteria are generally in the range of 0.5 to 5 μm in size. They are usually attached to dust or droplets in the air before they are inhaled by people into the respiratory tract. When the concentration of them as well as their derivatives in indoor air increases to a certain level, the incidence of some diseases tends to increase subsequently, such as cold, pneumonia, tuberculosis, SARS and other diseases.

• Pesticide products: There are moderate or low toxic substances in insecticidal aerosol, which will cause certain damage to human nervous system, respiratory system and cardiovascular system.

• Pets: In addition to the problem of cleaning pet hair all over the home, the space where people and pets live together also contains dust mites, pet dander, allergens, formaldehyde and other dust and air pollutants that are not visible to the naked eye, posing a threat to the health of people and pets.

• Tobacco smoking: Smoking burns lives, not only damaging the health of the smoker, but also causing health hazards to passive smokers around the smoker.

• Radon seeping in from the ground: As a major health risk radon is second only to tobacco. When radon or particles are inhaled into the lungs, some of them will accumulate and continue to emit radiation, which may increase the chance of lung cancer in patients.

2. Contaminants from Humans

• Contaminants from respiration: In addition to water and carbon dioxide, there are many harmful gases such as carbon monoxide, methanol, ethanol, hydrogen sulfide in the exhaled breath of human and other living organisms. It may also contain a variety of pathogenic microorganisms, especially pathogenic microorganisms emitted by the flying foam of patients with respiratory infectious diseases, which will cause indoor air pollution.

• Moisture from respiration: Breathing also increases the indoor humidity level.

• Skin particles: The skin includes hair, nails, sebaceous glands, sweat glands and other accessory organs. As the most important organ of the human body, it is also the largest source of pollution, with 271 types of waste and 151 types of sap excreted by it, including CO_2, CO, acetone, benzene, methane gas and dust. Discarded skin from the body can be unsettling in a confined space, and it is an important source of indoor dust.

• Odors: Smells from the body are disliked and have a pronounced effect on perceived air quality. In comparison, human odors are stronger in densely populated areas and crowded environments, and weaker in open areas or where there is better air circulation. Clearly, keeping the air clean and fresh will be beneficial to our human health.

• Heat: Bodies in a room add to the heating effect. At the same time and in the same place, a room with a lot of people will be relatively warmer. On the other hand the human body itself is a source of heat, and so many people together are one big heater.

3. Pollutants from Outdoor Air

• In all the regulations relating to ventilation, the assumption is that the outdoor air is clean and wholesome. However, many pollutants are present in the outdoor air, the majority of which are sulphur dioxide, nitrogen oxides, smoke and hydrogen sulphide (Figure 4-3). Once the opportunity is encountered, outdoor air can enter the house through doors, windows, holes or cracks in pipes. The main sources of these pollutants are dry industrial enterprises, means of transportation and a variety of small boilers around buildings, rubbish heaps and other sources of pollution from multiple seedlings.

(http://www.bbc.co.uk/news/world-asia-china-30826128)

(http://www.express.co.uk/news/nature/468085/Return-of-the-KILLER-SMOG-Worst-pollution-in-60-YEARS-to-strike-Britain-TOMORROW)

Figure 4-3 Outdoor air pollution

• The most important pollutants in ambient air are generally considered to be airborne particles (e.g. PM_{10}, $PM_{2.5}$), ozone, nitrogen dioxide, carbon monoxide and sulphur dioxide.

4.2 Indoor Air Quality Requirements

4.2.1 Indoor Air Quality Standards and Parameters

4.2.1.1 Standards

Indoor air quality (IAQ) refers to the quality of air within and around a building, which affects the health and comfort of people living in the building. It means that the air contains a constant value of each detected substance for a certain period of time and in a certain area (Table 4-3). It is an important indicator to indicate the health and livability of the environment. The main standards are oxygen content, formaldehyde content, water vapor content, particulate matter, etc. It is a comprehensive set of data that can fully reflect the air condition of a place.

Indoor air quality can be affected by gases (especially carbon monoxide, radon, VOCs), suspended particles, microorganisms (mold, bacteria) or other substances that can affect health conditions. The main ways to improve indoor air quality are source control, filtration, and ventilation to dilute pollutants. Residential units can further improve indoor air quality through regular carpet cleaning.

Confirming indoor air quality requires collecting air samples, monitoring people's exposure to pollutants, collecting samples from building surfaces, and building computer models of air movement within buildings.

The World Health Organization suggests that indoor air quality will be acceptable if

• Not more than 50% of occupants can detect any odor;

• Not more than 20% experience olfactory discomfort;

• Not more than 10% suffer from mucosal irritation;

• Not more than 5% experience annoyance, for less than 2% of the occupied time (This would equate to about 10 minutes in an eight-hour working day).

Table 4-3 Measurement unit of contaminant concentrations

Percentage	0.5%
Ratio	1:200
Volume ratio	0.005 m^3/m^3
Parts per million	5 000 ppm

Note: ① Volume ratio = Percentage/100;

② Volume ratio = Ratio (decimal);

③ Volume ratio = Parts per million × 10^{-6}.

4.2.1.2 Foreign Indoor Air Quality Requirements

WHO estimates indicate that approximately 7 million people die prematurely, primarily from non-communicable diseases, and that the diseases they acquire are the result of the combined effects of environmental and household air pollution. Global assessments of ambient air pollution alone indicate that hundreds of millions of healthy life are lost globally each year, with the greatest burden of diseases from ambient air pollution in low- and middle-income countries. Air pollution, in particular, increases morbidity and mortality from non-communicable cardiovascular and respiratory diseases, which are the leading causes of death globally; it also increases the disease burden of lower respiratory infections and increases premature births and other causes of death in children and infants, which remain the leading causes of disease burden in low- and middle-income countries.

In order to cope with this problem, the World Health Organization issued the "WHO Guideline for Indoor Air Quality" in 2010. Generally speaking, the level of pollutant limits set in the standard is related to the degree of development of the country or region. The more developed, the better the economic conditions of countries and regions, the more stringent the pollutant concentration limits required in the standard.

Recommended comfort criteria for specific applications are listed in Table 4-4.

Table 4-4 Recommended comfort criteria for specific applications (CIBSE Guide A)

Building/room type	Customary winter operative temperatures for stated activity and clothing levels			Customary summer operative temperatures (air conditioned building) for stated activity and clothing levels			Suggested air supply rate/(L · s^{-1} per person unless stated otherwise)	Filtration grade	Maintained illuminance/lx	Noise criterion		
	Temp. /℃	Activity /met	Clothing /clo	Temp. /℃	Activity /met	Clothing /clo				NR	dBA	dBC
Airport terminals:												
—baggage reclaim	12-19	1.8	1.2	21-25	1.3	0.6	10	F6-F7	200	45	50	75
—check-in areas	18-20	1.4	1.2	21-25	1.3	0.6	10	F6-F7	500	45	50	75
—concourse (no seats)	19-24	1.8	1.2	21-25	1.3	0.6	10	F6-F7	200	45	50	75
—customs area	18-20	1.4	1.2	21-25	1.3	0.6	10	F6-F7	500	45	50	75
—departure lounge	19-21	1.3	1.2	21-25	1.2	0.6	10	F6-F7	200	40	45	70
Art galleries—see Museum and art galleries												
Banks, building societies, post offices:												
—counters	19-21	1.4	1.0	21-25	1.3	0.6	10	F6-F7	500	35-40	40-45	65-70
—public areas	19-21	1.4	1.0	21-25	1.3	0.6	10	F5-F7	300	35-45	40-45	65-70

4.2.1.3 Domestic Indoor Air Quality Requirements

• To protect human health, prevent and control indoor air pollution, "Indoor Air Quality Standards" of China was approved for release on November 19, 2002 and officially implemented from March 1, 2003, jointly developed by the former General Administration of Quality Supervision, Inspection and Quarantine, the former State Environmental Protection Administration and the former Ministry of Health.

The standard harmonizes the implementation of factors including biological, physical, chemical and radiological limits and technical methods. These factors are closely related to the health of people living indoors as well as to standards for indoor air treatment and indoor air testing.

"Indoor Air Quality Standards" of China mainly refers to the standards of developed countries, and the experts has not yet had time to do a good research on China's indoor air pollutant composition, concentration levels and health hazards. In the past 10 years, China has carried out a large number of researches in this area, which provide mass data for China to revise the relevant standards. The new version (GB/T 18883-2022) of "Indoor Air Quality Standards" was revised to tighten the limits of many parameters, which means that the implementation of "Indoor Air Quality Standards" for the indoor air testing industry as well as indoor air purification industry will be more detailed and strict.

The following table shows the parameters of the specific control items in the "Indoor Air Quality Standards" including 19 indicators such as respirable particulate matter, formaldehyde, CO, CO_2, nitrogen oxides, benzo(a)pyrene, benzene, ammonia, radon, TVOC, O_3, total bacteria counts, toluene, xylene, temperature, relative humidity, air flow rate, noise and fresh air volume.

• With the continuous development of industrial enterprises, the air is entrained to varying degrees with a variety of pollutants. Usually in the natural ventilation of the open outdoors, air pollutants will not affect people's health, but with the improvement of people's living conditions, home decoration universal, and in order to save energy, indoor is usually in a closed state, resulting in the concentration of indoor pollutants is too high, and affect people's health. In order to regulate the quality of decorative materials, building materials and other standards to protect people's health, the state promulgated GB 50325-2010 "Code for Indoor Environmental Pollution Control of Civil Building Engineering" to put forward concentration limits for the five pollutants that have the most serious impact on human body in indoor air pollution (Table 4-5):

Table 4-5 Five pollutants proposed concentration limits

Contaminants	Ⅰ kind of civil buildings	Ⅱ kind of civil buildings
Radon (Bq/m^3)	≤200	≤400
Formaldehyde (mg/m^3)	≤0.08	≤0.10
Benzene (mg/m^3)	≤0.09	≤0.09
Ammonia (mg/m^3)	≤0.2	≤0.2
TVOC (mg/m^3)	≤0.5	≤0.6

Note: Ⅰ kind of civil buildings include residential buildings, hospitals, elderly buildings, kindergartens, school classrooms and other civil construction works; Ⅱ kind of civil buildings include office buildings, stores, hotels, cultural and entertainment venues, bookstores, exhibition halls, libraries, gymnasiums, public transportation yards waiting rooms, restaurants, barber stores and other civil building works.

4.2.2 Indoor Pollution Control Methods

In recent years, with the development of the economy, people's office and living space is getting bigger and bigger, and the decoration of office and living room is more and more frequent. However, a considerable part of the living and office buildings are in serious indoor pollution after disorderly decoration or due to negligence in environmental health management during the construction process.

At present, improving indoor air quality and preventing indoor pollution can be done from the aspects of indoor pollution source control, the use of green building materials, the use of nano-photocatalysts, ventilation, rational use of air conditioning, indoor air filters, interior virescence design and optimization design, etc.

4.2.2.1 Indoor Pollution Source Control

Source control is to avoid or reduce the production of pollutants from the source, or the use of barrier facilities to isolate pollutants from entering the indoor environment.

Source control can reduce or eliminate indoor air pollution from the source, is the most thorough way to ensure indoor air quality, and should be given priority. Indoor air pollutants mainly from construction and decoration materials, indoor supplies, human metabolism and human activities in the room, biological sources of pollution, ventilation and air conditioning systems and outdoor sources. Accordingly, the source of pollution control should also start from these aspects. Among them, building and decorative materials, indoor products involved in the release of pollutants are the most important factor affecting indoor air. Therefore, the use of environmentally friendly building and decoration materials is the basic solution to formaldehyde, benzene and ammonia pollution problems. For supplies that already exist indoors and release serious pollutants, it is advisable to permanently or temporarily move them out of the house. In addition, the use of energy is one of the main sources of air pollution in the home. For people using coal, firewood and straw as fuel, the choice of appropriate furnaces and supporting smoke exhaust facilities is also very important. Finally, the formation of good living habits, including not smoking and spitting, isolation of floor and kitchen leaks, timely cleaning of garbage, preventing livestock into the indoor space are effective source control measures.

4.2.2.2 Ventilation

Ventilation is the exchange of indoor and outdoor air. The higher the interchange rate , the higher the effect of reducing the pollutants produced indoors tends to be. Strengthening ventilation, using fresh outdoor air to dilute indoor air pollution, so that the concentration is reduced, is the most convenient and fast way to improve indoor air quality.

Opening windows is one of the most effective ways to ventilate and can always keep indoor with good air quality, which is critical to improve residential indoor air quality. In poorly ventilated dwellings, indoor smoke can exceed acceptable levels for small particles in outdoor air 100-fold. Studies have shown that the greater fresh air volume, the lower the risk of building syndrome. Even in the colder winter months, it is advisable to open some windows to allow fresh outdoor air to enter the home.

4.2.2.3 Rational Use of Nano-photocatalysts

As a new generation of environmental friendly catalyst, the nano-photocatalyst has a great potential application in air pollution control, especially in indoor air purification. The principle is that the nano-photocatalyst generates free electrons and holes through ultraviolet light catalysis, and then generates active oxygen with strong oxidation effect. The active oxygen can oxidize and decompose various organic compounds and some inorganic substances, making them decompose into harmless CO_2, H_2O, etc. , so as to achieve the purpose of air purification, and has the functions of deodorization, sterilization, anti-mildew, anti-fouling and anti-UV, etc.

4.2.2.4 Reasonable Use of Air Conditioning

The use of air conditioners when the fresh air volume is insufficient can cause a decline in indoor air quality. Therefore, the reasonable use of air conditioning is also one of the measures to prevent and control indoor pollution, which can be started from the following aspects.

Firstly, people should reasonably control the indoor temperature and humidity, the general temperature difference in summer maintained at 6 ℃ to 7 ℃ is more appropriate, and not too big temperature difference between the indoor and the outdoor is conducive to self-regulation of the human body. Winter air is dry, so it is best to be equipped with a humidifier to keep the air moist.

Secondly, most of the residences now use split-type air conditioners, through which the air is basically circulating indoors, so it is best to open a little window to introduce some fresh air when using the air conditioner.

Thirdly, the filter should be cleaned frequently, which can not only ensure the air volume of the coil, but also timely remove the dirt on the filter.

Fourthly, some indoor air treatment equipment should be used in conjunction with the air conditioner, such as dehumidifiers, humidifiers, filters, negative ion generators, etc.

4.2.2.5 Plant Purification

Plant purification is the metabolic action of plants to render pollutants entering the environment harmless, including the purification of atmospheric pollution by terrestrial plants and the purification of pollutants in water bodies by aquatic plants.

Plants purify the atmosphere mainly through the leaves, and their main functions are:

- Absorbing CO_2 and releasing O_2;
- Retaining and filtering dust and dust drift;
- Reducing the content of SO_2, HF, Cl_2, O_3 and other harmful gases in the air through absorption;
- Reducing photochemical smog pollution;
- Filtering bacteria or sterilization;

• Absorbing and purifying heavy metals from drift dust and particulate matter.

For example, tortoise bamboos have a strong ability to absorb carbon dioxide; moonflowers and roses can absorb hydrogen sulfide, hydrogen fluoride, phenol, ether and other harmful gases; hanging orchids and aloe vera can absorb formaldehyde; ironwood, chrysanthemum and ivy can absorb benzene and other volatile gases; plantain can absorb sulfur dioxide and fluorine; heliotrope can prevent dust and isolate sound; weeping plum, laurel pine, magnolia and cherry blossoms can reduce the mercury content in the air, etc.

4.2.2.6 Optimized Design and Reasonable Decoration

The building design should follow the design principle of ecological environment, which includes considering the reasonable planning of building general plan, the improvement of urban microclimate, the building materials to meet the "Indoor Air Quality Standards", and the use of natural energy as much as possible or with the least energy to achieve the comfortable environment that people need to live and work, which is also the fundamental measure to ensure the indoor air quality of the building.

4.2.3 Indoor Air Filter

4.2.3.1 Definition of Air Filter

Filters are inserted into the supply air system to stop pollutants being distributed around the building. Air filters are also known as products that can adsorb, decompose or transform various air pollutants (generally including $PM_{2.5}$, dust, pollen, odors, decoration pollution such as formaldehyde, bacteria, allergens) to effectively improve air cleanliness (Figure 4-4), mainly used in domestic, commercial, industrial and building applications.

Figure 4-4 Dirty filters

4.2.3.2 Operating Principle

The air filter is mainly composed of motor, fan and air filter system, and its working principle is: the motor and fan inside the machine make indoor air flow, the air filter inside the machine will remove or absorb various pollutants from the polluted air. Some models of air purifier will also be installed in the outlet of the negative ion generator (working in the negative ion generator in high voltage to produce DC negative high voltage), and the air will be continuously ionized, generating a large number of negative ions, which are sent out by the micro-fan and form negative ion airflow, so as to achieve the purpose of cleaning and purifying the air.

4.2.3.3 Application Places

- Recently renovated or refurbished homes.
- Homes with elderly people, children, pregnant women and newborns.
- Homes for people with asthma, allergic rhinitis and pollen allergies.
- Homes where pets and livestock are kept.
- Homes that are closed or affected by second-hand smoke.
- Hotels and public places.
- Housing for people who wish to enjoy a high quality of life.
- Hospitals, to reduce infection and prevent the spread of disease.

4.2.3.4 Suitable Crowd

(1) Pregnant women: Pregnant women who are exposed to high levels of air pollution may experience general discomfort, dizziness, sweating, dryness from smoke, chest tightness and vomiting, which can have a negative impact on the development of the foetus. They are three times more likely to suffer from heart disease than children born by pregnant women who breathe clean air.

(2) Children: Children's bodies are developing and their immune systems are weak, making them vulnerable to the dangers of indoor air pollution, leading to reduced immunity, delayed physical development, induced blood disorders, increased incidence of asthma, and greatly reduced intelligence.

(3) Office workers: Working in a high-end office building is an enviable career. However, in a constant and confined environment with poor air quality, it can easily lead to dizziness, chest tightness, fatigue, emotional ups and downs and other uncomfortable symptoms, affecting work efficiency, leading to various diseases, and in serious cases, can also cause cancer.

(4) Older people: Older people have reduced physical functions and are often plagued by many chronic diseases. Air pollution not only causes bronchitis, laryngitis, pneumonia and other respiratory diseases in the elderly. It can also induce hypertension, heart disease, cerebral hemorrhage and other cardiovascular diseases.

(5) Patients with respiratory diseases: Living in polluted air for long periods of time can lead to reduced respiratory function and increased respiratory symptoms, especially rhinitis, chronic bronchitis, bronchial asthma, emphysema and other diseases. The treatment effect is complementary and radical by breathing pure air.

(6) Drivers: Since the space in the car is relatively small and well sealed, it is difficult for

fresh air to enter the car and cause no air circulation; in addition, the automobile exhaust will cause serious pollution to the outside..

4.2.3.5 Selection Basis

Firstly, the environment and the effect to be achieved should be considered. General indoor air pollutants is: dust, viruses, bacteria, moulds and insect mites and other allergens; respirable organic volatile gases such as formaldehyde, benzene and ammonia; radioactive pollution caused by radon gas and its progeny released from the strata and building decoration materials. Therefore, the selection of air filter should be based on a comprehensive consideration of their functions and effects.

Secondly, The type of filter selected will depend on the application (quantity and size of pollutant particles to be removed) and the type of systems in which it is installed. The efficiency of filters can be varying from 20% to 90%, so the air filter capacity of the air purifier should be considered. If the room is large, you should choose an air purifier with a larger purifying air volume per unit time. Generally speaking, larger filters have a higher purification capacity, so when choosing one, you can refer to the description in the catalogue or instruction book.

Thirdly, consideration should be given to the service life of the filter and whether maintenance is convenient. For example, there are some products that use filtration, adsorption and catalytic principles of the filter; when the use of time increases, the air filter tends to saturate, it is important to check, clean or replace the filters periodically as the captured pollutants will decrease the filters' efficiency. Users should choose purification filters with regeneration capacity (including high-efficiency catalytic activated carbon) to extend the service life; there are also some electrostatic products that do not require replacement of the relevant module, and these products only need to be cleaned regularly.

Fourthly, the layout of the room should be taken into account to match the filter. There are 360-degree circular designs for the inlet and outlet air of air purifiers, as well as one-way.

Chapter 5

Building Service Engineering

5.1 Classification and Operation of Building Service System

5.1.1 Mechanical and Electrical Services

5.1.1.1 Definition of Building Service System

Building service system is concerned with the design, development, installation and maintenance of systems about convenience and comfort in buildings. It implies any system or equipment inside a building that makes this space comfortable and secure. Building services help to set up spaces in which people can live and work while having as little impact on the environment as possible. Building services are able to impart a special kind of life-promoting energy to intellectual and professional pursuits. Building services are responsible for the design, installation, operation and monitoring of various aspects in the building. Building services serve the electrical, HVAC, mechanical, plumbing, safety, sanitation and telecommunications systems of the building, which is the main source of supply and distribution throughout the building.

The Building Services Research and Information Association (BSRIA) says: "Building services are primarily used to help create a comfortable and safe living or working environment for people and processes, by providing warmth, light, water, power, sanitation, transport, communication, sound control, security & fire protection." "Building service engineering is all about making buildings meet the needs of the people who live and work in them." (Figure 5-1)

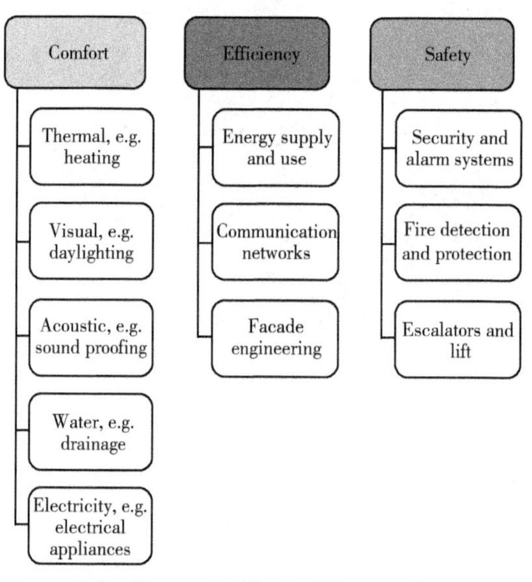

Figure 5-1 The goal of the building service system

(https://www.cibse.org/)

5.1.1.2 Building Service Systems in Accommodation

- Heating (radiators, warm air, floor heating);
- Ventilation (windows, toilet/kitchen extraction);
- Lighting (ceiling/wall lights);
- Power (computers, washing machine);
- Hot and cold water (toilets, sinks, baths, showers);
- Drainage (bathroom, kitchen);
- Telephone system;
- Fire alarm (ceiling battery device, wireless equipment);
- Cable digital TV, broadband;
- Security (motion sensors, PIR, window contactors, remote alarms).

5.1.1.3 Mechanical Services

Mechanical services work is the construction, installation, replacement, repair, alteration, maintenance, testing or commissioning of a mechanical heating, cooling or ventilation system in a building (Figure 5-2).

- Heating;
- Gas/Diesel/Wood fuel installations;
- Ventilation & air conditioning;
- Kitchen ventilation;
- Smoke & fume extraction;
- Controls/BMS;
- Hot & cold water;
- Drainage;
- Sprinklers;
- Compressed air, medical & industrial gases.

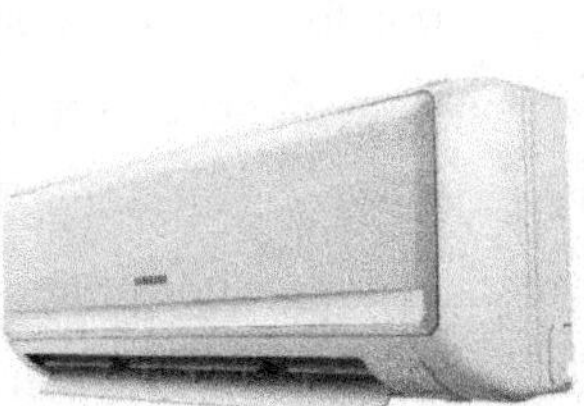

Figure 5-2 Mechanical services

5.1.1.4 Electrical Services

Electrical services provide electrical related services of BHS products including but not limited to installation, repairing, upgrading, evaluation, maintenance, machine relocation, performance improvement (Figure 5-3). The important role of electricity in our life and production cannot be ignored; it brings us great convenience and becomes an important source of energy in our production and life. The most critical factor for the normal operation and transmission of electricity in power plants is the electrical equipment.

- Power distribution;
- Power generation;
- Lighting;
- Vertical transportation;
- Communications;
- Fire alarms;
- Security;
- Public address;
- Lightning protection and earthing.

Figure 5-3 Electrical services

5.1.2 Duties of a Building Service Engineer

5.1.2.1 Definition of Building Service Engineer

Building service engineers design and install all of the functional components within a building from lighting to plumbing. Building service engineers combine their creativity with problem-solving and engineering skills to come up with innovative, ideal, eco-friendly interior system designs. Building service engineers must have strong communication skills and work as a team, as they collaborate with a variety of specialists, including architects, contractors, plumbers, electricians, surveyors, other engineers and construction workers. They must also have good computer and technical skills, as modern construction work relies significantly on automated design and other drafting and modeling software.

Building service engineers should consider human well-being and security, sustainability and cost in practically everything they do. As laws and regulations require more energy-efficient or zero-energy buildings, more and more building service engineers are turning to green and sustainable functions such as daylight-controlled lighting systems, water-saving devices, low-carbon techniques, and occupancy sensors. In particular, some building service engineers specialize in sustainable energy or renewable energy sources.

5.1.2.2 Typical Responsibilities of a Building Service Engineer

- Negotiation with builders and clients on their needs and budgets;
- Visit to construction sites;
- Contract negotiation;
- Project outline plan of work (Table 5-1);

Table 5-1 Project outline plan of work

Stage	Need identification	Start of design	Start of construction	Practical completion
Feasibility	Clients briefing Basic planning Rooms routes & risers - 3R's			
Preconstruction		Detailed design Sharing design Responsibility Design coordination		
Construction			Keeping time for commissioning Supervision & witnessing Installation drawing review	
After practical completion				Occupancy, energy and maintenance User feedback

- System selection to design unique systems for every building project (Figure 5-4);
- System designers—not component designers;
- Primary space planning;
- Deploying energy, plumbing, ventilation systems, etc.;
- Selecting appropriate building materials;
- Using CAD and modeling software to create plans, drawings, and diagrams;
- Reviewing contractor design and installing or supervising the installation of building components;
- Site supervision and witness testing to ensure they are working and make necessary adjustments;
- Maintaining, repairing and replacing systems as needed;

• Creating cost effective and energy efficient solutions for buildings;

• Ensuring all systems and equipment meet health, safety and environmental codes;

• Writing reports and giving presentations;

• Researching new technologies and systems.

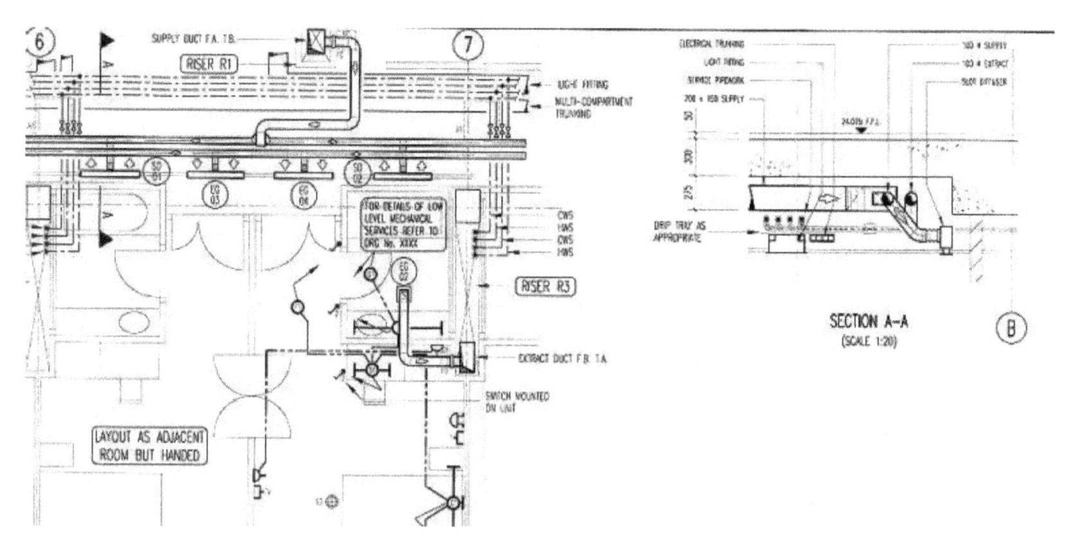

Figure 5-4 System design

5.1.2.3 Specific Work of a Building Service Engineer

One of the focuses of a building service engineer is to design and plan the electrical, lighting, and energy systems necessary for proper operation of a building. This can include laying out the wiring plans for electrical components and integrating other sources of energy, such as gas. Those specifically trained and hired for planning out energy systems typically have a background in electrical engineering.

The mechanical workings of a building also require attention. A building service engineer specializing in mechanical engineering focuses on designing and overseeing escalators, elevators, and industrial production equipment. They may also oversee maintenance on these systems once a building is completed.

5.1.2.4 Reasons for Service Problems

• Inexperienced designers and late changes to the brief;

• Insufficient space;

• Spatial/technical coordination/cooperation;

• Poor equipment performance;

• Poor or non-existent installation drawings;

• Inadequate commissioning;

• Lack of time;

• Wrong conditions;

• Poor services designer/contractor;

• Shared design duties poorly defined/understood.

5.1.3 Building Design: Key Steps & Considerations

5.1.3.1 Key Steps in Building Service Design

• Analyse the design project, e.g. client brief, environment, and legislation;

• Zone the building according to the functions of rooms;

• Calculate the cooling and heating loads of different zones;

• Select systems for different zones;

• Size the systems according to the calculated loads and selected system types, e.g. boilers, chillers, pumps and fans;

• Plan the system layout inside the building, e.g. plant room, pipework/ductwork distribution and terminal positions.

5.1.3.2 Key Considerations in Building Service Design

Considerations before selecting the system:

• Weather condition;

• Internal heat gains.

Considerations after selecting the system:

• System positioning;

• Pipework/ductwork distribution.

5.1.3.3 Professional Drawings

Schematic drawing: A schematic, or schematic diagram, is a representation of the elements of a system using abstract, graphic symbols rather than realistic pictures (Figure 5-5).

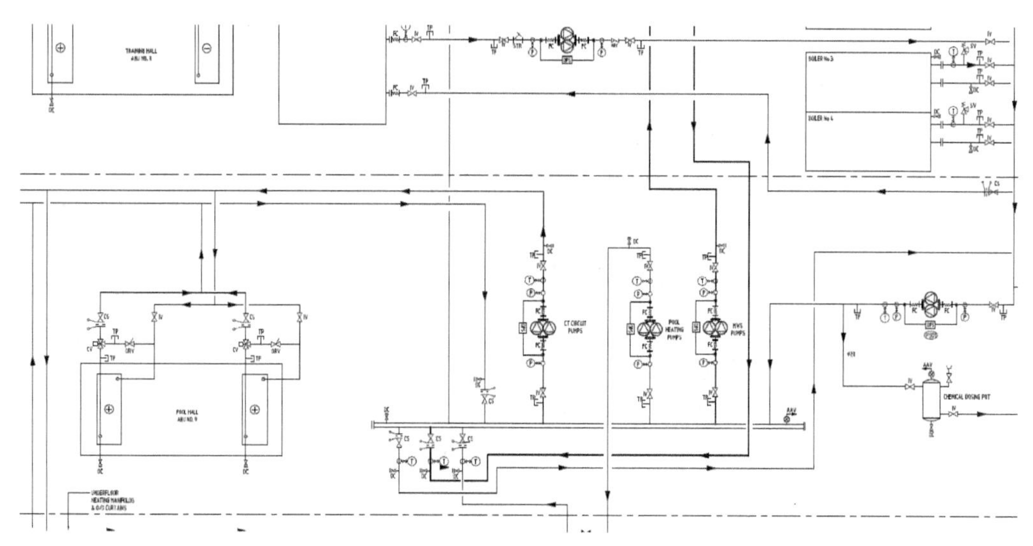

Figure 5-5 Schematic drawing

Layout drawing: Show the location of various system components in the building with indications in terms of space requirements (Figure 5-6).

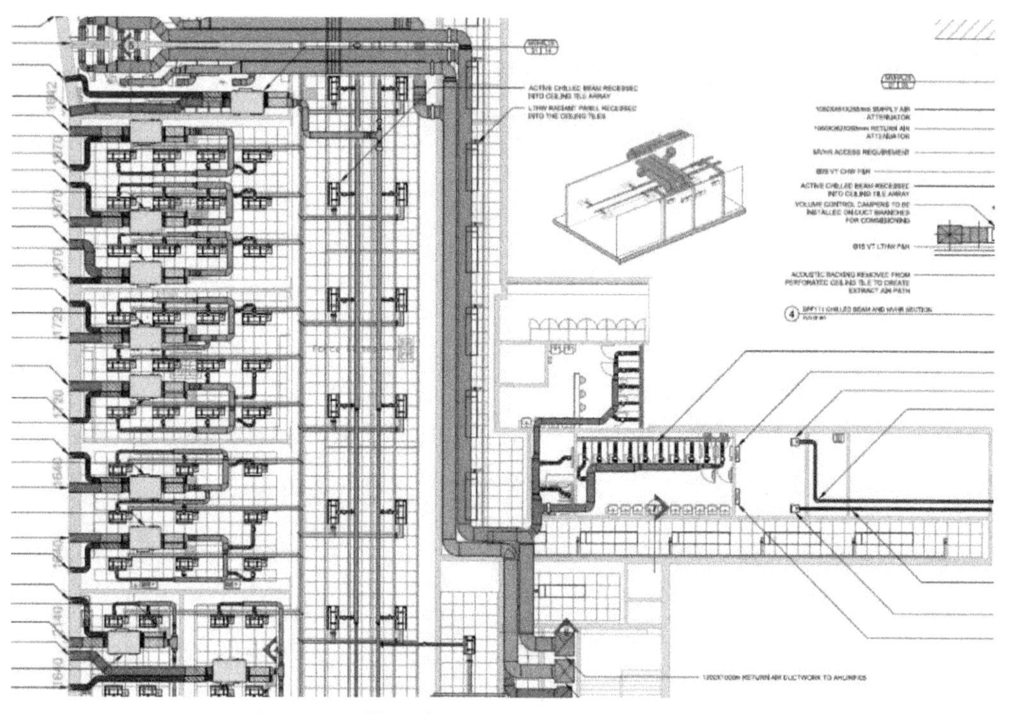

Figure 5-6 Layout drawing

Plant room drawing: Arrangement of primary plants in the plant room and pipework/ductwork to and from risers (Figure 5-7).

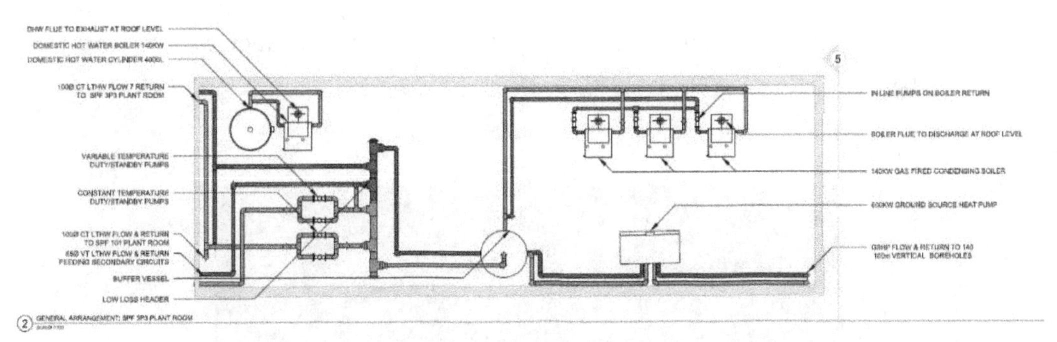

Figure 5-7 Plant room drawing

5.2 Appropriate Level of Equipment Design

5.2.1 Ventilation System

5.2.1.1 Necessity of Ventilation

"Ventilation is essential to the provision of a safe, healthy, productive and comfortable living or working environment."

—CIBSE Guide B2-Ventilation and Air Conditioning

A supply of outside air is required for one of the following purposes:

• Human respiration;

• Dilution and removal of odors, tobacco smoke, toxic and flammable gases and other contaminants;

• Control of internal humidity;

• Provision of air for fuel burning appliances;

• Maintaining occupants' thermal comfort.

5.2.1.2 Ventilation and Air Quality

Ventilation is the process of air exchange between the indoor and the outdoor, introducing fresh outdoor air into the room and exhausting indoor air to the outside for the maintenance or improvement of air quality. It is a convenient and efficient way for indoor air quality improvement by means of ventilation dilution or ventilation exclusion, diluting indoor air pollutants with outdoor fresh air, and thus reducing the concentration of pollutants.

Building indoor ventilation is an important factor affecting human health and comfort, directly affecting people through the physiological effects of fresh air and airflow, and indirectly acting on the human body through the effects on indoor temperature, humidity and internal surface temperature (Figure 5-8). Frequent opening of windows for ventilation can effectively use sunlight and ultraviolet light in the air to kill germs, and there is sufficient oxygen and negative ions in the fresh air, which can promote human metabolism.

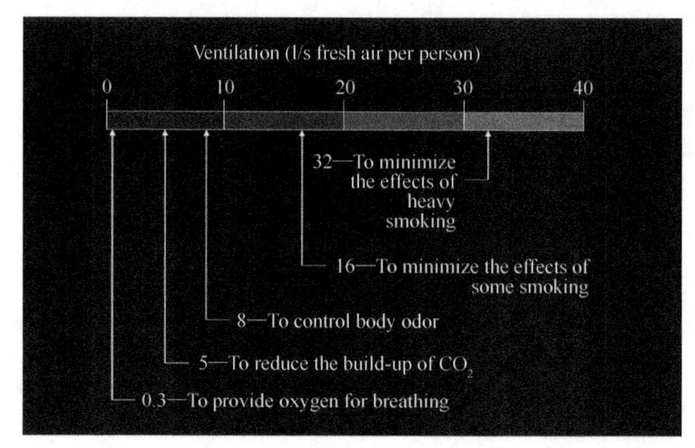

Figure 5-8 Ventilation levels

5.2.1.3 Types of Ventilation

• According to the classification of ventilation power: natural ventilation, mechanical ventilation (Figure 5-9).

• According to the scope of ventilation services: comprehensive ventilation, local ventilation.

• According to the classification of airflow direction: air supply (inlet air), air exhaust (smoke).

• According to the classification of ventilation purposes: general ventilation, hot air heating, detoxification and dust removal, accident ventilation, protective ventilation, building smoke exhaust, etc.

• According to the classification of power location: power centralized and power distributed.

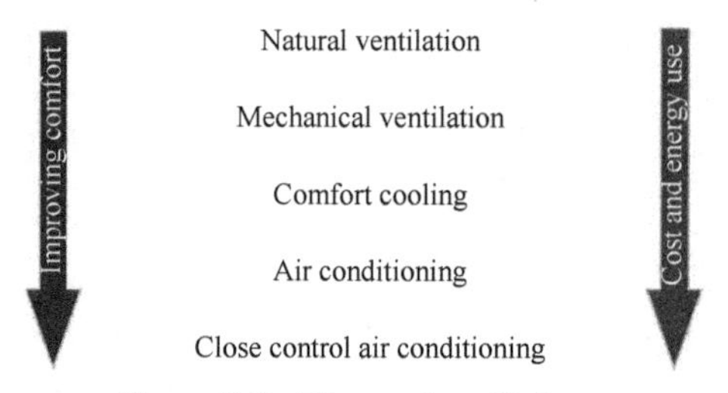

Figure 5-9 Types of ventilation

5.2.1.4 Definition of Natural Ventilation

"Natural ventilation may be defined as ventilation that relies on moving the air through a building under the natural forces of wind and buoyancy." "Natural ventilation is the airflow through a building resulting from the provision of specified routes such as openable windows, ventilator, ducts, shafts etc. driven by wind and density differences. ”

—CIBSE Guide B2-Ventilation and Air Conditioning

There are numerous ways of incorporating natural ventilation methods and paths into the building structure:

• Most simplistically, windows or ventilators on the facade of the structure;

• More complicatedly, the architectural integration of natural ventilation methodologies to utilize stack and wind driven ventilation.

Typically, natural ventilation alone can deal with around 20 W/m^2 of cooling load. When combined with additional methods, e.g. thermal mass, night purge, etc., this can rise to around 50 W/m^2.

5.2.1.5 Principles of Natural Ventilation

Natural ventilation uses natural forces to drive airflow in the space. There are two natural forces that can be used to drive air through buildings: wind and buoyancy (Figure 5-10).

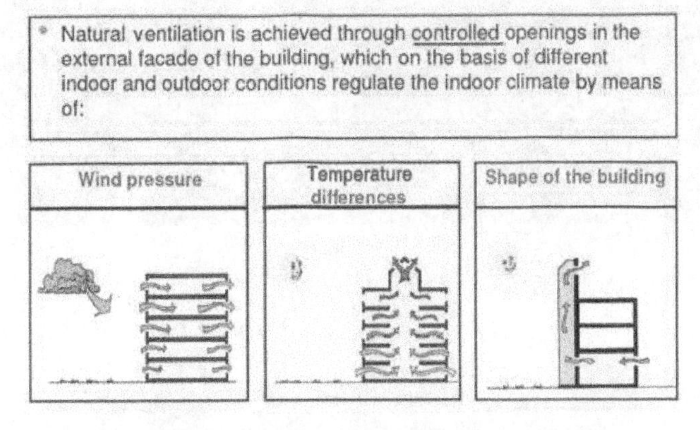

Figure 5-10 Primary drivers of natural ventilation

When the wind blows to the windward side, it creates a positive pressure on the outer wall. Similarly, when the wind flows away from the leeward side, it creates an area of lower pressure. If the windows on both the windward and leeward sides of a building are open, air will be forced through the building because of the pressure difference at the opening.

Stacked or buoyant ventilation is based on the difference in temperature between the air inside and outside the building to drive the airflow (Figure 5-11). Warmer air is lighter than colder air. When a building filled with warm air is exposed to cold air and two vertically spaced windows are opened, the lighter air will exit through the opening and the colder air will enter the building through the lower window (Figure 5-12).

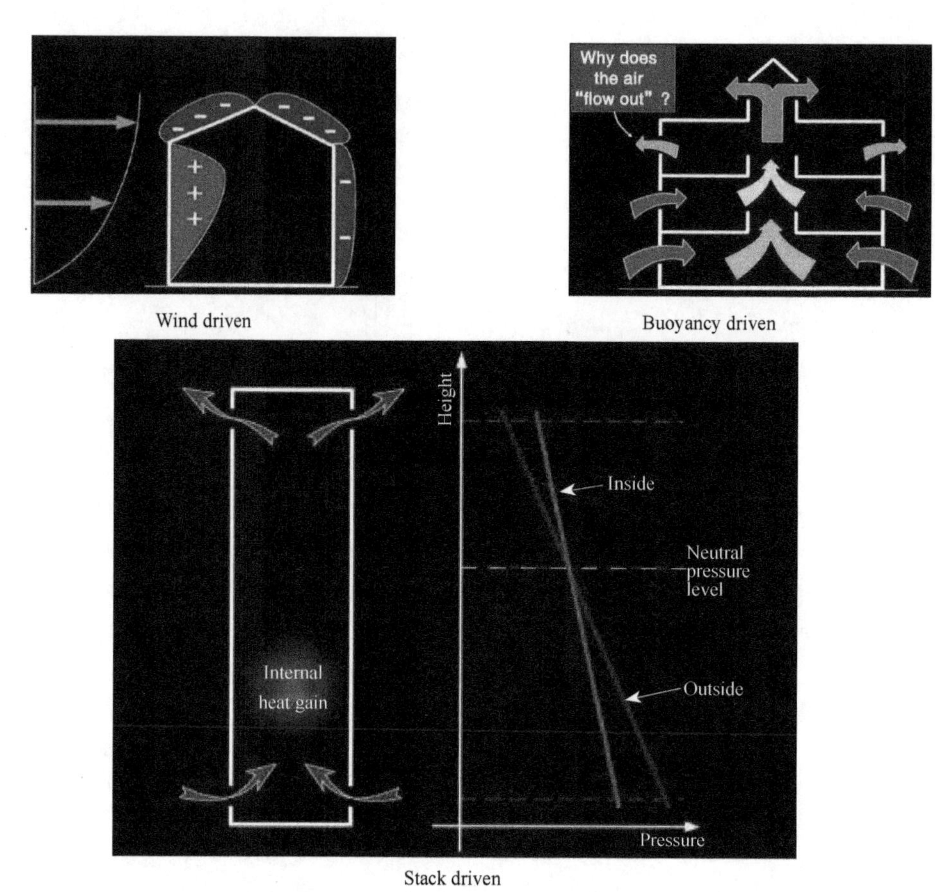

Wind driven

Buoyancy driven

Stack driven

Figure 5-11 Wind driven, buoyancy driven and stack driven natural ventilation

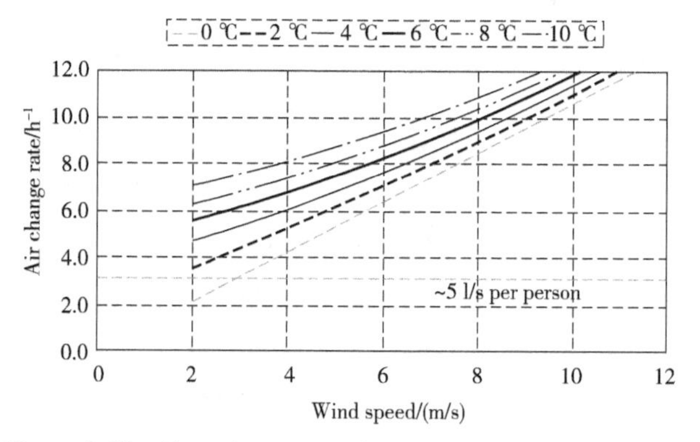

Figure 5-12 Air exchange rate by thermal buoyancy and wind

(https://www.windowmaster.com/)

Cross ventilation is the process of transferring fresh air from outside through the inlet to the outlet to create a steady breeze. A window is opened opposite the open door to generate a cross breeze. It pulls air from one end to the other, swiftly circulating and ventilating the entire room.

Chimney effect is the result of the joint action of the thermal pressure formed by the difference between indoor and outdoor temperatures and the outdoor wind pressure, which is usually dominated by the former, while the thermal pressure value is proportional to the difference in air density generated by the difference between indoor and outdoor temperatures and the difference in height of the air inlet and outlet. This indicates that the higher the indoor temperature is above the outdoor temperature, the higher the building is, and the more obvious the chimney effect is, and it also indicates that the chimney effect in civil buildings generally occurs only in winter. In the case of a building, theoretically, half of the height position of the building is regarded as the neutral surface, and the rooms below the neutral surface are considered to be infiltrated with air from outdoors, while the rooms above the neutral surface are infiltrated with air from indoors.

For the atrium, the open windows and the top of the building can make full use of the wind pressure to naturally ventilate the building; its structure through the building can form the temperature difference between the upper and lower ends, thus accelerating the indoor air flow and achieving the "chimney effect" in the true sense (Figure 5-13).

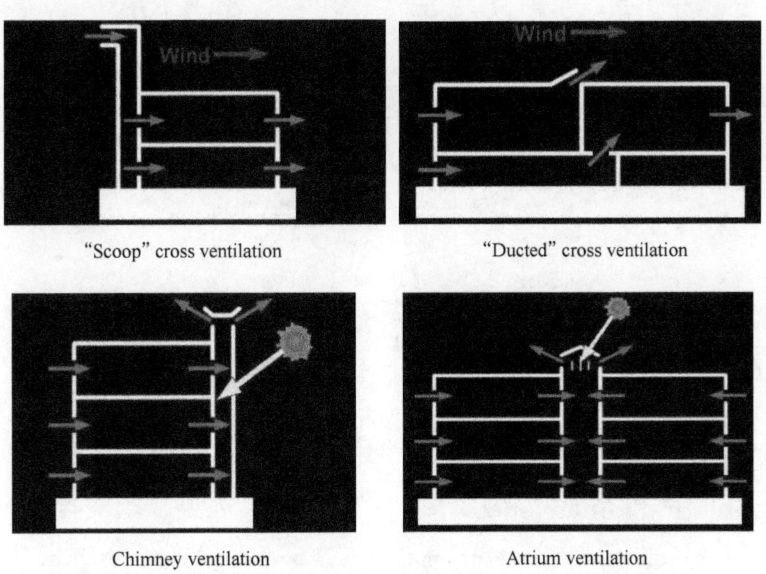

"Scoop" cross ventilation "Ducted" cross ventilation

Chimney ventilation Atrium ventilation

Figure 5-13 Various types of natural ventilation (CIBSE Guide B)

5.2.1.6 Natural Ventilation Type with General Constraints

The opening size and relative position of the room directly affects the air speed and inlet air volume. For large air inlet, the flow field is large; for small air inlet, although the flow rate increases, but the flow field shrinks. The relative position of the opening plays a decisive role in the route of airflow. Air inlet and outlet should be at relatively staggered positions, so that the airflow in the room will change direction and make indoor air more even, which ensures excellent ventilating effect.

The relative position and height of the openings will affect the airflow route. As shown in the figure, if the air inlet is more than one-half of the room height, the airflow will mostly move along the ceiling and cannot produce the desired wind speed at human body height (Figure 5-14). If low air inlet and low air outlet, or low air inlet and high air outlet, the airflow can act on the human body range of activities to play the role of ventilation and heat dissipation.

When ventilation is provided by a single opening on one side, the depth of the room should be less than or equal to twice the height of the ceiling.

When single-sided double-opening ventilation is used, the depth of the room should be less than or equal to twice the height of the ceiling.

When cross ventilation is used, the windows are generally located on opposite sides or connected to the roof vents, and the depth of the room should be less than or equal to five times the ceiling height.

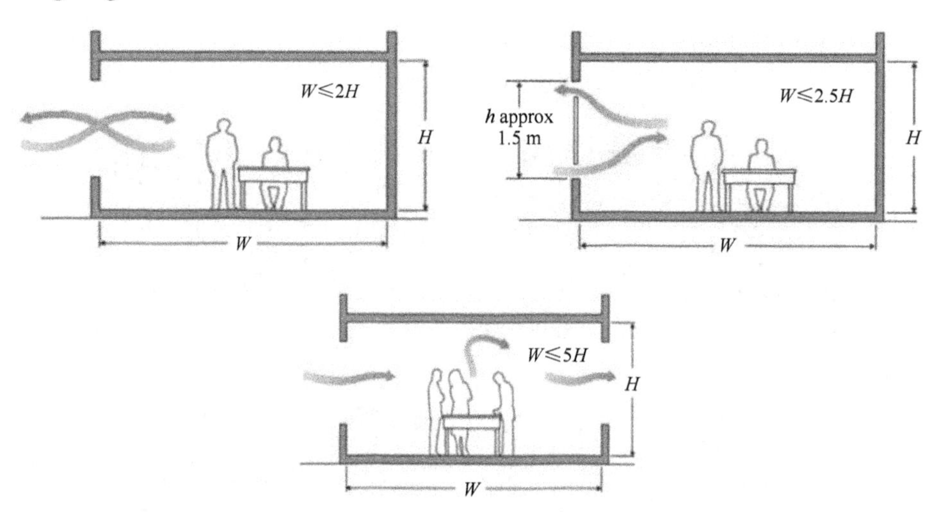

Figure 5-14 The opening size and relative position of the room

(https://www.bsria.com/uk/)

5.2.1.7 Ventilation Effectiveness

The effective ventilation of buildings has always been a primary design requirement (Figure 5-15). Ventilation effectiveness evaluation is divided into three aspects.

(1) Age of air: Age of air refers to the time taken by air particles from entering a room to arriving at a certain point indoors, which reflects the freshness of indoor air. It can comprehensively measure the ventilation effect of a room and is an important indicator for evaluating indoor air quality. When the pollution sources are evenly distributed in the room and the air supply is full fresh air, the smaller the air age at a point, the better the air quality at that point. It also reflects the room's ability to remove pollutants, the smaller the average air age of the room,

the stronger the ability to remove pollutants.

(2) Ventilation efficiency: Ventilation efficiency is the ratio of the average air age of the room to the piston flow under actual ventilation conditions (the value is no more than 1), which reflects the comparison between the speed of fresh air replacing the original air and the speed of piston replacing the original air under ventilation.

(3) Accessibility: Accessibility reflects the relative degree of air supply from every air outlet to every point in space within a certain time. After a long enough time for a single air outlet, the accessibility of every point in space is 1, while after a long enough time for multiple air outlets, the accessibility sum of every point in space is 1.

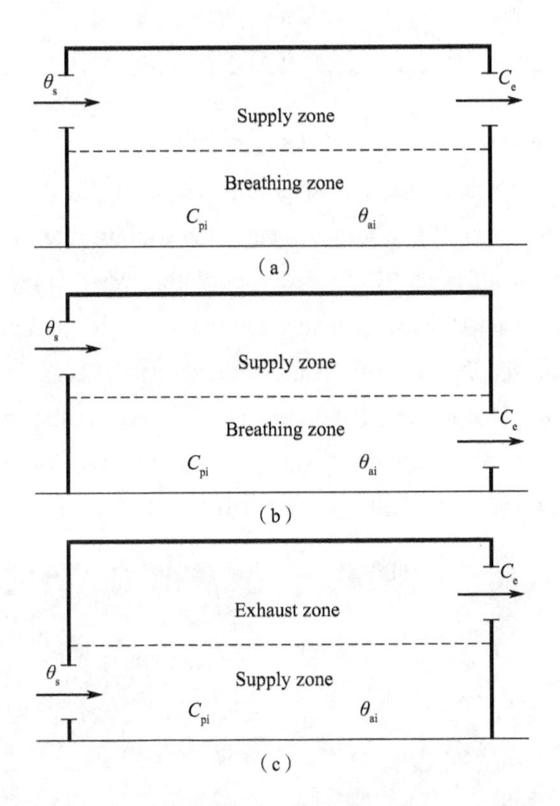

Figure 5-15 Supply/extract arrangements for ventilation

(a) Mixing, supply and exhaust at high level (b) Mixing, supply at high level, exhaust at low level (c) Displacement

Calculation of ventilation effectiveness takes into account pollution concentration in exhaust air, supply air and breathing zone indoor air (Table 5-2).

Table 5-2 Ventilation effectiveness for ventilation arrangements

Ventilation arrangement	Temp. difference between supply air and room air ($\theta_s - \theta_{ai}$)/K	Ventilation effectiveness, E_v
Mixing, high-level supply and exhaust (Figure 5-15(a))	< 0 0-2 2-5 > 5	0.9-1.0 0.9 0.8 0.4-0.7
Mixing, high-level supply, low-level exhaust (Figure 5-15(b))	< -5 (-5) -0 > 0	0.9 0.9-1.0 1.0
Displacement (Figure 5-15(c))	< 0 0-2 > 2	1.2-1.4 0.7-0.9 0.2-0.7

5.2.1.8 Manual Control vs. Automatic Control

Traditional windows: Traditional windows must be opened and closed manually to meet the requirements of ventilation. If you forget to close the window, the floor will get wet when it rains; or you will forget to open the window for ventilation when you are in a hurry to go for work; furthermore, you need to check whether the window is closed upstairs and downstairs when you go out. Obviously, this is a significant inconvenience.

Automatic windows: Due to the interconnection of devices in the smart home system, you can control the windows at home anytime and anywhere, and many times you can also achieve unexpected convenience and enjoyment (Figure 5-16).

Figure 5-16 Manual control vs. automatic control

5.2.1.9 Calculation of Air Change per Hour (ACH)

Air change per hour is a quantity to measure how quick the air is replaced in a defined space:

$$ACH=\frac{q\times 3\,600}{V}$$

where q is air volume flow rate (m^3/s); V is volume of the room (m^3).

5.2.1.10 Disadvantages of Natural Ventilation

Natural ventilation is highly susceptible to:

- Air pollution;
- Noise;
- Overheating.

And to a lesser extent:

- Rain;
- Poor security;
- Lack of control;
- No heat recovery (Figure 5-17).

The Problem with Natural Ventilation...

Natural Ventilation?

- Too hot!
- Too cold!
- Too stuffy!
- Draughty!
- High energy bills!
- Unpredictable results!

- Naturally ventilated BSF projects using 300% more energy than expected
- Manual Windows—do they really do the job?
- Good design, but poorly specified or value engineered
- Installed automated systems that can introduce more problems than they solve?
- Acoustics?
- Health?
- Security?
- Perceived performance improvement in SBEM systems...though at what cost?

The answer?
Well designed buildings and controls
Fine control....solution based, not product based

Figure 5-17 The problem with natural ventilation

Therefore, ventilation rates need a controlled condition (Figure 5-18):

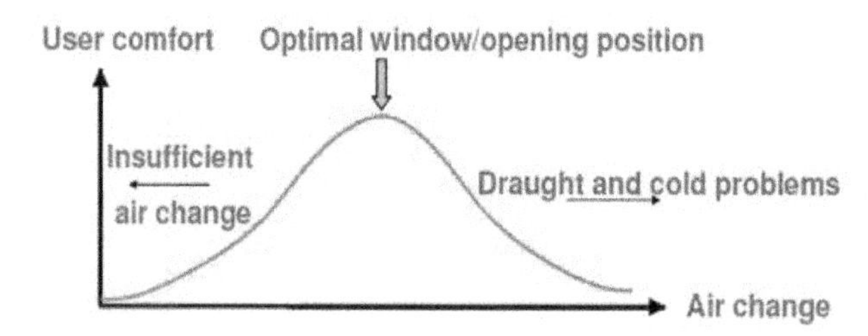

Figure 5-18 The relationship between air change and user comfort

(https://www.windowmaster.com/)

Natural ventilation saves energy and has low cost. It will not be affected by power outage and other emergencies, but it cannot be effectively controlled. If problems occur, manual intervention cannot be conducted in time, so the natural conditions of the building site are strictly required (Figure 5-19).

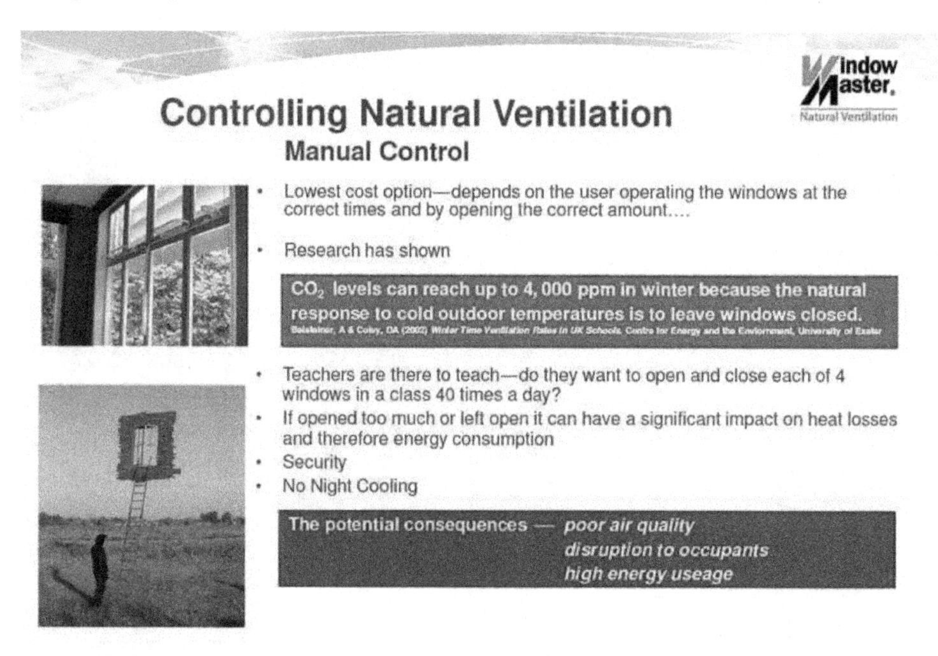

Figure 5-19 Controlling natural ventilation

5.2.1.11 Definition of Mechanical Ventilation

"Mechanical ventilation may be defined as the movement of air through a building using fan power; filtration and heating of the air may also take place."

—CIBSE Guide B2-Ventilation and Air Conditioning

Mechanical ventilation systems generally consist of fans, ducts, valves, and air supply and exhaust outlets. Mechanical ventilation can be divided into local ventilation and overall ventilation according to the distribution of harmful substances and the scope of the system action.

• Local ventilation: Local ventilation controls the spread of pollutants in local areas of the room or the concentration of pollutants in local areas to meet the health standards. Local ventilation is divided into local air exhaust and local air supply. Local exhaust is a type of local ventilation that directly removes pollutants from the source of pollution. When pollutants are concentrated in a certain place, local exhaust is the most effective way to deal with the harm of pollutants to the environment.

• Comprehensive ventilation: The principle of comprehensive ventilation is to send clean fresh air to a room, dilute the concentration of pollutants in the indoor air, and discharge the air containing pollutants to the outdoors, so as to make the concentration of pollutants in the indoor air meet the requirements of health standards.

Comprehensive ventilation is suitable for: the location of harmful substances is not fixed; large area or local ventilation devices affect the operation; a room or section in which the diffusion of harmful substances is not restricted. This is to allow harmful substances into the indoor, while introducing outdoor fresh air to dilute the concentration of harmful substances, so that they are reduced to meet the health requirements of the allowable concentration range, and then discharged from the indoor.

General ventilation includes general air supply and exhaust, both of which can be used simultaneously or separately. When used alone, it needs to be combined with natural air supply and exhaust.

5.2.1.12 Necessity of Mechanical Ventilation

• To provide adequate background fresh air ventilation or compensate for natural means when they are inadequate for occupant well being;

• To provide adequate fresh air ventilation for fume/contamination control, when a fixed rate would normally be applied;

• To cool the building when the outside air is at an appropriate temperature.

5.2.1.13 Grille vs. Diffuser

A grille is a device for supplying or extracting air vertically without any deflection.

A diffuser normally has profiled blades to direct the air at an angle as it leaves the unit into the space (Figure 5-20).

Figure 5-20 Grille and diffuser

5.2.1.14 Domestic and Non-domestic Ventilation

The NuAire XS window fan kit is a representative device for domestic ventilation (Figure 5-21), which is available in 6, 9 and 12 inch impeller sizes suitable for supply or extract and can be the heart of a room's automatic ventilation system.

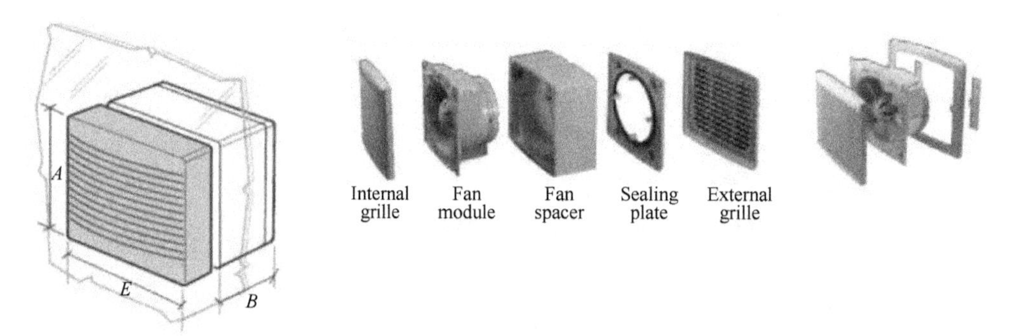

Figure 5-21 Composition of the XS (GL) window fan kit

• Domestic ventilation requirement is shown in Table 5-3 and Table 5-4.

Table 5-3 Extract ventilation rates (UK Building Regulations Part L)

Room	Intermittent extract	Continuous extract	
	Minimum rate	Minimum high rate	Minimum low rate
Kitchen	30 l/s adjacent to hob; or 60 l/s elsewhere	13 l/s	Total extract rate should be at least the ***whole dwelling ventilation*** rate given in Table **5.4**
Utility room	30 l/s	8 l/s	
Bathroom	15 l/s	8 l/s	
Sanitary accommodation	6 l/s	6 l/s	

Table 5-4 Whole dwelling ventilation rates (UK Building Regulations Part L)

	Number of bedrooms in dwelling				
	1	2	3	4	5
Whole dwelling ventilation rate [a, b]/(l/s)	13	17	21	25	29

Notes:

a. In addition, the minimum ventilation rate should be not less than 0.3 l/s per m^2 of internal floor area. (This includes all floors, e.g. for a two-storey building add the ground and first floor areas.)

b. This is based on two occupants in the main bedroom and a single occupant in all other bedrooms. This should be used as the default value. If a greater level of occupancy is expected add 4 l/s per occupant.

• Non-domestic ventilation requirement is shown in Table 5-5 and Table 5-6.

Table 5-5 Extract ventilation rates (UK Building Regulations Part L)

Room	Extract rate
Rooms containing printers and photocopiers in substantial use (greater than 30 minutes per hour)	Air extract rate of 20 l/s per machine during use. Note that, if the operators are in the room continuously, use the greater of the ***extract*** and ***whole building ventilation*** rates
Office sanitary accommodation and washrooms	***Intermittent*** air extract rate of: 15 l/s per shower/bath 6 l/s per WC/urinal
Food and beverage preparation areas (not commercial kitchens)	***Intermittent*** air extract rate of: 15 l/s with microwave and beverages only 30 l/s adjacent to the hob with cooker(s) 60 l/s elsewhere with cooker(s) All to operate while food and beverage preparation is in progress
Specialist buildings/spaces (e.g. commercial kitchens, sports centres)	See Table 6.3

Table 5-6 Whole building ventilation rate for air supply to offices (UK Building Regulations Part L)

	Air supply rate
Total outdoor air supply rate for offices (no smoking and no significant pollutant sources)	10 l/s per person

5.2.1.15 Different Systems of Mechanical Ventilation

Different systems of mechanical ventilation are shown in Figure 5-22 to Figure 5-24.

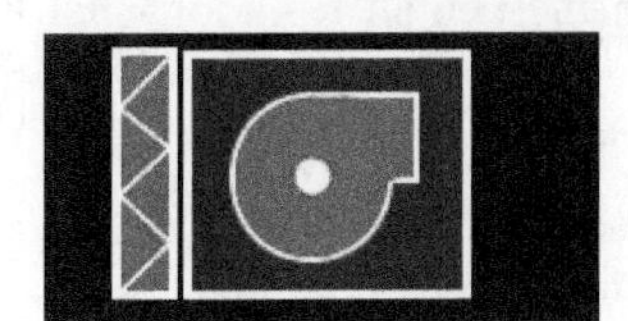

Figure 5-22 Mechanical ventilation: filtration only

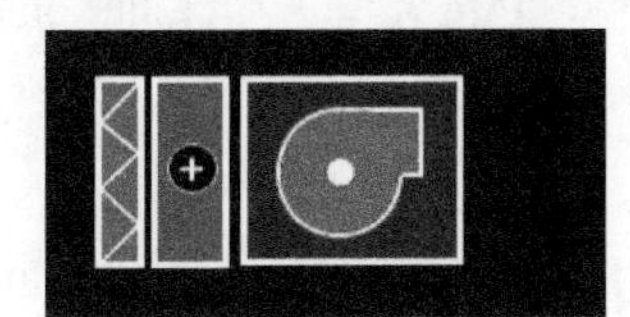

Figure 5-23 Mechanical Ventilation: filtration + heating

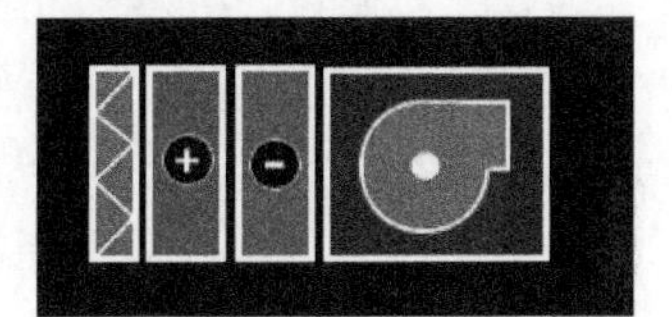

Figure 5-24 Comfort cooling: filtration + heating & cooling

5.2.1.16 Mechanical System Design

1. Ductwork Design

Suitable conditions for ductwork (Figure 5-25):

Figure 5-25 Ductwork

• Delivery to and from spaces;

• From central plant to rooms;

• Ductwork reduces leakage and space required;

• Required air volume (based on room loads, etc.) will be delivered in full to where it is needed.

Factors influencing ductwork design:

• Duct size;

• Volume flow rate (m^3/s) = Velocity (m/s) × Duct cross sectional area (m^2);

• Effect of velocity;

• Particles in air stream;

• Friction by square law;

• Turbulence and noise at fittings—changes of direction & branches.

The basis of size determination:

• Smaller ductwork is cheaper.

BUT:

• Friction in the ductwork increases fan power;

• Fan power (W) = Pressure (Pa) × Flow rate (m^3/s);

• Operating costs;

• Part L requirements (specific fan power).

• Part L refers to a specific section of building regulations in the United Kingdom that is focused on the conservation of fuel and power.

Process Requirements:

• Process applications (Figure 5-26) can require that certain velocities be maintained;

• Kitchen extraction;

• Dust extraction;

• Chemical extraction.

Figure 5-26 Process applications

Acoustic considerations:

Noise is probably the main reason. The noise mainly comes from regenerating noise due to the velocity and fan is the primary noise source. It can be prevented in two ways—selecting fans carefully and using attenuation, but they will increase the pressure drop (and hence fan noise and power).

Duct sizing (three main duct sizing methods):

• Static regain—Most suitable for long lengths of high velocity ducting;

• Equal pressure drop—Duct size reduces through the system and lowers installed cost;

• Velocity method—Uses standard velocities for applications, which is the most commonly used approach today.

(1) Static regain method (Figure 5-27):

• Velocity pressure + Static pressure = Total pressure;

• Considers a "perfect" expander.

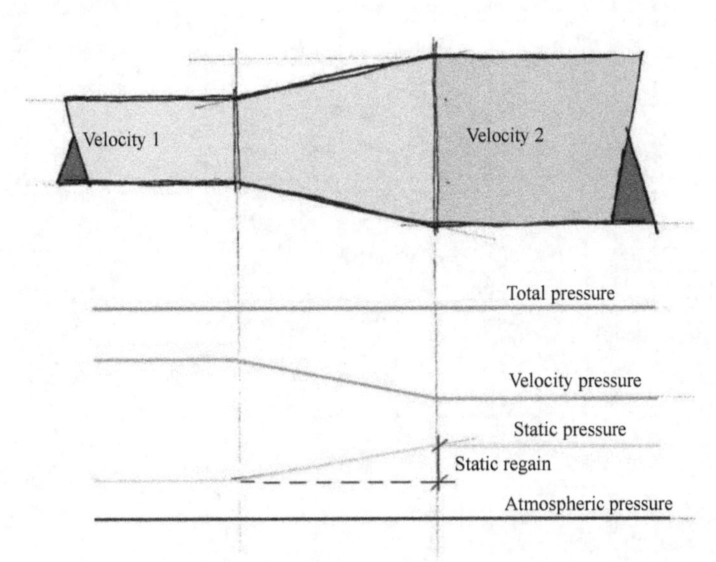

Figure 5-27 Static regain method

Advantages:

• Provides a size for all ductwork in the system;

• The system is virtually self balancing;

• Energy efficient;

• Good for medium/high pressure systems.

Disadvantages:

• Larger duct sizes at the end of branches;

• Trial and error approach required, i.e. an iterative approach;

• Not applicable to extract systems.

(2) Equal pressure drop method:

• Flow rate 0.5 m^3/s (Figure 5-28);

- Size for 1 Pa/m;
- Check velocity—5.5 m/s;
- Duct diameter—350 mm;

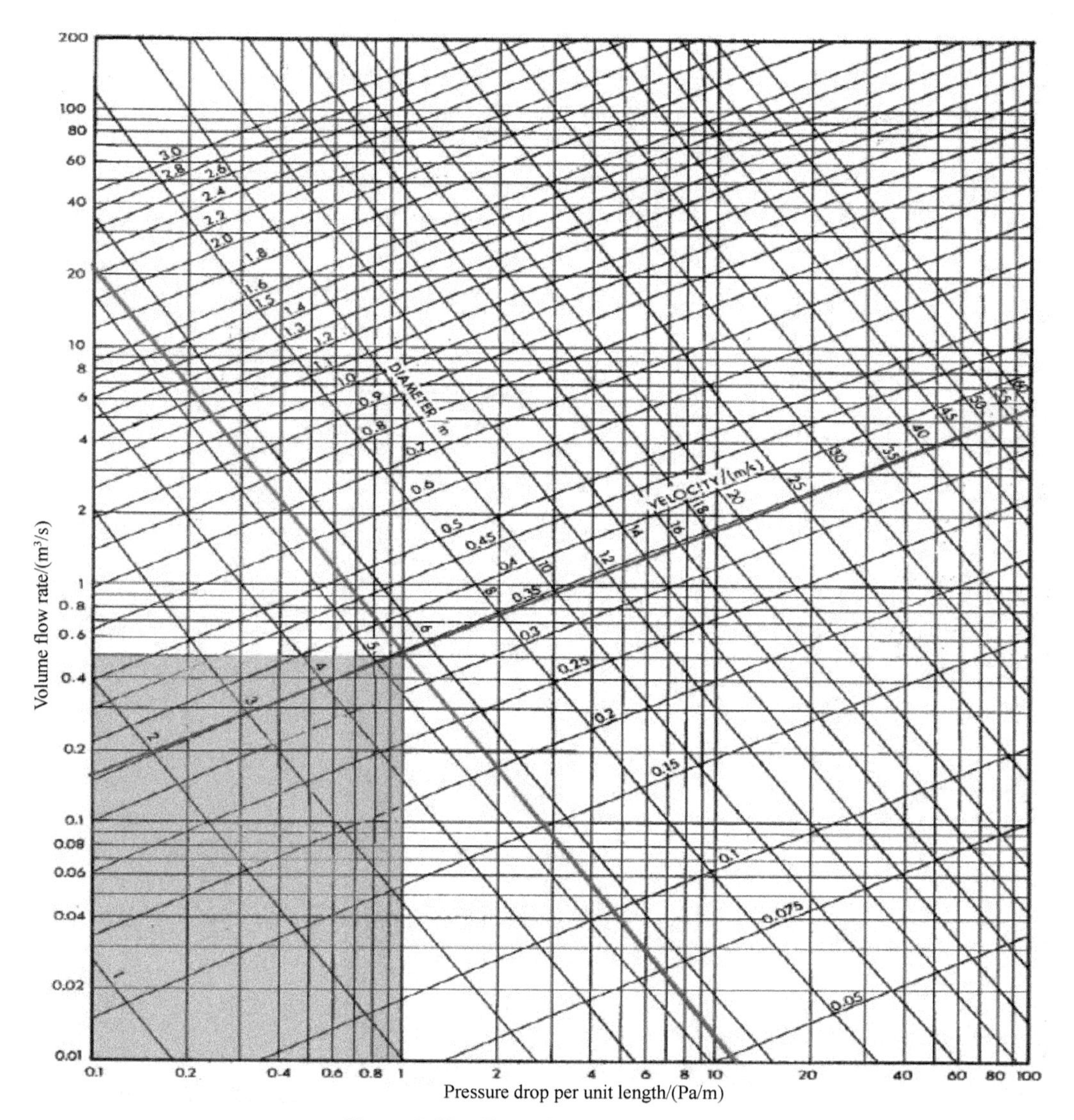

Figure 5-28 Flow of air in round ducts

- Duct—350 × 300 (Table 5-7).

Table 5-7 Equal volume flow rate, pressure drop and surface roughness

Dimen. of side, b	Dimension of side of duct, a																				Dimen. of side, b
	100	125	150	175	200	225	250	300	350	400	450	500	550	600	650	700	750	800	850	900	
100	110	123	134	145	154	163	171	185	199	211	222	232	242	251	260	268	276	284	291	298	100
125	347	138	151	162	173	183	192	209	225	239	251	263	275	285	295	305	314	323	331	339	125
150	385	394	165	178	190	202	212	231	248	264	278	291	304	316	327	338	348	358	368	377	150
175	421	430	440	193	206	218	230	251	269	287	303	317	331	344	357	369	380	391	401	411	175
200	454	464	474	484	220	233	246	269	289	308	325	341	356	371	384	397	409	421	433	444	200
225	485	496	507	517	527	248	261	286	308	328	346	364	380	395	410	424	437	450	462	474	225
250	515	527	538	549	560	570	275	301	325	346	366	385	402	419	434	449	463	477	490	503	250
300	570	583	596	608	620	632	643	330	357	381	403	424	443	462	479	496	512	527	542	556	300
350	620	635	649	662	676	689	701	714	385	412	436	459	481	501	520	539	556	573	589	605	350
400	667	683	698	713	727	741	755	768	794	441	467	492	515	537	558	578	597	616	633	650	400
450	710	727	744	760	776	791	806	820	848	874	496	522	547	571	594	615	636	655	674	693	450
500	751	770	787	804	821	837	853	869	898	927	954	551	577	603	627	650	672	693	713	733	500
550	790	810	828	847	864	882	898	915	946	976	1005	1033	606	633	658	682	706	728	749	770	550
600	827	848	867	887	905	924	941	959	992	1024	1054	1084	1112	661	688	713	738	761	784	806	600
650	862	884	905	925	945	964	982	1001	1036	1069	1101	1132	1162	1191	716	743	769	793	817	840	650
700	896	919	940	962	982	1002	1022	1041	1077	1113	1146	1179	1210	1240	1269	771	798	824	849	873	700
750	928	952	975	997	1018	1039	1059	1079	1118	1154	1190	1223	1256	1287	1318	1347	826	853	879	904	750
800	959	984	1008	1031	1053	1075	1096	1117	1157	1195	1231	1267	1301	1333	1365	1396	1426	881	908	934	800
850	989	1015	1039	1063	1086	1109	1131	1152	1194	1234	1272	1308	1344	1378	1411	1443	1474	1504	936	963	850
900	1018	1044	1070	1095	1119	1142	1165	1187	1230	1271	1311	1349	1385	1421	1455	1488	1520	1551	1582	991	900
950	1046	1073	1100	1125	1150	1174	1198	1221	1265	1308	1349	1388	1426	1462	1498	1532	1565	1597	1629	1659	950
1000		1101	1128	1155	1180	1205	1230	1254	1299	1343	1385	1426	1465	1503	1539	1575	1609	1642	1675	1706	1000
1050			1156	1184	1210	1236	1261	1285	1332	1378	1421	1463	1503	1542	1580	1616	1652	1686	1719	1752	1050
1100				1211	1239	1265	1291	1316	1365	1411	1456	1499	1540	1580	1619	1657	1693	1729	1763	1797	1100
1150					1266	1294	1320	1346	1396	1444	1490	1534	1577	1618	1658	1696	1734	1770	1806	1840	1150
1200						1322	1349	1375	1427	1476	1523	1568	1612	1654	1695	1735	1773	1811	1847	1882	1200
1250							1377	1404	1456	1507	1555	1602	1646	1690	1732	1772	1812	1850	1888	1924	1250
1300								1432	1486	1537	1587	1634	1680	1725	1768	1809	1850	1889	1927	1965	1300
1400									1542	1596	1648	1697	1745	1792	1837	1881	1923	1964	2004	2043	1400
1500										1652	1706	1758	1808	1857	1904	1949	1993	2036	2078	2119	1500
1600											1762	1816	1868	1919	1968	2015	2061	2106	2149	2192	1600
1700												1872	1926	1979	2029	2078	2126	2173	2218	2262	1700
1800													1982	2036	2089	2140	2189	2237	2284	2330	1800
1900														2092	2147	2199	2250	2300	2348	2396	1900
2000															2203	2257	2310	2361	2411	2459	2000
2100																2313	2367	2420	2471	2521	2100
2200																	2423	2477	2530	2582	2200
2300																		2533	2587	2640	2300
2400																			2643	2697	2400
2500																				2753	2500
	950	1000	1050	1100	1150	1200	1250	1300	1400	1500	1600	1700	1800	1900	2000	2100	2200	2300	2400	2500	
Dimen. of side, b	Dimension of side of duct, a																				Dimen. of side, b

$$d = 1{\cdot}265\left[\frac{(ab)^3}{a+b}\right]^{0.2}$$

Read diameters above stepped line up to top scale and diameters below stepped line down to bottom scale.

Advantages:

• Quick and easy to use;

• Gives size reduction at branches;

• More suitable for commercial buildings with complex duct routes.

Disadvantages:

• The system will use more energy than a system sized using the static regain method.

(3) Velocity method:

• Standard velocities—empirical values for different applications (Table 5-8);

Table 5-8 Velocity between splitters in primary attenuator

NR	Duct velocity/(m/s)			Velocity between splitters in primary attenuator/(m/s)
	Main	Branch	Runout	
20	4.5	3.5	<2.0	5
25	5.0	4.5	2.5	7
30	6.5	5.5	3.0	9
35	7.5	6.0	3.5	10
40	9.0	7.0	4.5	12

- Widely used;
- Gives simple approach to sizing;
- Avoids many potential noise issues;
- Needs to be combined with maximum pressure drops, particularly in main runs.

2. System Pressure

Extract mechanical ventilation is shown in Figure 5-29.

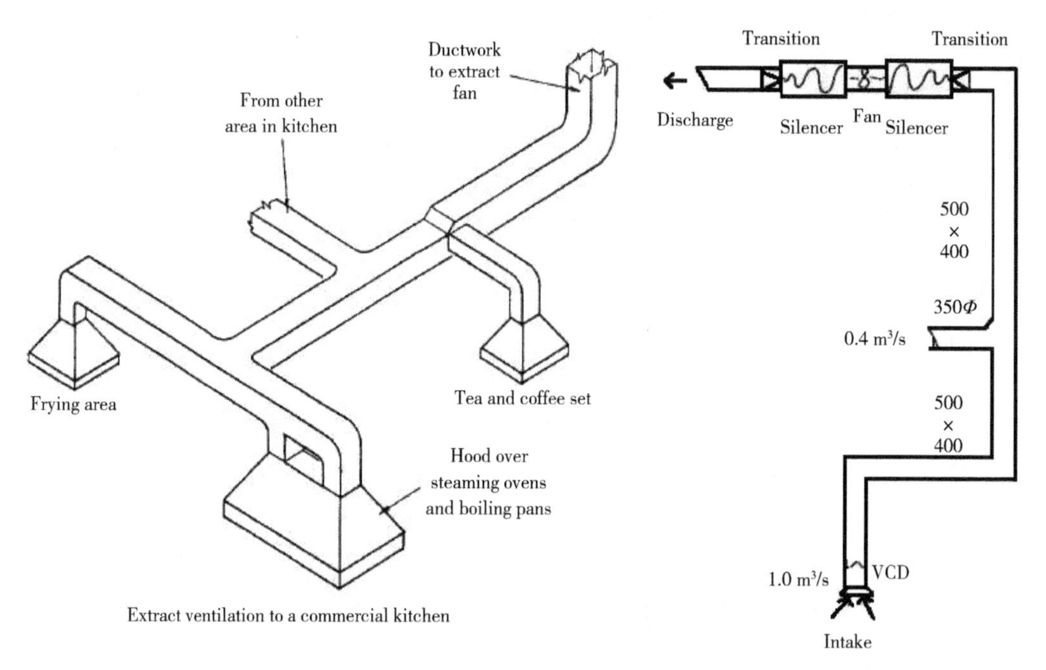

Figure 5-29 Extract mechanical ventilation

System resistance-effects from bends can be seen from Figure 5-30.

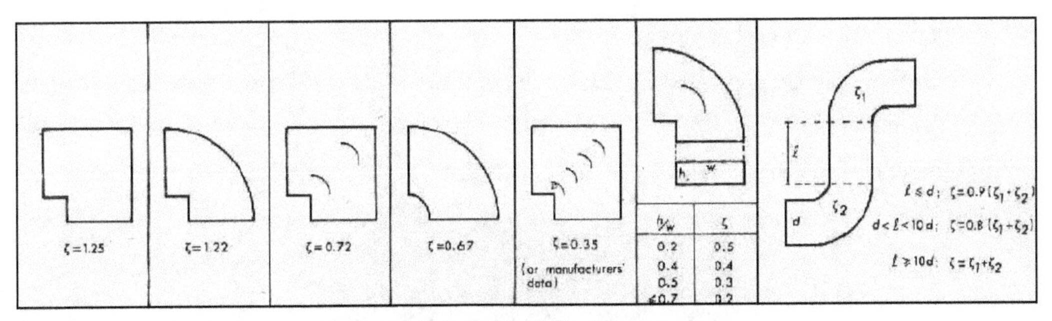

Figure 5-30 Miter bends

Pressure drop calculation is shown in Table 5-9.

Table 5-9 Pressure drop calculation sheet

Pressure Drop Calculation Sheet - Equal Volume Flow Rate, Pressure Drop and Surface Roughness method. (CIBSE p.C4-59) Enter data in Purple Cells

Location	Section Description	Volume Flow Rate (m^3/s)	Duct Width (mm)	Duct Height (mm)	Hydraulic Diameter (mm)	Velocity (m/s)	Dynamic Pressure (Pa)	Pressure Drop (Pa/m)	Section Length (m)	Straight Section Pressure Drop (Pa)	Section K Factor	Fittings Pressure Drop (Pa)	Total Section Pressure Drop (Pa)	Total Pressure Drop (Pa)
Intake	bellmouth	1	500	400	492	5.3	16.6	0.6		0.0	0.5	8.3	8.3	8.3
	VCD	1	500	400			Based on manufacturer's data					5.0	5.0	13.3
	2 mitre bends	1	500	400	492	5.3	16.6	0.6		0.0	0.7	11.6	11.6	24.9
	Straight Duct	1	500	400	492	5.3	16.6	0.6	3	1.8		0.0	1.8	26.8
	Through Branch	1.4	500	400	492	7.4	32.6	1.1		0.0	0.15	4.9	4.9	31.6
	Straight Duct	1.4	500	400	492	7.4	32.6	1.1	8	9.2		0.0	9.2	40.8
	mitre bend	1.4	500	400	492	7.4	32.6	1.1		0.0	0.35	11.4	11.4	52.2
	transition	1.4	500	400	492	7.4	32.6	1.1		0.0	0.2	6.5	6.5	58.8
	silencer	1.4	500				This is based on manufacturers data					15.0	15.0	73.8
	silencer	1.4	500				This is based on manufacturers data					15.0	15.0	88.8
	transition	1.4	500	500	551	5.9	20.7	0.7		0.0	0.2	4.1	4.1	92.9
	Straight Duct	1.4	500	500	551	5.9	20.7	0.7	2	1.3		0.0	1.3	94.2
Discharge	mesh opening	1.4	500	500	551	5.9	20.7	0.7		0.0	0.3	6.2	6.2	100.4
												Total	(+10%)	110

3. Fans (Figure 5-31)

There are three main types of fan used in building service applications.

- Backward curved centrifugal fan: Used for lower volume higher pressure systems.
- Forward curved centrifugal fan: Used for higher volume systems.
- Axial fan: Used for high volume low pressure systems.

Figure 5-31 Fans

(1) Backward curved centrifugal fan (Figure 5-32):

- Highest efficiency of all fans;
- Low volume flow rate/high pressure;

• Steep pressure/flow characteristic;

• Therefore changes in system resistance have little effect on volume flow rate, good for variable volume systems;

• Requires greater manufacturing time, skill & cost.

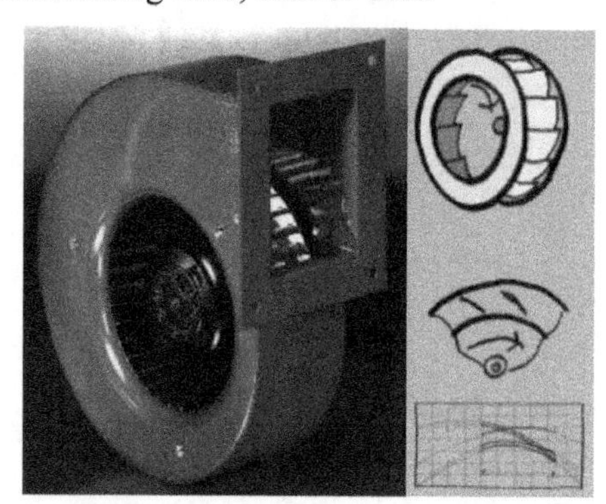

Figure 5-32 Backward curved centrifugal fan

(2) Forward curved centrifugal fan (Figure 5-33):

• In relation to fan size, considerable volume flow rate;

• Lower efficiency and lower noise levels than backward curved centrifugal fan;

• Flow rate increases substantially as pressure drops;

• If fan runs with lower system resistance than design, motor may overload;

• Cheaper to manufacture.

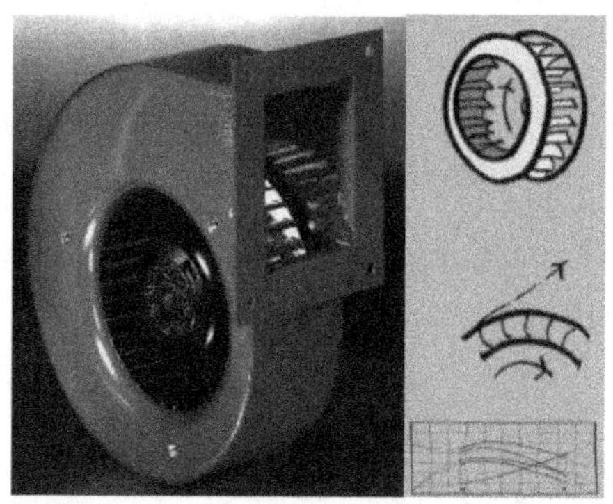

Figure 5-33 Forward curved centrifugal fan

(3) Axial fan (Figure 5-34):

• Compact;

• Low pressure/high volume flow rate;

• Risk of stalling;

• Alters blade pitch angle;

• Higher noise levels than centrifugal fans—at frequencies that are difficult to attenuate adequately;

• Easily fire rated;

• Efficiency and noise can be improved by more accurate control of tolerances—adding cost.

Figure 5-34 Axial fan

Different axial fan selections for the duty: 1.4 m^3/s @ 110 Pa.

3 selections:

• 40JM at 2 840 rpm;

• 50JM at 1 420 rpm;

• 63JM at 900 rpm.

Criteria considered:

• Noise;

• Efficiency;

• Cost.

Fan Selection No.1:

• Duty 1.4 m^3/s @ 110 Pa;

• Size 400 mm;

• Speed 2 840 rpm;

• Efficiency 58%;

• Noise Level 88 dB;

• Cheapest selection, BUT noisiest!

Fan Selection No.2:

• Duty 1.4 m^3/s @ 110 Pa;

• Size 500 mm;

• Speed 1 420 rpm;

• Efficiency 72%;

• Noise Level 81 dB;

• Most efficient selection!

Fan Selection No.3:

• Duty 1.4 m^3/s @ 110 Pa;

• Size 630 mm;

• Speed 900 rpm;

• Efficiency 64%;

• Noise Level 78 dB;

• Quietest selection, BUT most expensive!

Fan selection summary: 1.4 m^3/s @ 110 Pa (Table 5-10).

Table 5-10 Fan selection summary

	No.1 fan	No.2 fan	No.3 fan
Diameter/mm	400	500	630
Speed/rpm	2 840	1 420	900
Efficiency/%	58	72	64
Noise/dB	88	81	78
Power/W	600	300	350
Cost/£	560	750	900
Attenuation?	Yes + Yes	Yes	No

4. Attenuator Selection

An attenuator is also needed, and it adds pressure, increases fan noise and power (Figure 5-35). The following factors should be considered when selecting an attenuator:

• Circular or rectangular;

• Cost;

• Performance;

• Duct configuration;

• Pressure drop.

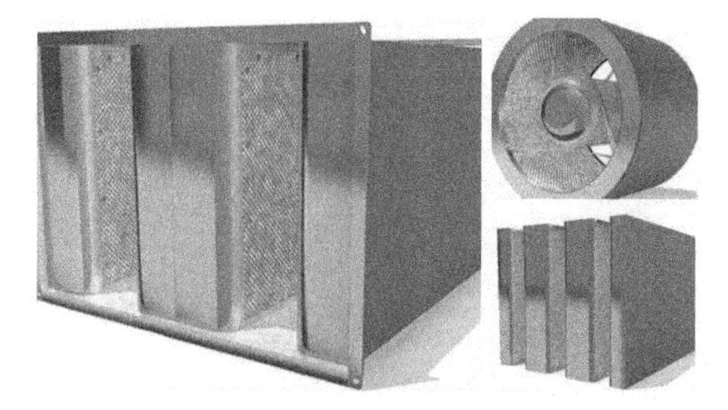

Figure 5-35 Attenuator

(http://www.ductacmind.com/en/2015/11/28/temper-1/attenuatormontage/)

5. Ductwork Shape

Circular ductwork (Figure 5-36):

• The most efficient in materials terms;

• Cheapest for smaller sizes;

• Looks neat;

• Thinner material, less weight & cost.

Rectangular ductwork (Figure 5-37):

• Offers greatest flexibility in aspect ratio;

• Thicker material, more weight & cost labor;

• Bracing & stiffening needed.

Flat oval ductwork:

• Offers benefit of reduced depth and also looks good;

• More expensive than rectangular ductwork.

Shape for application:

• Risers are normally most efficient in rectangular ductwork;

• Buildings mostly square;

• Fit to equipment;

• Structural openings;

Figure 5-36 Circular ductwork 1

Figure 5-37 Rectangular ductwork

• Limited ceiling voids and high air volumes give large ducts with high aspect ratios (Figure 5-38);

• Fire dampers, access doors, flexible connections.

Figure 5-38 Limited ceiling voids and high air volumes

Final runs in circular ductwork (Figure 5-39):

• Normally plenty of space by this point;

• Cheap and straightforward installation (Figure 5-40);

Figure 5-39 Circular ductwork 2

Figure 5-40 Exposed ductwork leads to other solutions

6. Ductwork Routing (Figure 5-41, Figure 5-42)

• Integrated into multi-disciplinary design;

• Avoids the structure;

• Fire zones;

• Ceiling voids;

• Commissioning and balancing.

Figure 5-41 Ductwork routing

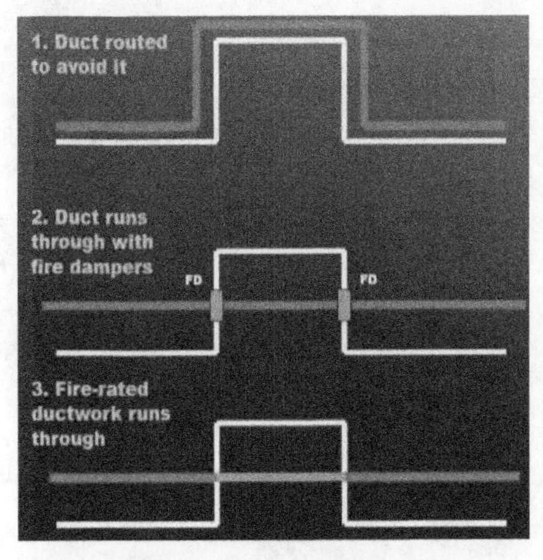

Impact of fire compartment

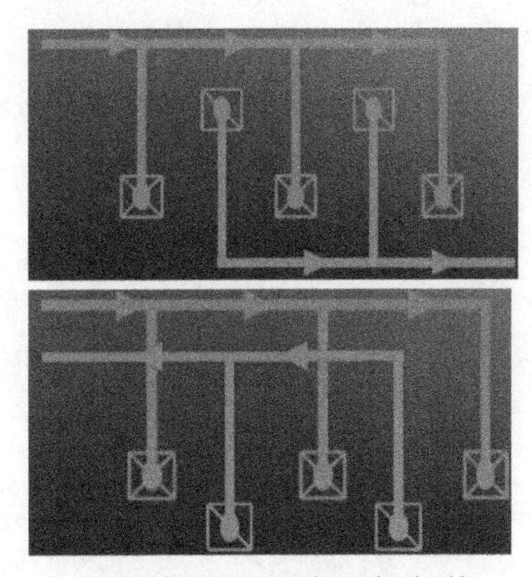

Avoiding cross-overs in restricted voids

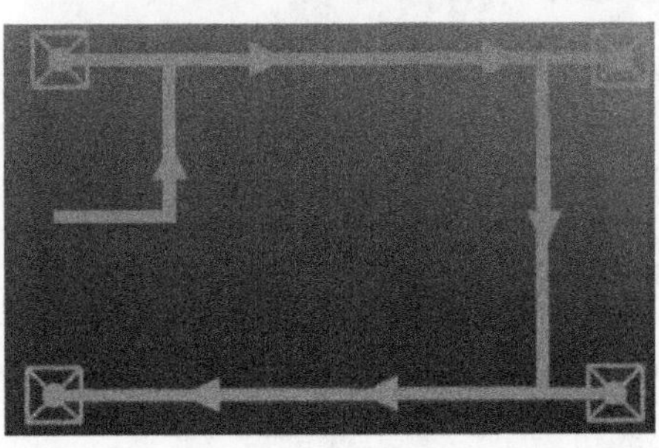

Bad ductwork design

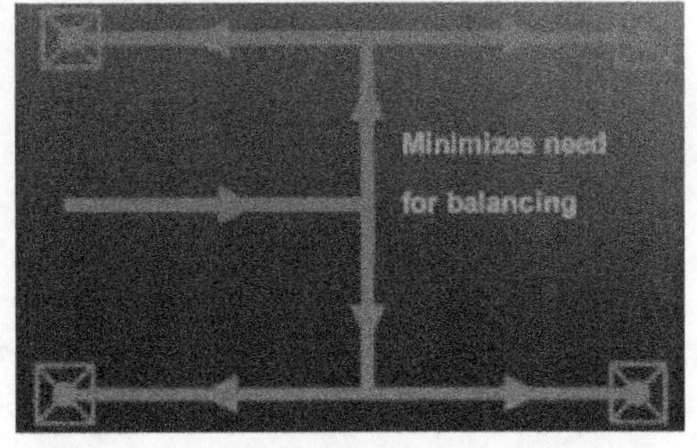

Self-balancing system

Figure 5-42 Ductwork design

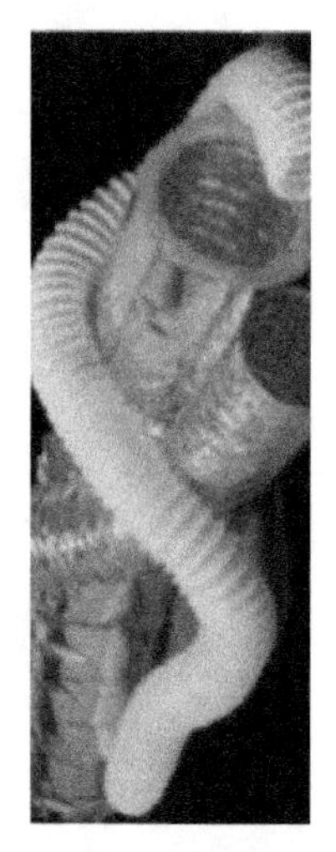
Figure 5-43 Plastic duct

7. Duct Materials

• Hot dip galvanized steel (normal);

• Plastic (corrosive environments, domestic) (Figure 5-43);

• Stainless steel (special applications);

• Pre-coated steel (special applications);

• Aluminium (not where fire rating is required);

• Flexible ductwork (terminal connections only);

• Fabric ductwork (supply only).

8. Ductwork Insulation

Thermal:

• To reduce heat loss or gain to surrounding area;

• Condensation protection.

Fire:

• To maintain compartment separation;

• To prevent fire and smoke spreading from source.

Noise:

• External insulation can reduce break-in/out (Figure 5-44);

• Internal lining can reduce noise levels, but it has installation and longevity issues.

Figure 5-44 External insulation

9. Fixing

Vibration isolation (Figure 5-45):

• Spring hangers;

• Flexible connections.

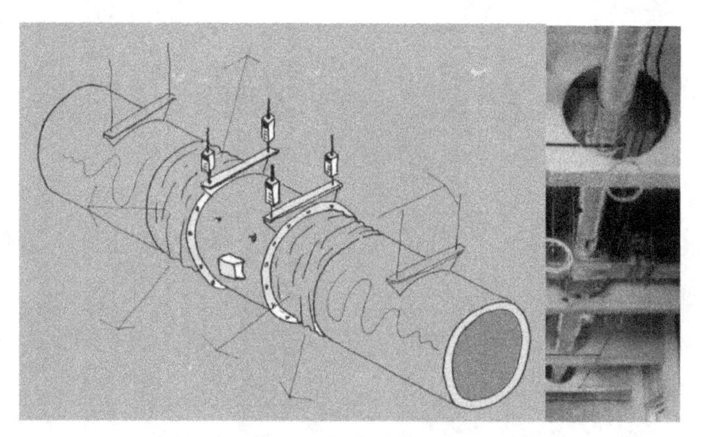
Figure 5-45 Vibration isolation

10. Other Considerations

Fittings:

- Volume control dampers or VCDs, self acting;
- Fire dampers;
- Bends and transitions (Figure 5-46);
- In-duct coils;
- Silencers;
- Terminal filters—grease, coarse, HEPA (high efficiency particulate absorber).

Figure 5-46 Bends and transitions

Access is required for:

- Inspection;
- Cleaning;
- Maintenance;
- Drain points;
- Commissioning.

5.2.1.17 Mechanical Ventilation with Heat Recovery (MVHR)

MVHR is the solution to the ventilation needs of energy efficient buildings, is also called as heat recovery ventilation (HRV) or comfort ventilation. A heat recovery ventilation system properly fitted into a house provides a constant supply of fresh filtered air maintaining the air quality whilst being practically imperceptible (Figure 5-47).

Figure 5-47 Equipment mechanical ventilation with heat recovery

MVHR works quite simply by extracting the air from the polluted sources e.g. kitchen, bathroom, toilets and utility rooms and supplying air to the "living" rooms e.g. bedrooms, living rooms, studies (Figure 5-48). The extracted air is taken through a central heat exchanger (Figure 5-49) and the heat recovered into the supply air. This works both ways, if the air temperature inside the building is lower than the outside air temperature then the coolth is maintained in the building.

1) Benefits of heat exchanger

(1) Air quality:

• Continuous supply of fresh air to provide good indoor air quality;

• No drastic CO_2 peaks;

• No build-up of air pollution, e.g. from carpets and furniture or radioactive radon;

• Elimination of bad odors;

• Filtered air, for example, pollen filter is a great advantage for allergic people;

• Keeps the midges out of the house.

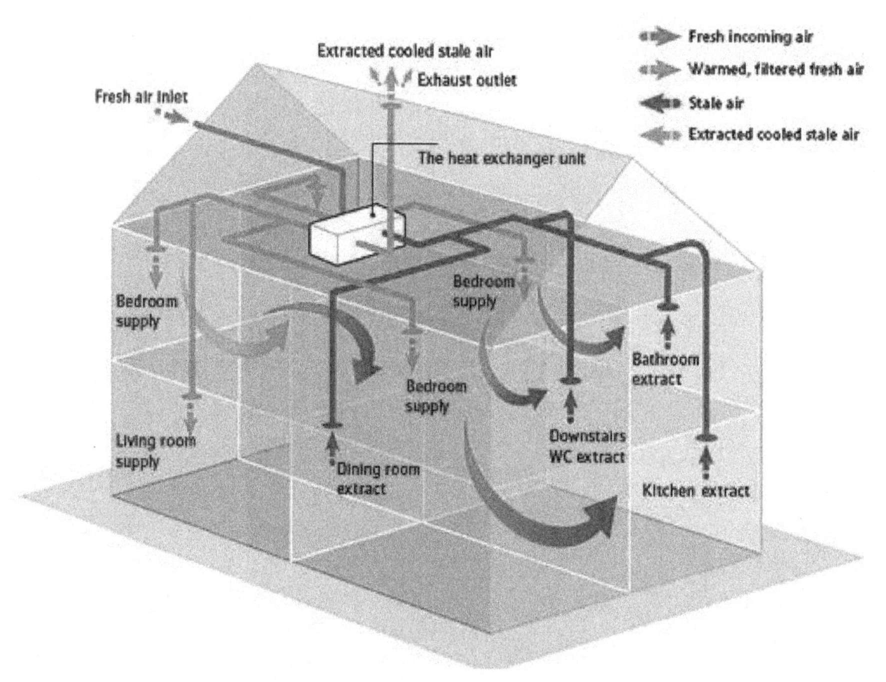

Figure 5-48 The working process of MVHR

(http://www.acarchitects.biz/self-build-blog-mvhr-systems/)

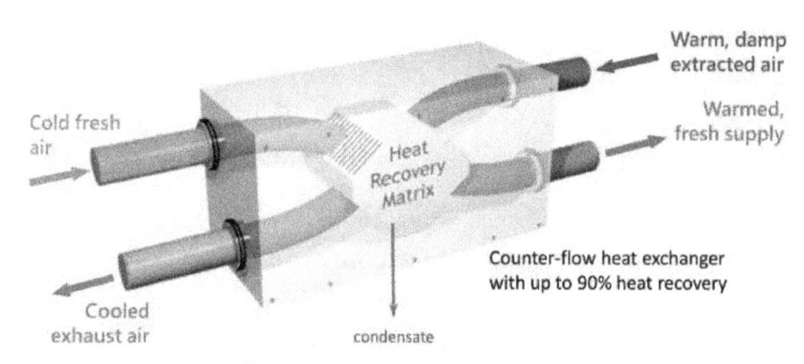

Figure 5-49 Heat exchanger

(http://www.acarchitects.biz/self-build-blog-mvhr-systems/)

(2) Humidity control:

• Preservation of the building fabric through steady ventilation;

• Keeps mould, fungus, dust mites in check;

• Active dehumidification in the cold season.

(3) Comfort:

• Less noise inside (windows can remain closed), undisturbed sleep;

• No drafts (in conjunction with an airtight building fabric);

• Good indoor climate.

(4) Energy saving:

• Approximately 30% of the heating energy can be saved in airtight buildings with highly efficient MVHR systems compared to naturally (uncontrolled) ventilated buildings (Figure 5-50).

Figure 5-50 The role of heat exchanger

(https://www.youtube.com/watch?v=dwXGYlFm8Ko)

2) Design example (Figure 5-51)

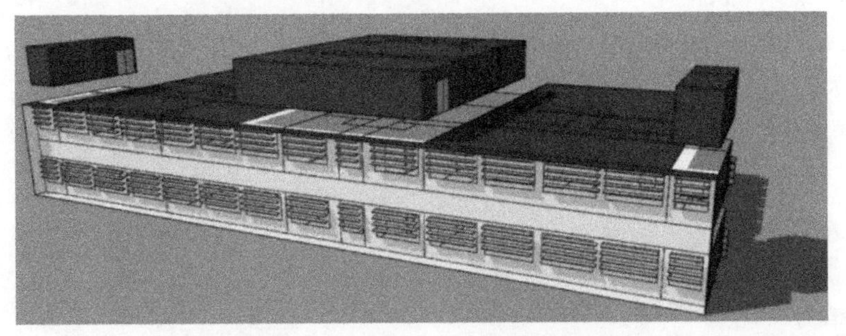

Figure 5-51 A common dirty extract system to serve all the WCs over both floors of the case study building

3) Design steps

Step 1: Ventilation rate determination (Table 5-11).

Table 5-11 Extract ventilation rates

Room	Extract rate
Office sanitary accommodation and washrooms	***Intermittent*** air extract rate of: 15 l/s per shower/bath 6 l/s per WC/urinal

Required minimum extract ventilation rates = 6 l/s per WC.

5 toilet bowls were designed to install in each toilet from the layout CAD drawing in brief document.

Thus, required minimum extract ventilation rates for each toilet = 6 l/s per WC × 5= 30 l/s = 0.03 m^3/s.

Step 2: Grille Selection (Figure 5-52).

A Gilberts Series PE Extract Grilles was selected as it can provide a comprehensive performance and established range of flexible size.

The size range is from 100 × 100 up to 1 200 × 1 200 in 1mm increments as standard.

Extract grille size:	250 x 250
Extract air volume flow rate:	0.05 m^3/s
Jet velocity:	4 m/s
Pressure drop:	15 Pa

Recommended comfort criteria for specific applications — *continued*

Building/room type	Winter operative temp. range for stated activity and clothing levels			Summer operative temp. range (air conditioned buildings†) for stated activity and clothing levels			Suggested air supply rate / ($L \cdot s^{-1}$ per person) unless stated otherwise	Filtration grade	Maintained illuminance / lux	Noise rating§ (NR)
	Temp. / °C	Activity / met	Clothing / clo	Temp. / °C	Activity / met	Clothing / clo				
General building areas:										
— toilets	19–21	1.4	1.0	21–23	1.4	0.65	>5 ACH	G4–G5	200	35–45

Figure 5-52 Grille selection

Step 3: Ductwork Design (Figure 5-53).

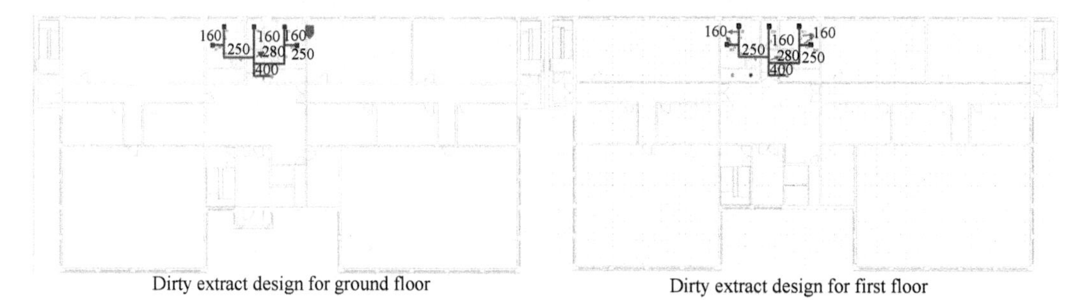

	System resistance /(N/m^2)	Total flow rate /(m^3/s)	On floor
Dirty extract	51.25	0.5	1

Figure 5-53 Ductwork Design

Step 4: Fan Selection (Figure 5-54).

The Nuaire Air-Volve Twin Fan External outdoor unit was selected. The low noise and SFP's were guaranteed by manufacturer to meet the latest current legislation and building regulations with a local constant pressure controller. The detail catalogue was attached in Appendix C.

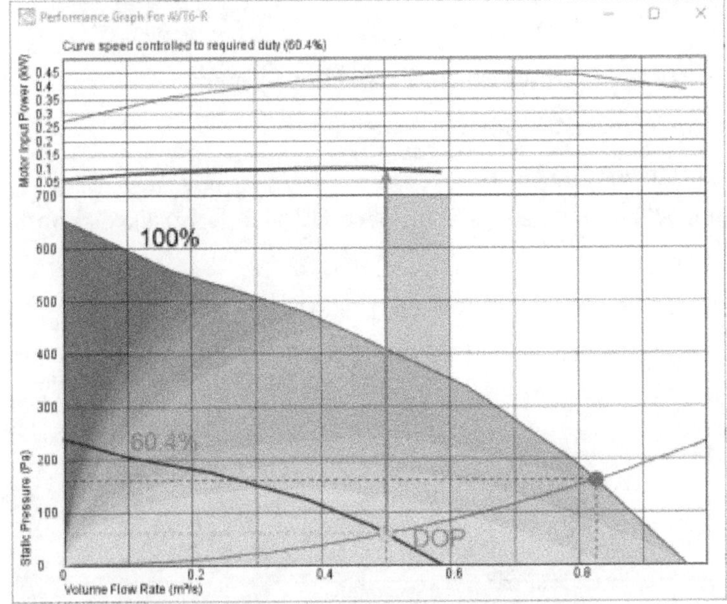

Option 2: when the fan was operated at 60.4% of full speed to provide the design condition, the power input was 0.1 kW.

The *Specific Fan Power (SPF)* = $\frac{Design\ Power\ Input}{Design\ air\ flow\ rate} = \frac{0.1\ kW}{0.5\ m^3/s} = 0.2$

The capital cost of option 2 may be higher than option 1 as the fan operating capacity was bigger but option 2 was provided a smaller specific fan power and the dirty extract system would be operating at design condition constantly, the energy consumption would be reduced thus for the operating cost. Therefore, option 2 Nuaire AVT6-R was determined to serve the dirty extract system.

Figure 5-54 Fan selection

5.2.1.18 Pro's & Con's of Mechanical Ventilation

Advantages of mechanical ventilation:

• Control of air-flow;

• Control of air conditions;

• Possibility of heat recovery;

• Potential to reduce energy consumption and running costs;

• Control of noise;

• Control of pollution.

Disadvantages of mechanical ventilation:

• Additional capital cost;

• The inappropriate installation of a mechanical ventilation system may increase the maintenance costs dramatically.

• Additional space requirements;

• Additional maintenance requirements.

5.2.2 Heating System

5.2.2.1 Composition of Conventional Heating System

• Fuel (e.g. gas, coal, oil and wood);

• Heat source (e.g. boilers and solar energy);

• Heat transfer medium (e.g. water, air, oil, electricity and steam) and pipework;

• Power (e.g. pumps);

• Heat emitters (e.g. radiators and convectors);

• Control devices (optional) (Figure 5-55).

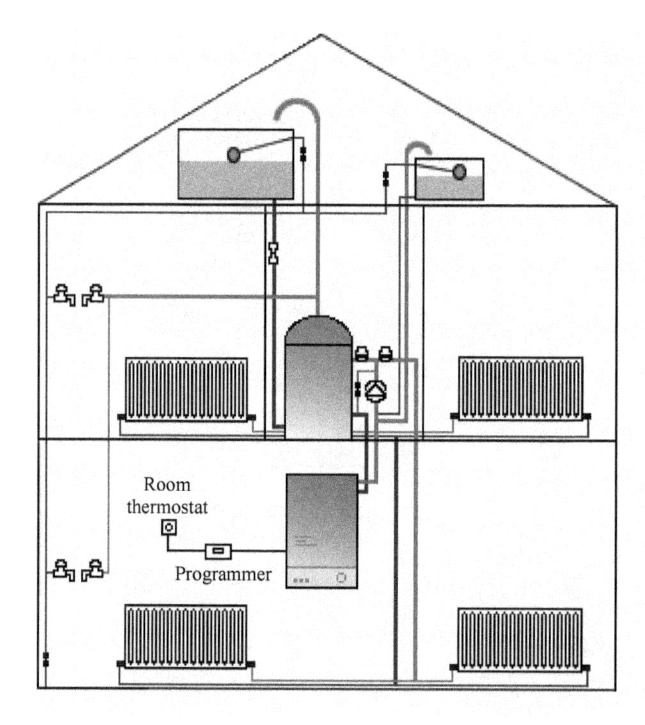

Figure 5-55 Composition of the conventional heating system
(http://www.ukgasspares.co.uk/ConventionalTraditionalBoilerSystem.htm)

5.2.2.2 Boilers

Boiler is a kind of energy conversion equipment. The energy input to the boiler is chemical energy in fuel, electrical energy, and the boiler outputs with a certain amount of thermal energy steam, high temperature water or organic heat carrier (Figure 5-56).

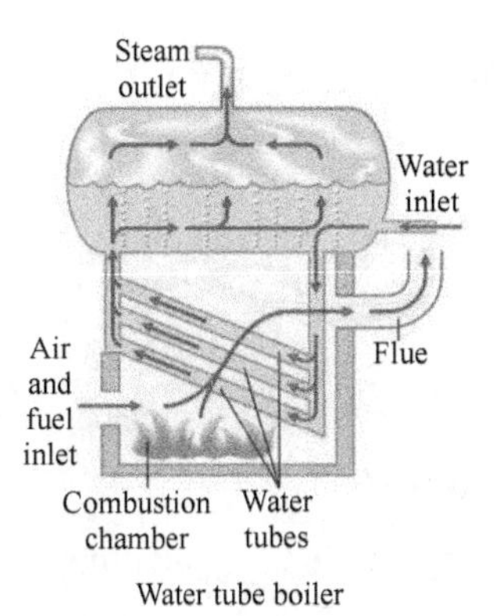

Water tube boiler

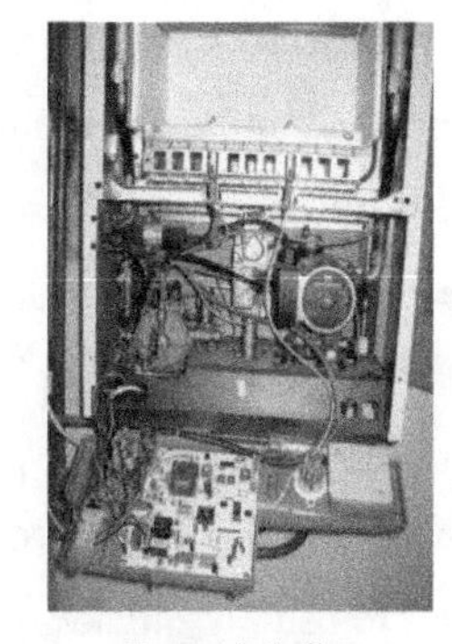

Domestic boiler

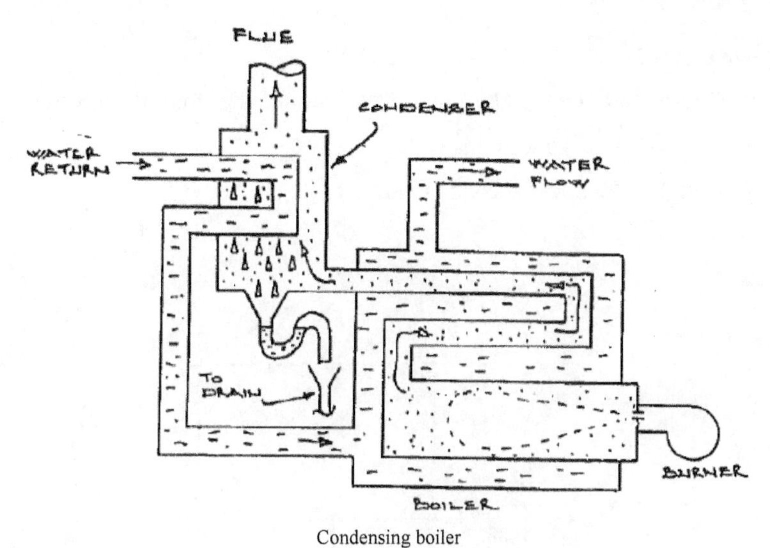

Condensing boiler

Figure 5-56 Types of boiler

Types of boiler systems are as follows.

1) Conventional boiler systems (Figure 5-57)

- People also call them regular or conventional boiler systems;
- They send hot water to your radiators and a hot water cylinder;
- Requiring a cold-water storage tank, normally located in the loft.

Figure 5-57 Conventional boiler systems

2) Combi boiler systems (Figure 5-58)

• Combi or combination boilers, are the UK's most popular type of boiler.

(1) Advantages:

• You can get unlimited hot water whenever you need it because it heats water straight from the mains.

• You don't need a cold water tank or hot water cylinder, so you save space. That means they work particularly well if you live in a smaller house or flat.

• They're usually quicker to install because they've got less boiler parts than other heating systems.

(2) Disadvantages:

• If you use more than one tap at a time, it will reduce the flow of hot water. But that does also depend on the strength of your mains water pressure.

• Your water might take a few seconds to heat up.

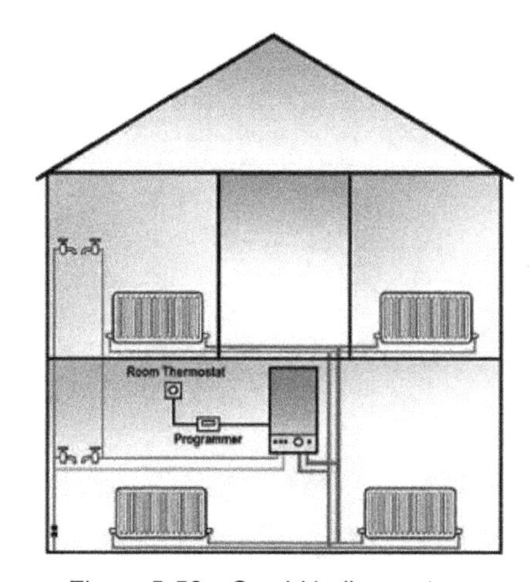

Figure 5-58 Combi boiler systems

3) System boiler systems

• A system boiler directly heats your central heating system and produces hot water for a storage cylinder.

• They take their water supply directly from the mains, so no cold-water storage tank is needed.

Advantages of system boiler systems :

• Easier installation compared to conventional boiler systems;

• Great for high hot water demand;

• Stronger water pressure;

• Fast and economical with a pump to shorten response time;

• No need for a cold water feed tank, hence saving space;

• Compatible with solar thermal heater.

5.2.2.3 Pumps (Figure 5-59 and Figure 5-60)

A pump is a machine that conveys a fluid or pressurizes a fluid. It transmits mechanical energy from the prime mover or other external energy to the fluid to increase the energy of the fluid.

Single pump (pump curve):

$$H=p/\rho g$$

where H is hydraulic head (m); p is pressure (MPa); ρ is density (kg/m^3); g is acceleration due to gravity (m/s^2).

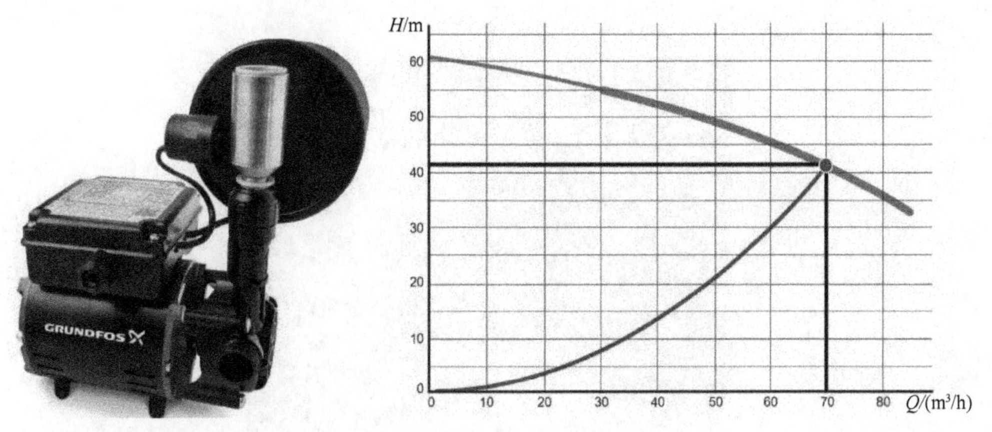

Figure 5-59 Pump curve

(http://www.snhtradecenter.co.uk/shower-pumps/grundfos-2bar-universal-single-shower-pump-ssr2-cn.html)

When two (or more) pumps are arranged in series, their resulting pump performance curve is obtained by adding their heads at the same flow rate as indicated in Figure 5-60(a).

When two or more pumps are arranged in parallel, their resulting performance curve is obtained by adding the pumps flow rates at the same head as indicated in Figure 5-60(b).

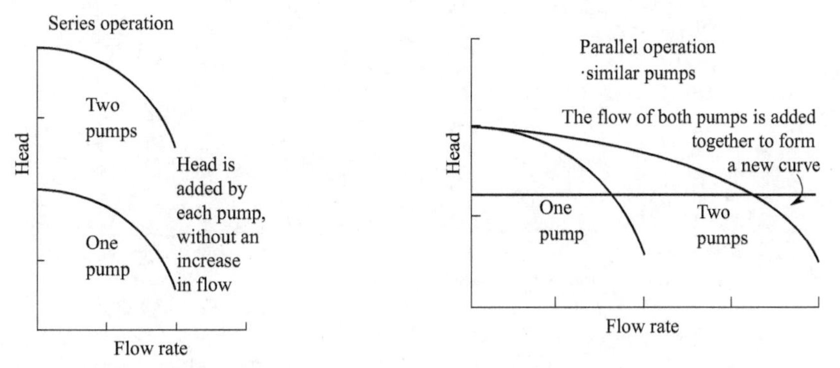

Pumps in series—each pump operates at the same flow rate

Pumps in parallel—each pump operates at the same pressure

Figure 5-60 Different working processes of pumps

(http://webwormcpt.blogspot.co.uk/2008/04/series-parallet-pump-calculator.html)

5.2.2.4 Expansion Tank

The expansion tank (Figure 5-61) is an important part of the HVAC system that provides space in which the liquid can expand or contract as its volume changes with temperature. Expansion tanks are generally made of steel plates and usually round or rectangular in shape. The expansion tank can also serve to stabilize the pressure of the system and remove the air released by the water in the heating process.

Three basic configurations (Figure 5-62):

- An open tank;
- A closed tank;
- A diaphragm tank.

Figure 5-61 Expansion tank

(https://terrylove.com/forums/index.php?threads/water-heater-expansion-tank.54062/)

Closed tank (contains a capture volume of compressed air and water)

p_1

h

x

$p_x = p_1 + \rho_w h$

Diaphragm tank (a flexible membrane is inserted between the air and water)

p_a

h

x

$p_x = p_a + \rho_w h$

Open tank (open to the atmosphere)

x

h

p_1

$p_x = p_1 - \rho_w h$

Figure 5-62 Three basic configurations

5.2.2.5 Piping Arrangement

According to piping arrangement type, there are two types of pipe systems (Figure 5-63).

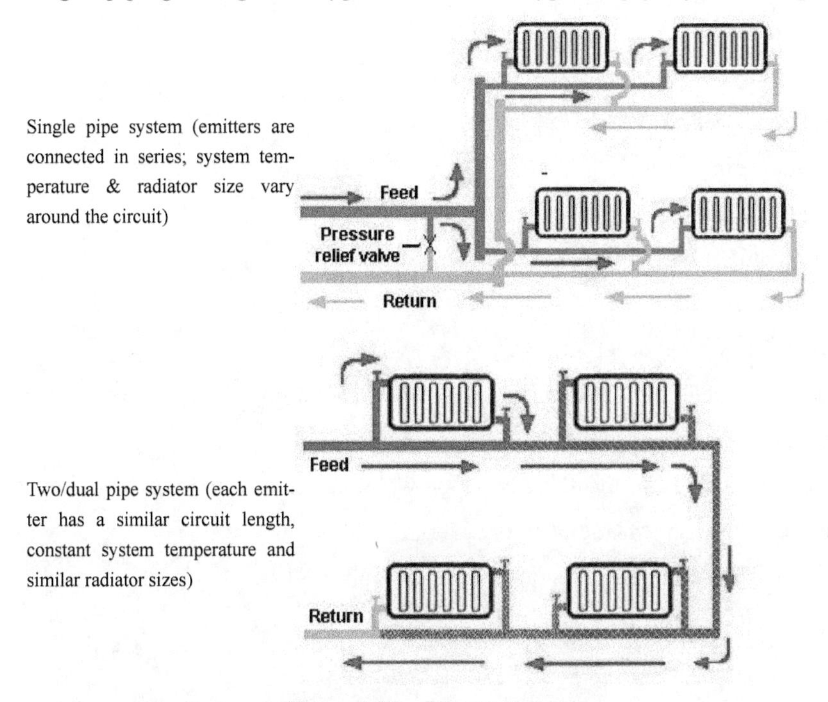

Figure 5-63 Pipe systems

(http://www.traderadiators.com/blog/What-are-the-different-types-of-pipework-installations-used-in-central-heating-systems)

5.2.2.6 Heat Transfer Modes

There are three heat transfer modes (Figure 5-64).

• Conduction: involves the transfer of heat from one material to another through direct molecular contact.

• Convection: involves the transfer of heat from one place to another by the motion of a gas or a liquid across the heated surface.

• Radiation: involves the transfer of heat through radiation due to temperature difference.

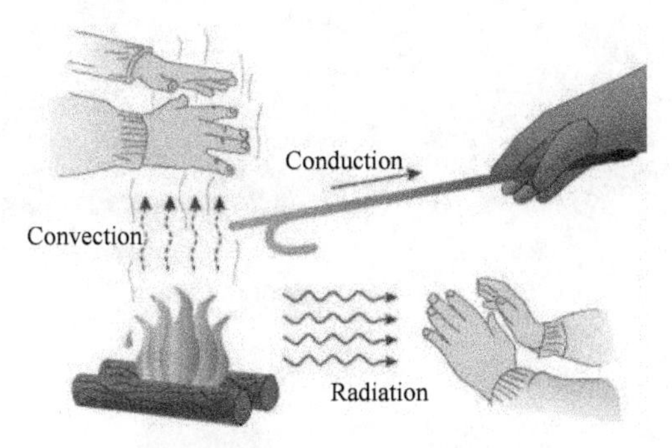

Figure 5-64 Heat transfer modes

(http://spot.pcc.edu/-lkidoguc/Aquatics/AqEx/Water_Temp.htm)

5.2.2.7 Heat Emitters

Types of heat emitters are listed below:

- Radiators;
- Convectors;
- Underfloor heating system;
- Radiant heating system;
- Warm air heaters;
- Fan coil units;
- Air handling units.

1. Radiators

- Heat output includes 60% convection and 40% radiation (Figure 5-65);
- Generally used in naturally ventilated buildings;
- Typically located around the perimeter walls & under windows to offset cold down-draughts and prevent condensation (Figure 5-66);
- Simple with good local temperature control (Table 5-12).

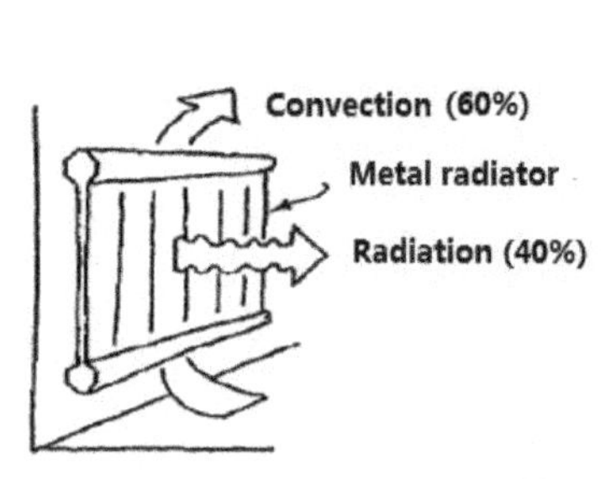

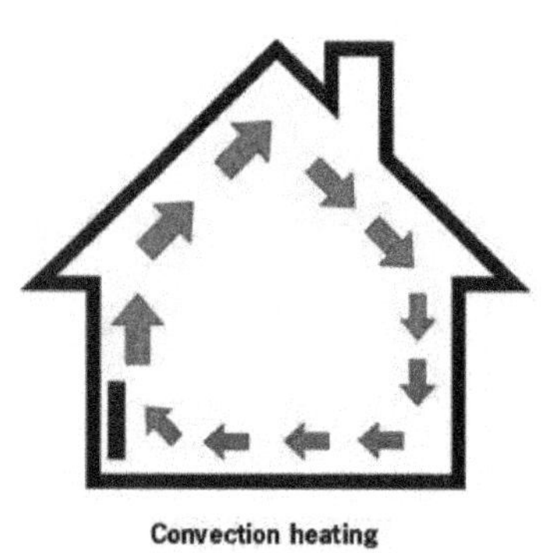

Figure 5-65 Heat output

Cast iron radiator Towel rail Storage heater Panel radiator Electric tube radiator

Figure 5-66 Types of radiators

Table 5-12 Radiators (Pro's & Con's)

Benefits	Limitations
Simple & compact	Slow thermal response
Low maintenance	Impact on floor layout
Familiar	Restrain re-decoration
Good temperature control	
Variety in design	

Radiator control is conducted by thermostatic radiator valves (Figure 5-67).

(https://www.greatrads.co.uk/products/the-radiator-company-ancona-column-radiator)

(https://www.screwfix.com/p/danfoss-ras-c-white-chrome-angled-trv-15 mm-x-15 mm/70871)

Figure 5-67 Thermostatic radiator valves (TRV)

2. Convectors

The surface area of the heating element in the convector is made up of fins attached to hot water heating pipes (finned pipe), which is used where the heating load would require too large an area of radiators. Convectors have two types: natural and fan convectors (Figure 5-68).

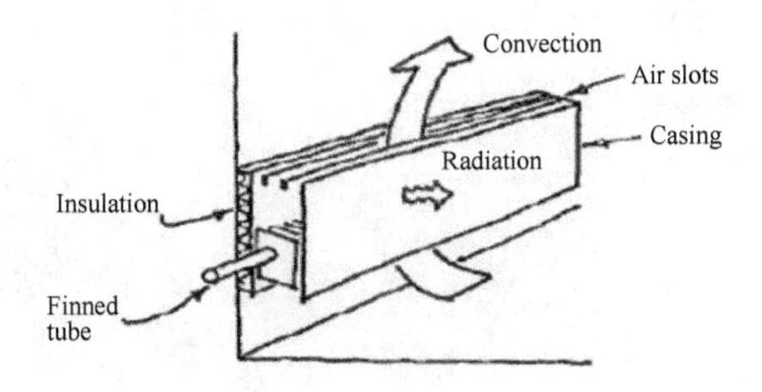

Figure 5-68 Convector

1) Natural convectors

• The convector consists of a casing enclosing the finned pipe through which air is drawn by buoyancy forces (Figure 5-69);

• For local control a damper is sometimes provided;

• A form of natural convector is often used to protect large areas of glazing by placing the heating element (fined pipe) at low level below the window.

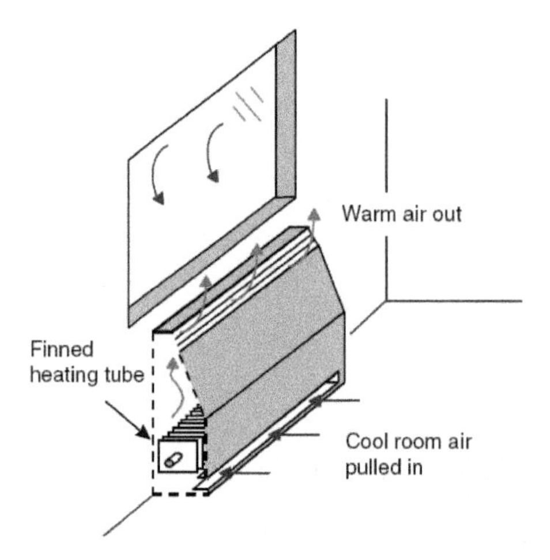

Figure 5-69 Internal view of a natural convector

2) Fan (or forced) convectors (Figure 5-70, Figure 5-71 and Table 5-13)

• In fan convectors the air is forced through the emitter by fans (mechanical device);

• This forced circulation allows more heat to be transferred to the air from the water;

• There is an increase in heat output per unit size;

• The location of mechanical convectors improves the circulation of air in a room (Figure 5-71);

• Time taken to heat a room from cold is reduced;

• Fan speed can be controlled to suit different requirements.

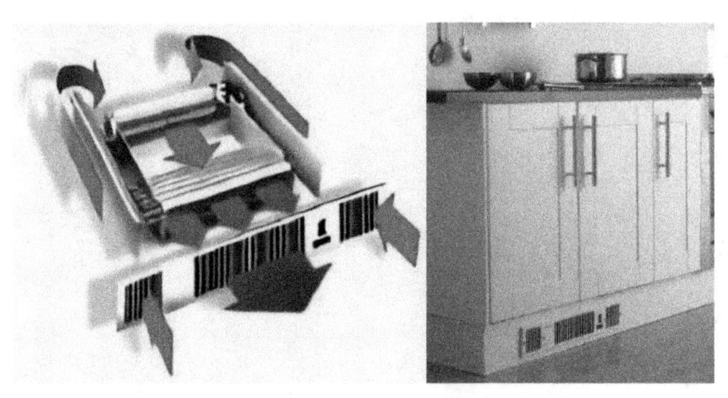

Figure 5-70 Fan convector

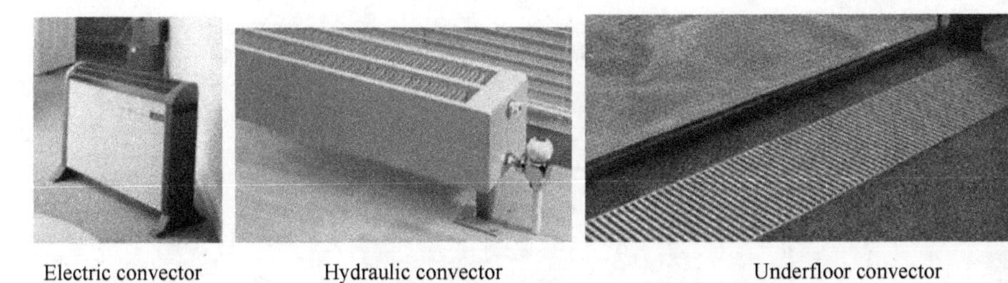

Figure 5-71 Convector examples

Table 5-13 Convectors (Pro's & Con's)

Benefits	Limitations
Faster warm up time compared to radiators	Dust collects in casing which requires periodic cleaning
High output per unit	Fan convectors require more maintenance
Flexibility in casing design	Fan convectors require a power supply
Good air movement	Fan convectors get noisier with age
Casings reduce risk of burning for occupants	

3. Underfloor Heating System

Underfloor heating system (Figure 5-72) comprises of plastic pipes or heating cables embedded between a top layer of screed and the insulation, which has two basic forms: water and electric. Floor temperature is normally around 40 ℃, lower than radiator systems. Floor heating pipe arrangement needs to calculate the pipe spacing, laying shape; if the pipe arrangement is not reasonable, it will cause uneven heating (Figure 5-73). The choice of insulation material also requires more effort; if the insulation material is not good, the heat will be distributed to the downstairs, and in addition to affecting the experience, will also cause the use of the cost to increase, and even the release of harmful substances. This kind of heat emitter dispense with holes in the floor. Benefits and limitations of underfloor heating system are listed in Table 5-14.

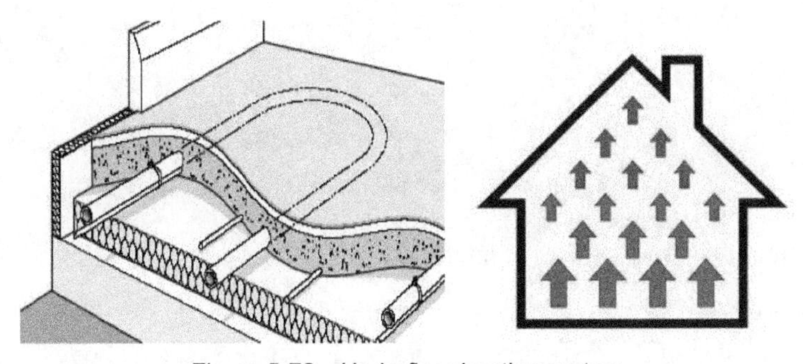

Figure 5-72 Underfloor heating system

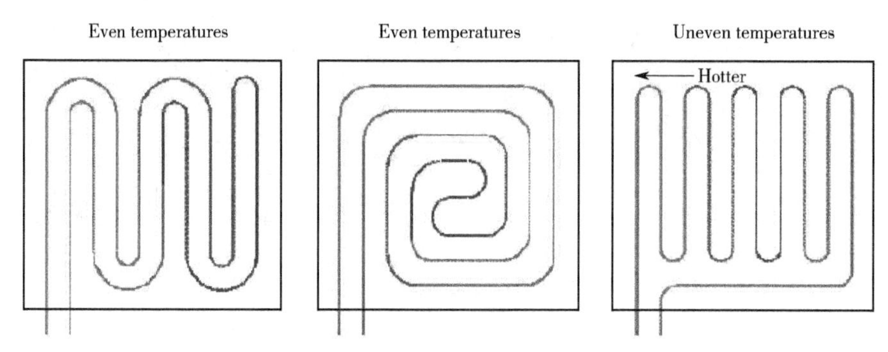

Figure 5-73 Underfloor piping layout

(http://www.heatweb. com/techtips/Underfloor/underfloorheating.html)

Table 5-14 Underfloor heating system (Pro's & Con's)

Benefits	Limitations
Invisible heating system	Heat output is limited
Relatively even temperature distribution throughout the space	Slow response to changes in temperature setting
Suited to public spaces e.g. foyers, churches, shopping complexes, nursery schools	Not applicable for buildings with raised floors for under floor services e.g. office power and data systems

4. Radiant Heating System

Radiant heating system (Figure 5-74) generally provides radiant heat downwards into the occupied area that warms people directly without heating the air within the space and suited to tall buildings with large openings e.g. warehouses, industrial units and sports halls. There are four types of radiant panel or heater.

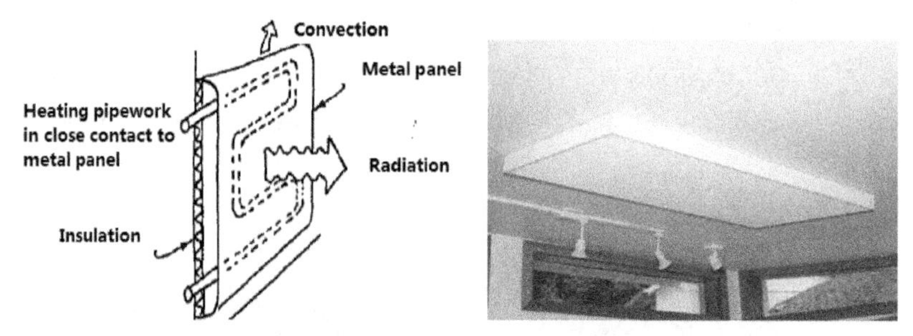

Figure 5-74 Radiant heating system

1) Gas fired radiant tube heater (Figure 5-75)

Gas is burnt in a long metal tube, the surface temperature of the tube becomes hot, and then the heat is radiated to the occupants. A reflector is usually located above the tube. The system requires a gas supply and an electrical supply to power a fan and is normally provided with a flue outlet. It is a very efficient system with over 80%-90% efficiency achieved in the process.

2) Gas fired plaque heater (Figure 5-76)

Gas is burnt in close proximity to ceramic plaques mounted in a metallic reflector, the plaques glow red hot and thereby produce a radiant heat output. The radiant heat is directed into the occupied space below. In these systems the products of combustion usually pass directly into the building and therefore sufficient ventilation must be provided. They are not a popular system.

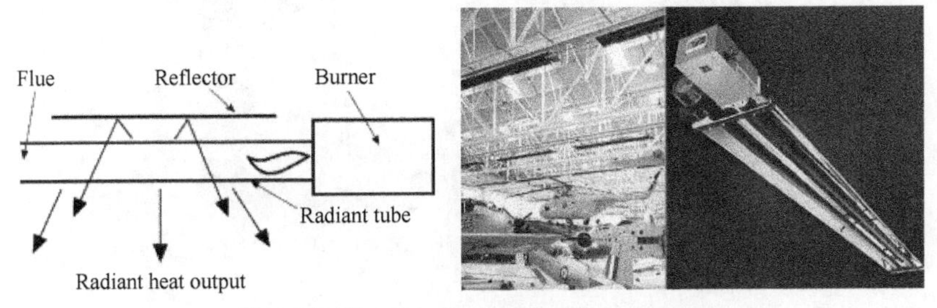

Figure 5-75 Gas fired radiant tube heater

Figure 5-76 Gas fired plaque heater

3) Electric quartz heater (Figure 5-77)

Electric quartz heaters have replaced gas fired plaque heaters and are reasonably inexpensive. They provide an easily controllable (on/off) system, and electric quartz lamps produce radiant heat output in front of metallic reflectors that direct the output to the space below. Heat output is produced very quickly once the unit is switched on. In addition, the maximum mounting height is approximately 5 m.

Figure 5-77 Electric quartz heater

4) Hot water radiant panel (Figure 5-78)

Heating pipework is attached to a plate to provide a greater surface area for the emission of radiant heat. Pressed steel radiators may also be used in lower pressure systems. These systems are often used in commercial/factory type buildings. Benefits and limitations of hot water radiant panel are listed in Table 5-15.

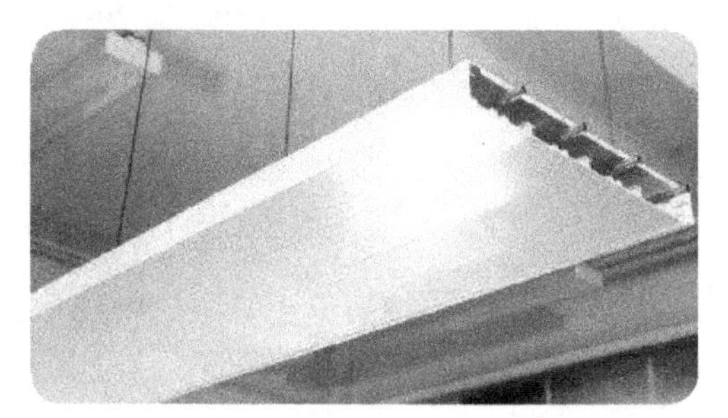

Figure 5-78 Hot water radiant panel

Table 5-15 Radiant Panel (Pro's & Con's)

Benefits	Limitations
A building can be pre-heated more rapidly than a warm air system	Radiant heat can cause some materials to become discolored
Air movement is not required to distribute heat within the space	Gas burners require regular checks
High level mounting frees up floor space	Not readily accepted in arces other than tall spaces

5. Warm Air Heaters (Figure 5-79)

This kind of heat emitter is typically used in industrial applications and has two basic types:

• Unflued unit, where the heat and products of combustion pass directly into the space. It is energy efficient but requires adequate ventilation.

• Flued unit, which incorporate a heat exchanger that draws air from the space to be heated indirectly with the flue gases vented to outside.

Benefits and limitations of warm air heater are listed in Table 5-16.

Figure 5-79 Warm air heater

Table 5-16 Warm air heaters (Pro's & Con's)

Benefits	Limitations
Quick and simple installation	Gas burners require regular checks
Free standing units have good access for servicing	Unflued units need good ventilation
	Noise may be an issue in quiet enviroments

6. Air Handling Unit (AHU)

An air handling unit, commonly called an AHU, is the composition of elements mounted in large, accessible box-shaped units called modules, which can meet the appropriate ventilation requirements for purifying, air-conditioning or renewing the indoor air in a building or premises (Figure 5-80). AHUs are usually installed on the roof of buildings and, through ducts, the air is circulated to reach each of the rooms in the building. Benefits and limitations of AHU are listed in Table 5-18.

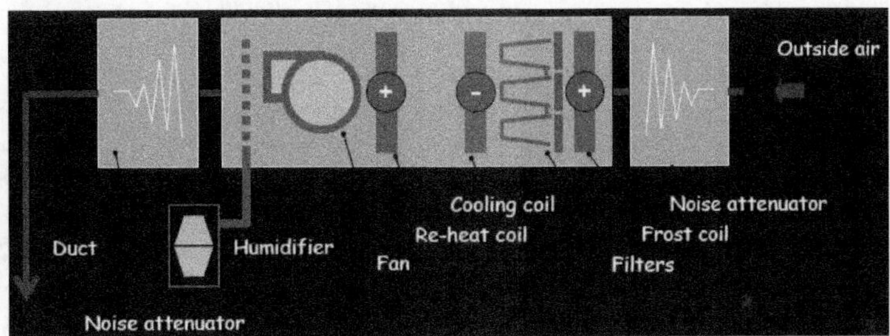

Figure 5-80 Typical layout of AHU

Table 5-17 AHU (Pro's & Con's)

Benefits	Limitations
Constant rate of ventilation	Fans consume a significant amount of energy
Heating coil can provide the main source of heat for the building	Occupants have less control over their environment
Recirculation can be used to cut down the amount of fresh air to be heated	The AHU and ductwork take up space and require maintenance
Heat recovery device can be used, which transfers heat from the exhaust air to the supply air	
Allows nighttime cooling	
Air movement within the space can be controlled	
Security and noise problems associated with openable windows are avoided	

7. Fan Coil Unit (FCU)

Fan coil unit is a relatively small piece of equipment that consists of a fan, a coil, and other components that are used to cool or heat the air recirculating system within a room (Figure 5-81). Some will also add fresh air to the space. There are three types of FCU (Figure 5-82). The units are quite common and can be found in most commercial or residential applications. Benefits and limitation of FCU are listed in Table 5-18.

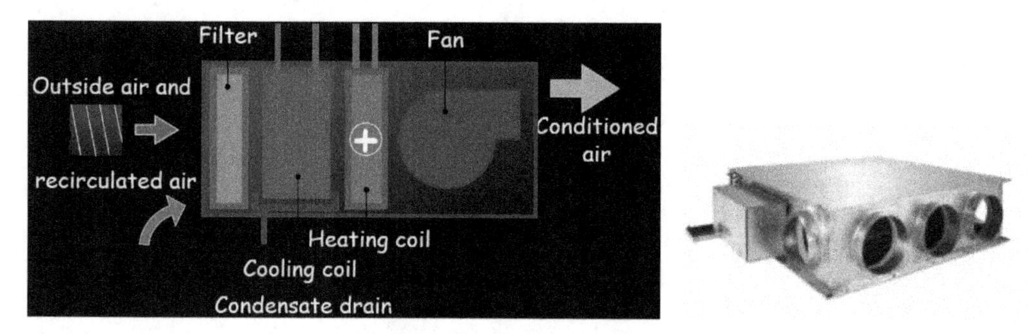

Figure 5-81 Typical layout of FCU

4-pipe system
Supply side: FlowCon ABS
Return side: FlowCon AB with
SME and electrical actuator

2-pipe heating system
Supply side: FlowCon ABS
Return side: FlowCon AB with
SME and electrical actuator

2-pipe cooling system
Supply side: FlowCon ABS
Return side: FlowCon AB with
SME and electrical actuator

Figure 5-82 Types of FCU

(https://ipaper.ipapercms.dk/FlowconInternational/WW/Brochure/FlowConApplicationGuide/?page=8)

Table 5-18 FCU (Pro's & Con's)

Benefits	Limitations
Constant rate of ventilation	Fans consume a significant amount of energy
Heating coil can provide the main source of heat for the area served	Fan and air noise; High risk of leaks from multiple pipe connections
Outside air can be ducted to the FCU or separately to the room	The outside AHU and duckwork take up space and require maintenance
Heat recovery device can be used, which transfers heat from the exhaust air to the supply air	The FCU needs a power supply for the fan and the control value

Continued

Benefits	Limitations
Allows cooling to be added by a separate coil	The filter needs cleaning/replacement regularly
Air movement within the space can be controlled	The fan and motor bearings need maintenance
Security and noise problems associated with openable windows are avoided	All maintenance needs to be carried out at high level over work places

5.2.2.8 Heating System Design and Drawings

1. Key Steps

Step 1: Calculating room heating load (Table 5-19).

• CIBSE steady state method;

• Dynamic building simulation;

• Steady state load (SSL) vs. Intermittent load (IL) (IL should not be more than 2 times of SSL).

$$Q_{heating} = Q_{fabric} + Q_{ventilation} + Q_{infiltration}$$

Table 5-19 Calculation of room heating load

Room name	Floor area/m²	Average height/m	Volume/m³	SS[①] heating load/kW	INT[②] heating load/kW	F_3 value
GF_R001 Office	86.59	2.85	246.78	1.791 2	3.441 1	1.92
GF_R005 Office	65.17	2.85	185.74	1.328 6	2.569 8	1.93

Note: ① SS means single switch; ② INT means induction.

$$F_3 = INT/SS < 2$$

Step 2: Determining location and number of radiators.

• Typically located around the perimeter walls & under windows to offset cold down-draughts and prevent condensation;

• The number of radiators depends on the number of windows or the size/layout of the room.

Step 3: Selecting radiators.

• Determining the heating load of each radiator based on the total room heating load and the number of radiators;

• Correcting the heating out put required from each radiator (see following slides);

• Selecting radiators from the manufacture database.

Correcting the heating output required from each radiator:

• The nominated heating output of radiators from manufacture is based on a certain temperature difference between the average temperature of the radiator (i.e. average between the flow and return temperatures) and the indoor design temperature, which may be different from the actual design condition (e.g. flow temperature = 50 ℃ ; return temperature = 30 ℃ and indoor temperature = 21 ℃);

• A higher temperature difference gives more heat, hence requiring a smaller radiator, and a lower temperature difference gives less heat, hence requiring a larger radiator (Figure 5-83).

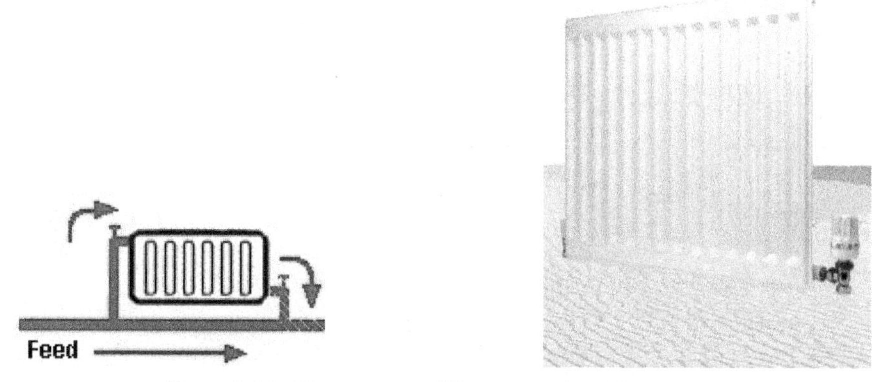

Figure 5-83 Temperature difference and radiators

The following information is taken from the Myson Premier HE documentation, which is available from www.myson.co.uk.

• The heat output provided in the database is based on a mean water to air temperature difference of 50 ℃ (Table 5-20). When the temperature difference is not 50 ℃ , the output should be multiplied by the appropriate correction factor within Table 5-20. People can select radiator from the manufacture database (Table 5-21).

Table 5-20 The appropriate correction factor

Centigrade	Factor	Fahrenheit
20 ℃	0.30	36 ℉
25 ℃	0.41	45 ℉
30 ℃	0.51	54 ℉
35 ℃	0.63	63 ℉
40 ℃	0.75	72 ℉
45 ℃	0.87	81 ℉
50 ℃	1.00	90 ℉
55 ℃	1.13	99 ℉
60 ℃	1.27	108 ℉
65 ℃	1.41	117 ℉
70 ℃	1.55	126 ℉

Table 5-21 Selecting radiators from the manufacture database

Output/W	Output/(Btu/h)	Order code	Output/W	Output/(Btu/h)	Order code	Output/W	Output/(Btu/h)	Order code
365	1 244	21 SC 17*	539	1 839	21 DPX 17*	688	2 347	21 DC 17*
449	1 532	21 SC 21*	664	2 265	21 DPX 21*	847	2 891	21 DC 21*
534	1 821	21 SC 25*	789	2 692	21 DPX 25*	1 007	3 436	21 DC 25*
618	2 109	21 SC 29*	914	3 118	21 DPX 29*	1 166	3 980	21 DC 29*
703	2 397	21 SC 33*	1 039	3 544	21 DPX 33*	1 326	4 524	21 DC 33*
787	2 686	21 SC 37*	1 164	3 971	21 DPX 37*	1 485	5 086	21 DC 37*
872	2 974	21 SC 41*	1 289	4 397	21 DPX 41*	1 645	5 613	21 DC 41*
956	3 263	21 SC 45*	1 414	4 824	21 DPX 45*	1 804	6 157	21 DC 45*
1 041	3 551	21 SC 49*	1 539	5 250	21 DPX 49*	1 964	6 701	21 DC 49*
1 125	3 840	21 SC 53*	1 664	5 676	21 DPX 53*	2 124	7 245	21 DC 53*
1 210	4 128	21 SC 57*	1 789	6 103	21 DPX 57	2 283	7 790	21 DC 57
1 294	4 416	21 SC 61*	1 914	6 529	21 DPX 61	2 443	8 334	21 DC 61
1 463	4 993	21 SC 69*	2 164	7 382	21 DPX 69	2 762	9 422	21 DC 69
1 633	5 570	21 SC 77	2 413	8 235	21 DPX 77	3 081	10 511	21 DC 77
1 802	6 147	21 SC 85				3 400	11 599	21 DC 85

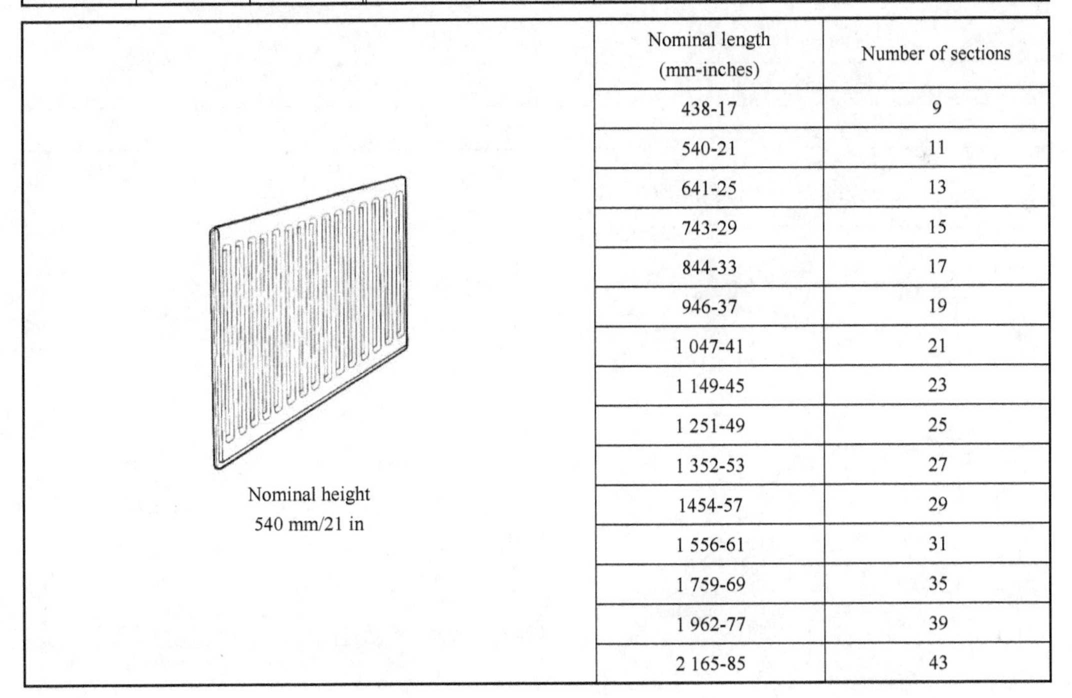

Nominal length (mm-inches)	Number of sections
438-17	9
540-21	11
641-25	13
743-29	15
844-33	17
946-37	19
1 047-41	21
1 149-45	23
1 251-49	25
1 352-53	27
1454-57	29
1 556-61	31
1 759-69	35
1 962-77	39
2 165-85	43

Step 4: Designing pipework (Figure 5-84) and calculating system resistance.

- Manual calculation (e.g. CIBSE Guide C);
- Computer software (e.g. Bentley Hevacomp) (Table 5-22).

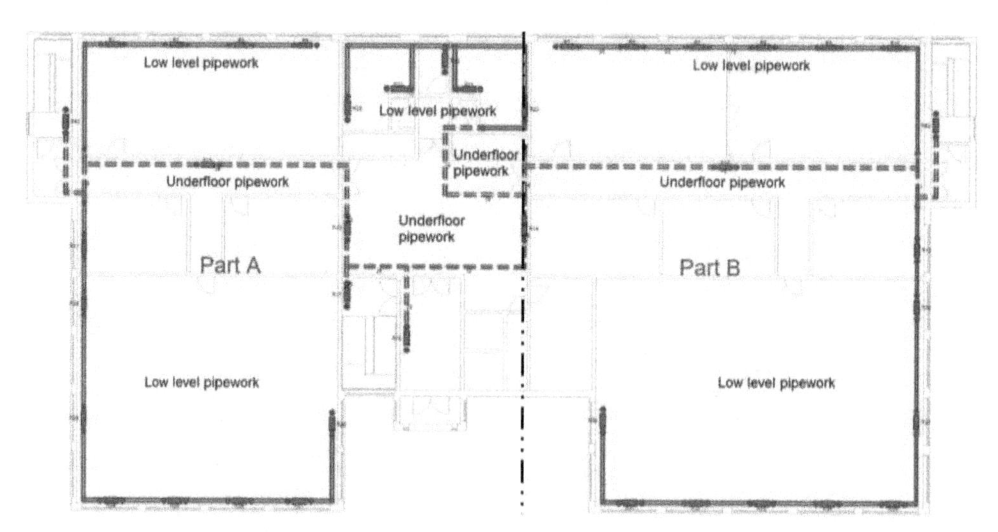

Figure 5-84 Ground floor pipework layout design

Table 5-22 Index run result

Heating system roof: Index run	
System resistance	24.17 kN/m²
Total flow rate	1.378 l/s
Index run to emitter	R30
on floor	0
Water content	982.33 liters

Step 5: Selecting pumps (Figure 5-85).

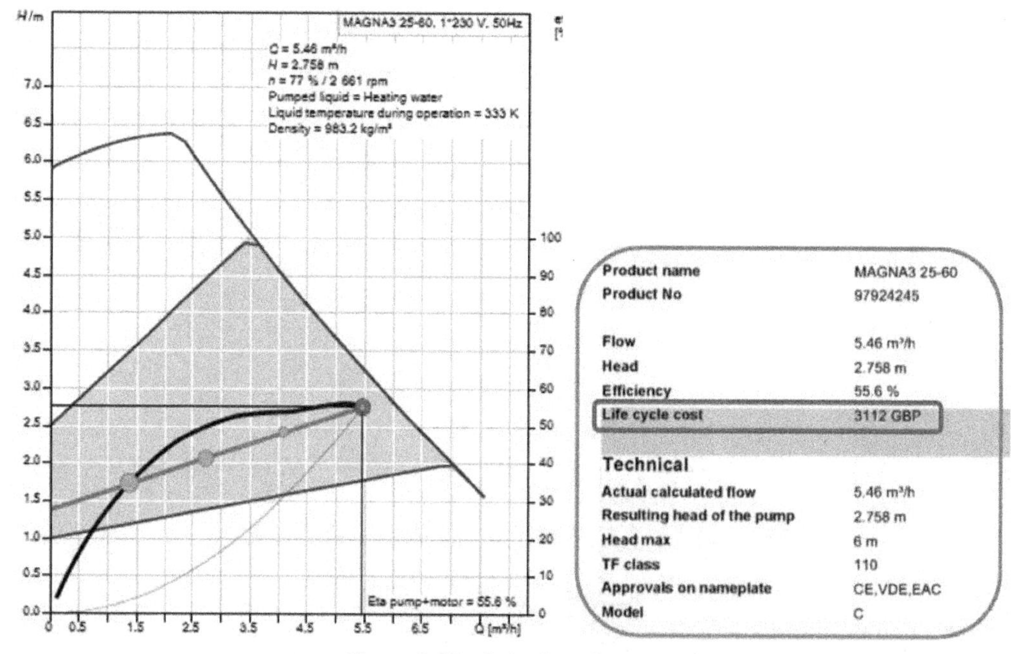

Figure 5-85 Selection of pumps

Step 6: Selecting boilers (Table 5-23).

• Total boiler duty

Q_{boiler} = 125 kW with 10% design margin

Table 5-23 Paramount four boiler selection range

Performance	Fuel Type		Paramount four 30	Paramount four 40	Paramount four 60	Paramount four 80	Paramount four 95	Paramount four 115
Output @ 80/60°C	NG	kW	29.2	36.9	56.3	74.7	92.2	111.7
Output @ 50/30°C	NG	kW	31.3	39.0	60.1	79.7	98.1	118.6
Input (Gross) Maximum	NG	kW	33.3	42.2	64.4	85.5	105.5	127.6
Output @ 80/60°C	LPG	kW	29.2	36.8	56.3	74.7	92.2	111.7
Output @ 50/30°C	LPG	kW	31.7	39.3	60.1	79.7	98.1	118.6
Input (Gross) Maximum	LPG	kW	33.3	42.2	64.4	85.5	105.5	127.6
Efficiency	Value (%)							
Efficiency @ 80/60°C - 100% Load	% Gross		87.6	87.7	87.7	87.7	87.8	87.6
Efficiency @ 50/30°C - 30% Load	% Gross		97.9	97.8	97.8	97.8	98	97.8
Current Building Regulations - Part L2 Seasonal Efficiency	% Gross		96.3	95.3	95.7	96.2	95.8	95.7
ErP efficiency rating			A	A	A	N/A	N/A	N/A

• Option 1: Two boilers with output of 79.7 kW each (minimum controllable load: 10.6%; fails: 63.76%) of the design; boiler coverage when one.

• Option 2: Three boilers with output of 60.1 kW each (minimum controllable load: 8.02%; fails: 96.16%) of the design; boiler coverage when one.

2. Heating System Drawings

• Heating layout (Figure 5-86);

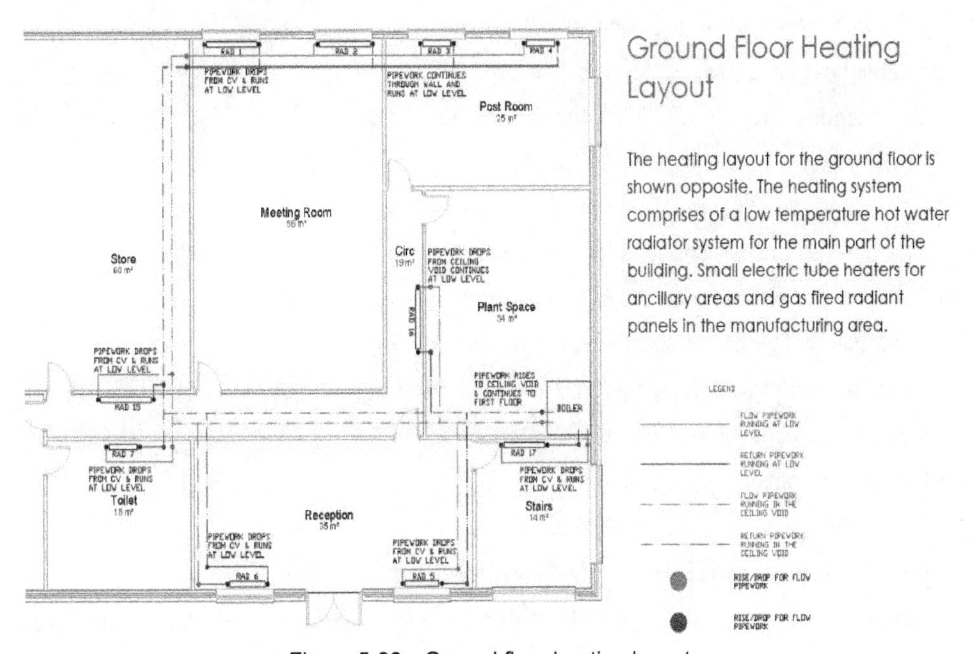

Figure 5-86 Ground floor heating layout

- Major system components;
- Pipework distribution with level of height;
- Clear legend.

5.2.3 Air Conditioning System

5.2.3.1 Definition of Air Conditioning

Air conditioning is a process of air treatment to filter, heat, cool, humidify/de-humidify the room air and provide clean outside air, and to match the requirements of the room. Air conditioning plant comprises of a number of pieces of equipment combined to perform this task, including filters, heating coil, cooling coil and humidifier/de-humidifier.

5.2.3.2 Purpose of Air Conditioning

- Ventilation, temperature control, humidity control and air movement need to be met.
- Benevolent employers care about the comfort of their staff.
- Employers know from independent research that good comfort standards improve staff productivity by up to 15%, thus allowing significant reductions in staff numbers & costs.
- Some industrial processes such as electronic and pharmaceutical manufacture demand close control over temperature and humidity for product reliability and quality control.
- Museums and galleries require stable humidity conditions to keep ancient artifacts in good condition for future generations.

5.2.3.3 Reasons for using air conditioning

Sensible heat gain/loss:

- Conduction heat gain/loss—via the fabric;
- Infiltration heat gain/loss—via leaks & cracks;
- Body heat gain—convection & radiation;
- Lighting heat gain—convection & radiation;
- Equipment heat gain—convection & radiation;
- Solar heat gain—direct short wave, indirect long wave.

Latent heat gain/loss (change of state of water from liquid to vapor or vice versa):

- Body gain (respiration & perspiration);
- Infiltration (outside humid air).

Total Heat = Sensible heat + Latent heat

5.2.3.4 Working Principle of Air Conditioning

Second law of thermodynamic: Heat always flows spontaneously from hotter to colder bodies, and never the reverse, unless external work is performed on the system (Figure 5-87).

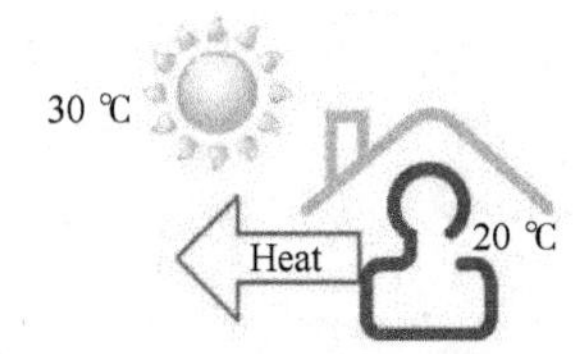

Figure 5-87 Heat flows in the presence of temperature differences

Refrigerant is a compound used in a heat cycle that undergoes a phase change from gas to liquid and vice versa. Refrigerants usually have high thermal conductivity, high latent heat and moderate boiling point.

Heat pump/refrigeration cycle refers to a machine that moves heat from one location at a lower temperature to another location at a higher temperature using mechanical work or a high temperature heat source (Figure 5-88).

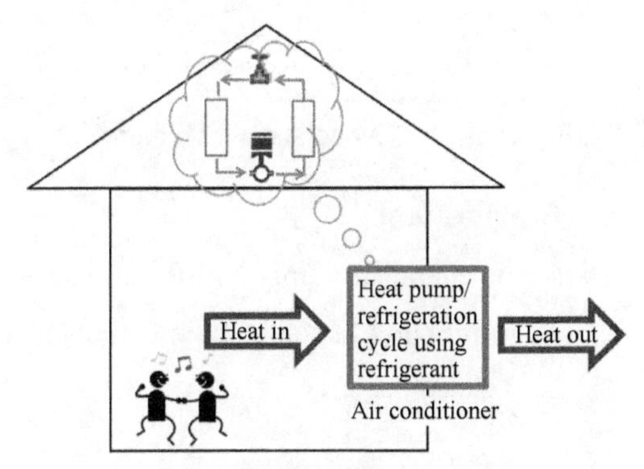

Figure 5-88 Heat pump/refrigeration cycle

5.2.3.5 Refrigeration Cycle Theory and Components

Composition of the air conditioner is as follows (Figure 5-89).

- Compressor: Compress the refrigerant to increase its temperature and pressure.
- Condenser: Release heat to the ambient environment.
- Expansion valve: Expand the refrigerant to decrease its temperature and pressure.
- Evaporator: Absorb heat from the indoor environment.

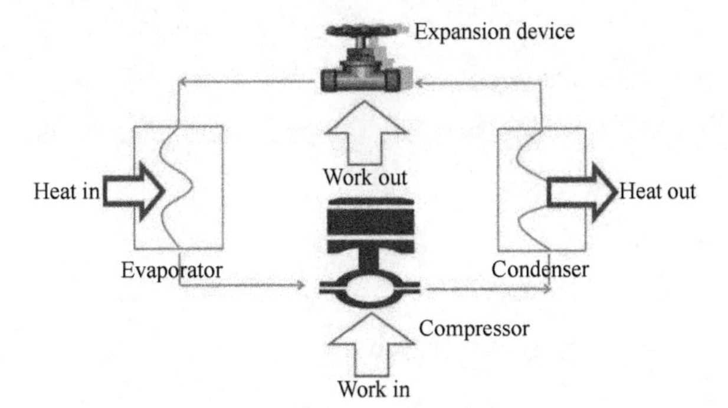

Figure 5-89 Composition of the air conditioner

Coefficient of performance (CoP) is a ratio of useful heating or cooling provided to the external work required by the system. Higher CoPs equate to lower operating costs and higher efficiency. CoP calculations should include energy consumption of all power consuming auxiliaries. An air conditioning system is more efficient when the room temperature is closer to the outside temperature (Figure 5-90).

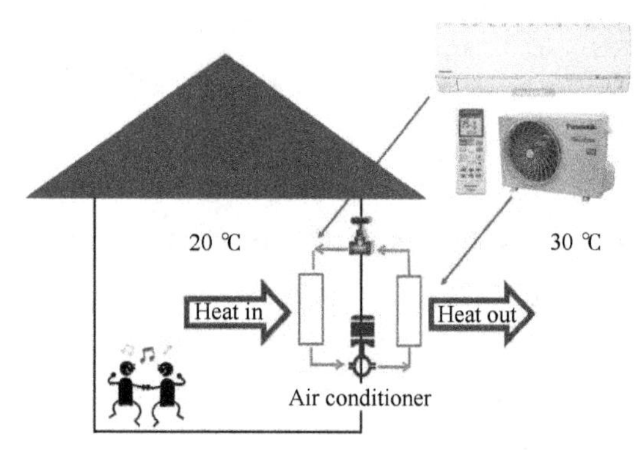

Figure 5-90 Refrigeration cycle theory

5.2.3.6 Working Fluid—Refrigerant

A refrigerant is a substance or chemical compound that is circulated by the compressor within the refrigeration cycle. The following factors should be considered when selecting a refrigerant:

- Safety;
- Non-toxic, non-irritating;
- Non-flammable, free from risk of explosion;
- Thermodynamic properties;
- Low specific volume;
- High latent heat;
- High thermal conductivity, good wettability;
- Low viscosity;
- Moderate boiling point (somehow below the target temperature);
- Relatively high density in gaseous form;
- As boiling point and gas vary with the pressure, the operating pressure should also be considered in the refrigerant selecting process;
- Mechanical component;
- Inert;
- Non-reactivity with compressor lubricants;
- Environmental;

• Not harmful to the environment.

Refrigerants are categorized into the following four groups.

(1) Inorganic (700 series numbering system): R-717 (ammonia).

(2) Organic (2-digit numbers following "R" are methane (CH_4) derived; 3-digit numbers after "R" starting with 1 are ethane (C_2H_6) derived): R-22 (chlorodifluoromethane, banned by developed countries, e.g. R-22 Legislation in the UK).

(3) Zeotropes (400 series numbering system): Formed from mixtures of halocarbon compounds, e.g. R-407 C (R-407 C is an HFC (hydrofluorocarbons) blend designed to have similar properties to R-22 in air conditioning systems), but with temperature glide.

(4) Azeotropes (500 series numbering system): Formed from mixtures of halocarbon compounds, but free from temperature glide, e.g. R-507.

Common refrigerants and their numbers are listed in Table 5-24.

Table 5-24 Common refrigerants and their numbers

Refrigerants	R No.	ODP
Trichlorofluoromethane (CCl_3F)	R-11	1
1,1,1,2-Tetrafluoroethane	R-134a	0.000 015
Chlorodifluoromethane ($CClF_2$-H)	R-22	0.05
Chlorotrifluoromethane ($CClF_2$-F)	R-13	1
Dichlorodifluoromethane ($CClF_2$-Cl)	R-12	1.00
Bromochlorodifluoromethane ($CClF_2$-Br)		7.9
Carbon tetrachloride (CCl_4)		0.82
Nitrous oxide (N_2O)		0.017
Alkanes (propane, isobutane, etc)		0
Ammonia (NH_3)	R-717	0
Carbon dioxide (CO_2)	R-744	0
Nitrogen (N_2)	R-728	0

(https://en.wikipedia.org/wiki/Ozone_depletion_potential)

Chlorine separates from CFCs (chloro-fluro-carbons) and HCFCs (hydro-chloro-fluoro-carbon) when released to the atmosphere, and reacts with sunlight, resulting in the breakdown of ozone (O_3) to oxygen, i.e. thinning the ozone layer.

Ozone depletion potential (ODP) of a refrigerant represents its effect on atmospheric ozone, and the CFC R-11 is assigned an ODP index of 1; other fluids such as HCFCs ranked against this (e.g. R-22 has ODP of 0.05). Ozone hole is shown in Figure 5-91.

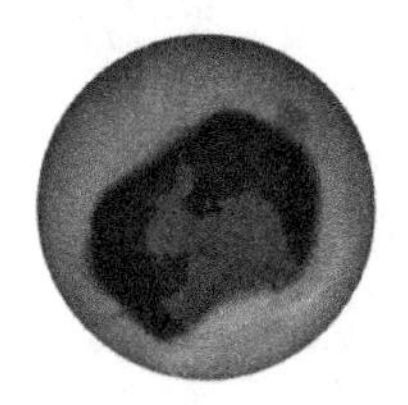

Figure 5-91 Ozone hole

Global warming index (GWI) expresses global warming potential of fluids with reference to CO_2 (which is assigned reference GWI of 1).

Methane/ethane-derived refrigerants contain carbon, which, when oxidized upon release to the atmosphere, contributes to global warming. HFC fluids are regarded as transitional refrigerants, eventually to be replaced by zero GWP fluids such as propane and butane.

Common global warming indices are listed in Table 5-25.

Table 5-25 Common global warming indices

No.	ODP	GWI
R-11 (CFC, banned)	1	4 000
R-22 (HCFC, to be banned)	0.05	1 700
R-134A (HFC)	0	1 300
R-404A (HFC)	0	3 748
R-407A (HFC)	0	1 610
R-717 (ammonia)	0	0
R-744 (CO_2)	0	1

CFCs were banned under 1987 Montreal Protocol—this included R-11, R-12, R-502—the predominant air conditioning fluids used:

- Supply of CFCs to the market is banned from 1 Jan., 2000;
- Servicing and topping up existing CFCs system is banned from 1 Jan., 2001;
- Supply of HCFCs to all new systems is banned from 1 July, 2002;
- Recycled HCFCs for topping up existing system is banned from 2015.

5.2.3.7 Air Handling Unit(AHU)

Components of AHU (Figure 5-92) include:

- Filters;
- Heating coil;
- Cooling coil;
- Humidifier/De-humidifier.

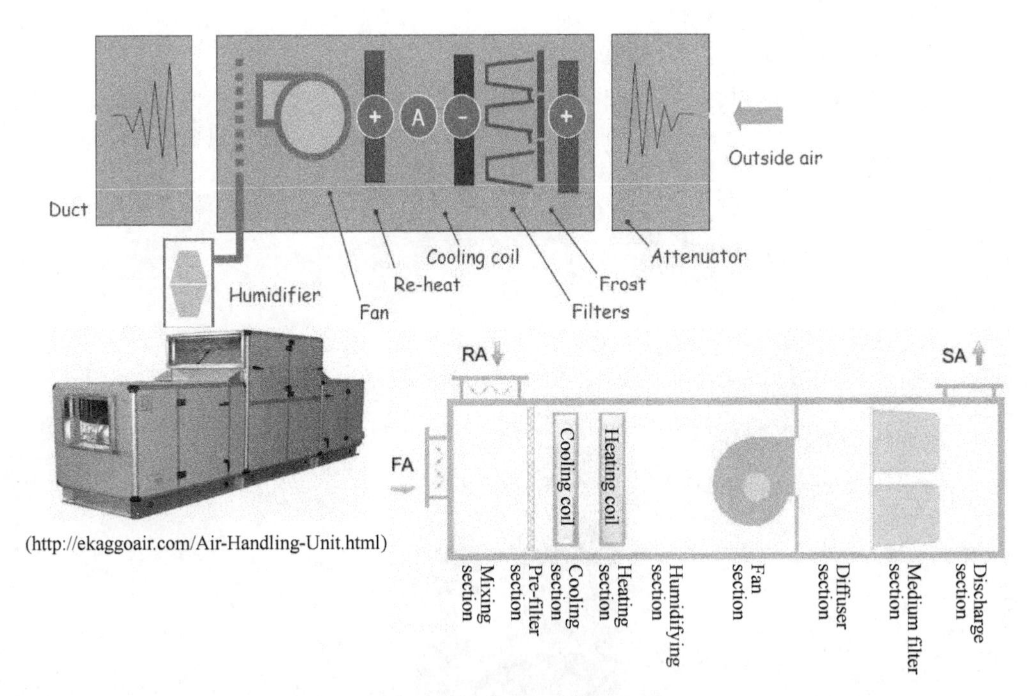

Figure 5-92 Components of AHU

1. Heating Coil (Figure 5-93)

It raises the temperature of air at constant moisture content (sensible heating).

Figure 5-93 Heating coil

2. Cooling Coil (Figure 5-94)

- Sensible cooling reduces the temperature of air at constant moisture content.
- Latent cooling reduces the temperature lower to the dew point to reduce moisture content of air.

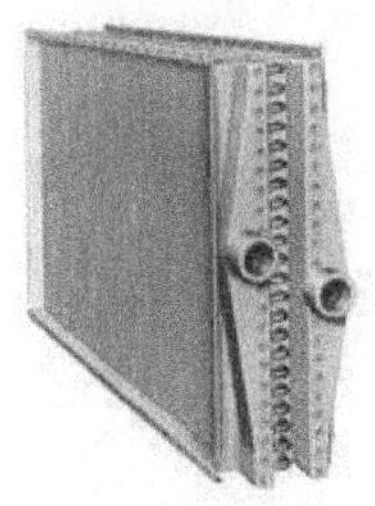

Figure 5-94 Cooling coil

3. Humidifier (Figure 5-95)

Humidifier adds moisture content to the air by either steam or water.

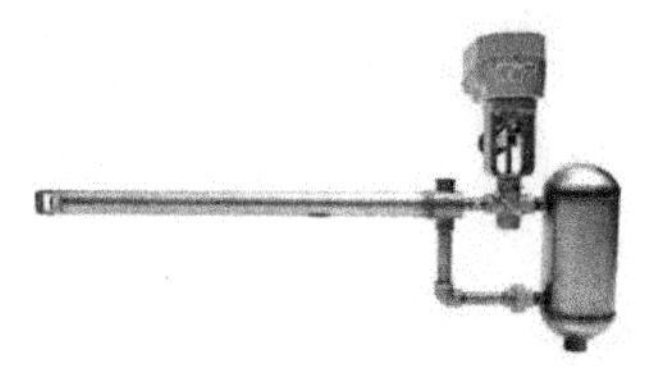

Figure 5-95 Humidifier

(http: //www.humidification.ca/isothermic/steam-injection-humidifier-four.php)

4. Fan (Figure 5-96)

Fan moves the air in the HVAC system.

Figure 5-96 Fan

5. Attenuator (Figure 5-97)

It reduces noise produced by fan units and other equipment within heating, ventilating and air conditioning (HVAC) systems.

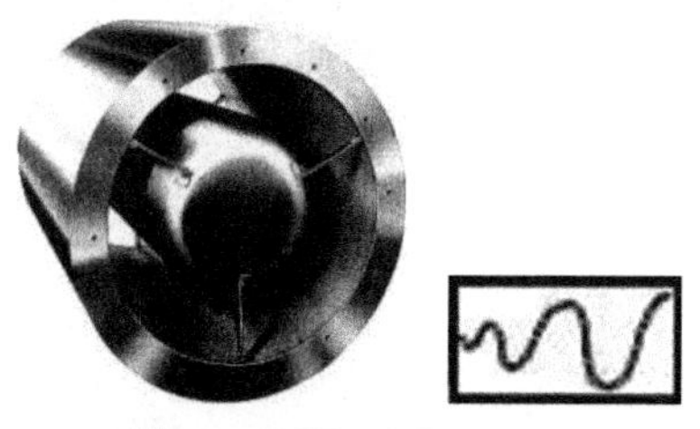

Figure 5-97 Attenuator

6. Filter (Figure 5-98)

- It removes allergens and pollutants to improve indoor air quality;
- Bag filters;
- Panel/Cartridge filters.

Figure 5-98 Filter

7. Boiler (Figure 5-99) and Boiler Plant (Figure 5-100)

1) Boiler

• It is only used for heating.

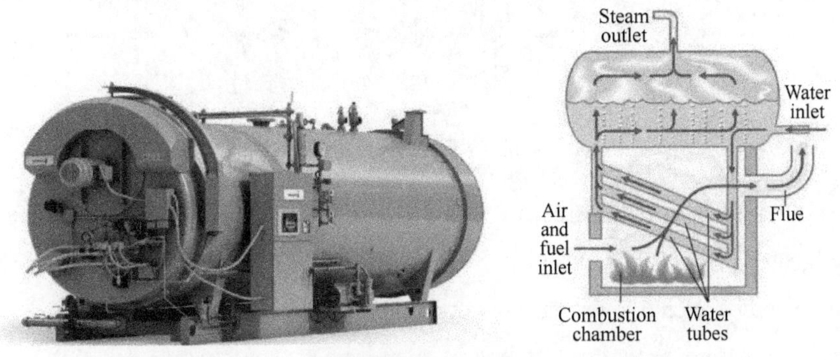

Figure 5-99 Boiler

2) Boiler plant

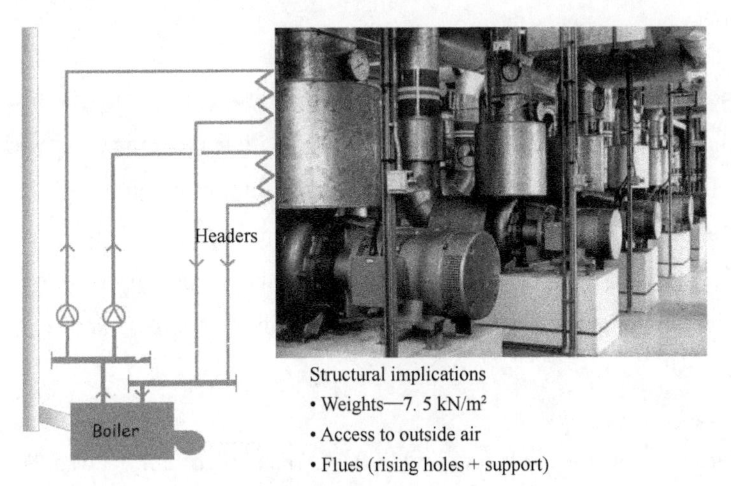

Figure 5-100 Boiler plant

• Boiler plant—1% GIA(gross internal area);

• Flues must discharge at roof level—at least 3 m above the level of any adjacent area to which there is access;

• Boilers can be either basement or roof located (Figure 5-101).

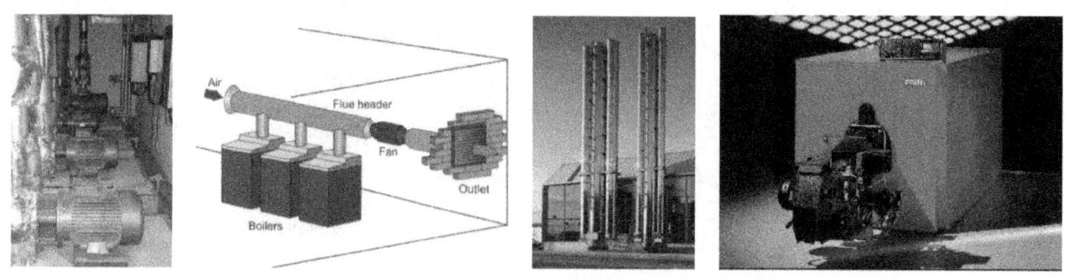

Figure 5-101 Different locations of the boiler

Summary of air conditioning component structure (Figure 5-102):

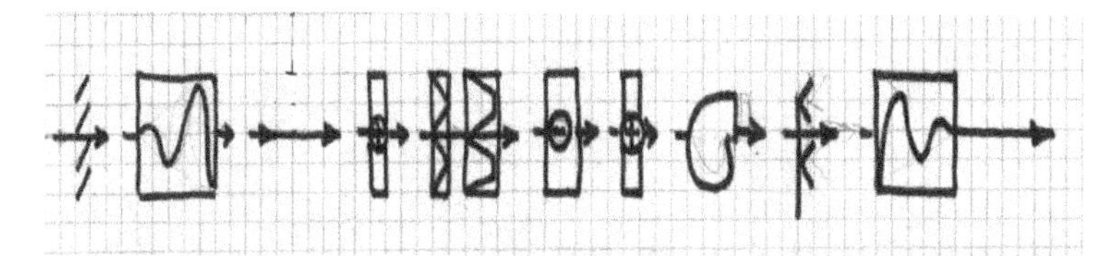

Figure 5-102 All components of air AHU

5.2.3.8 Primary Plant Selection

Key objectives:

• Maximized efficiency over seasonal operation;

• Good control (i.e. accurate matching of plant capacity to demand);

• Standby capacity to meet essential load requirements when one unit is off line as a result of break-down.

Implementation ways:

• Avoidance of gratuitous over-sizing, ensuring for instance that any standby capacity specified will be idle in normal plant operation (but rotated with active units to ensure equal wear on plant) i.e.duty and standby pumps—twin head pumps or duplicate pump sets.

• Limit standby capacity to a fraction of design on the grounds that a plant break-down during peak usage would be a rare event. Question if standby capacity is required at all. It may be acceptable just to spread the heating load over a number of boilers so that the effect of one failing is not critical.

Case example:

• A building has fan coil loads totalling 250 kW (design cooling load) and 175 kW (design heating load).

• A primary air handling unit (AHU) delivers 1.6 m^3/s of full fresh air, which enters the unit at 29 ℃ (dry bulb), 26 ℃ (wet bulb) in summer and −3 ℃ in winter. The AHU is to deliver air tempered to 18 ℃ all year round.

• Chilled water at 5 ℃ (flow), 10 ℃ (return) and hot water at 80 ℃ (flow) and 70 ℃ (return) are to be generated by suitable primary plant.

• Size the boiler plant to give 25% preheat capacity.

1. Chillers Selection

Chillers are typically applied in non-process/non-critical applications (e.g. comfort cooling) (Figure 5-103); some practitioners do not specify standby capacity—this is reasonable provided that it is possible to obtain good natural ventilation in peak summer.

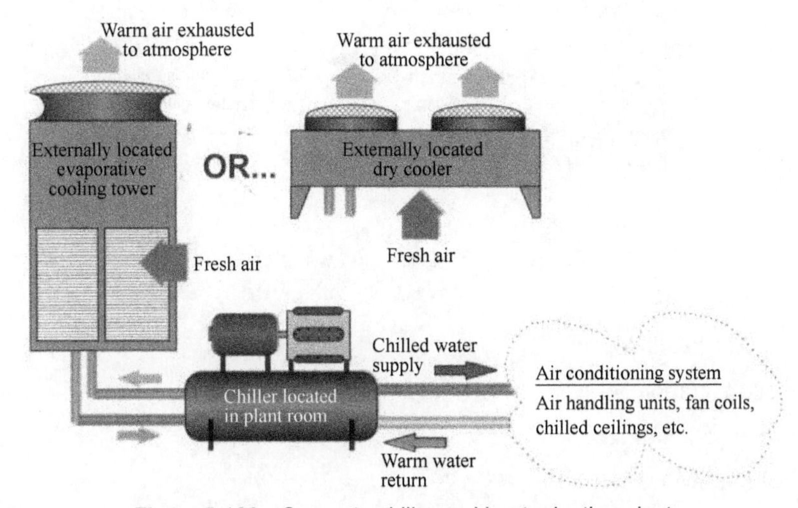

Figure 5-103 Separate chiller and heat rejection plant

Select units of plant with better operational performance i.e. capable of operating at near-design efficiency/coefficient of performance at light demand (typically to 10% of design) (Figure 5-104). Specifying a number of small units as opposed to one large unit can satisfy this in many cases though the use of multiple small units adds to the capital cost of the project.

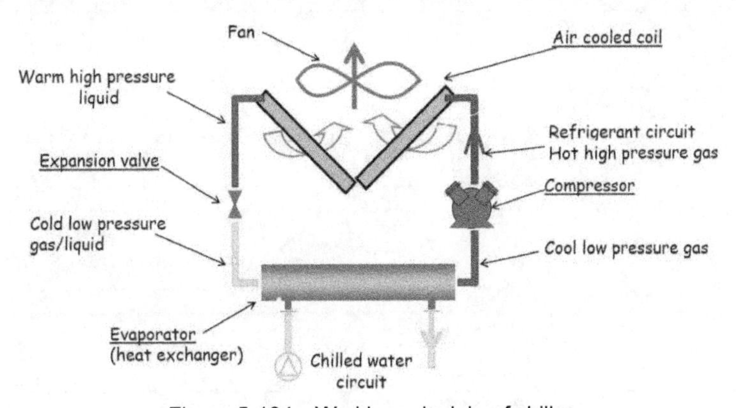

Figure 5-104 Working principle of chiller

Packaged water chiller uses the evaporator to make the heat exchange between water and refrigerant. After the refrigerant system absorbs the heat load in the water and cools the water to produce cold water, the heat is brought to the shell and tube condenser by the compressor, and the heat exchange between the refrigerant and water is made so that the water absorbs the heat and then the heat is dissipated through the water pipe to the external cooling tower (Figure 5-105).

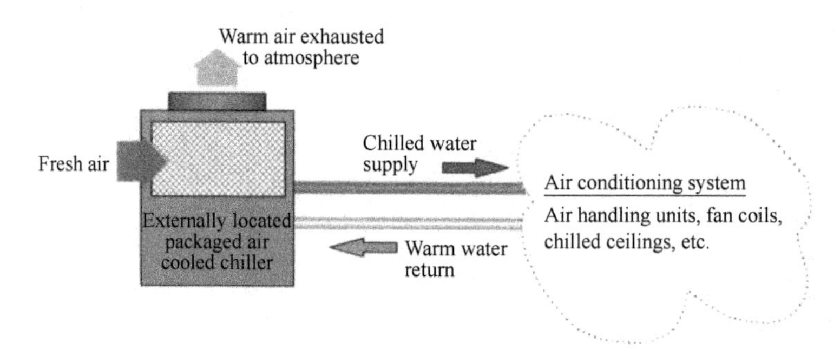

Figure 5-105 Packaged water chiller

Structural implications;

• Good access to outside air;

• Noise and vibration;

• Weights—7.5 kN/m^2.

Electrical implications:

• At CoP of 3.0, 150 kW of cooling will require 50 kW of electrical power (Table 5-26).

• Note: CoP varies with load!

Table 5-26 Chiller data

30RH		040	050	060	070	080	090	100	120	140	160	200	240
Net nominal cooling capacity, single pump*	kW	38.2	44.5	54	66	72	83	93	109	133	143	180	213
Net nominal cooling capacity, dual pump*	kW	37	43.4	53	65	71	83	92	107	132	142	180	213
Net nominal heating capacity, single pump**	kW	39.3	47.3	58	66	80	86	97	116	133	160	192	227
Net nominal heating capacity, dual pump**	kW	40.4	48.3	59	67	80	87	98	117	134	160	192	227
Operating weight, with hydronic module	kg												
Single pump		566	624	647	661	691	1183	1196	1238	1312	1368	2233	2405
Dual pump	kg	646	704	727	741	768	1260	1273	1338	1412	1468	2321	2493
Unit without hydronic module		542	600	623	637	665	1152	1165	1200	1274	1330	2086	2258
Refrigerant charge	kg	R-407C											
Compressors		Hermetic scroll compressor, 48.3 r/s											
Quantity, circuit A		1	2	2	2	2	1	1	2	2	2	2	3
Quantity, circuit B		-	-	-	-	-	2	2	2	2	2	3	3
No. of capacity steps		1	2	2	2	2	3	3	4	4	4	5	6
Control type		PRO-DIALOG Plus											
Air heat exchangers		Grooved copper tubes, aluminium fins											
Fans		Axial Flying Bird II fans with rotating shroud											
Quantity		1	1	1	1	1	2	2	2	2	2	4	4
Total air flow (high speed)	l/s	3870	3660	4080	5600	5600	7350	7950	8160	11200	11200	17343	20908
Water heat exchangers		Direct-expansion welded plate heat exchanger											
Hydronic module		Pump, screen filter, safety valve, expansion tank, pressure gauge(s), purge valves (water and air), flow switch and flow control throttle valve											
Pump		Single centrifugal monocell pump, 48.3 r/s											

* Standard Eurovent conditions: water heat exchanger entering/leaving temperature 12°C/7°C, outdoor air temperature 35°C.

**Standard Eurovent conditions: air heat exchanger entering/leaving temperature 40°C/45°C, outdoor air dry bulb temperature 7°C with 87% relative humidity.

Solution: Chiller

Note that Carrier stipulates a maximum oversizing margin equivalent to 15% of the design cooling load (Table 5-27).

Design chiller load is 250 kW (fan coils) + 58.37 kW (primary AHU cooling coil) = 308.37 kW.

There are two chiller possibilities:

• 3 machines (Carrier 30RH-120), each giving 109 kW of cooling; (327 kW cooling combined)

• 2 machines (Carrier 30RH-200), each giving 180 kW of cooling. (360 kW cooling combined)

Table 5-27 Net nominal cooling capacity, single pump

30RH	040	050	060	070	080	090	100	120	140	160	200
Net nominal cooling capacity, single pump/kW	38.2	44.5	54	66	72	83	93	109	133	143	180

• 3 chillers: The margin of design-oversizing is (3 × 109−308.37)/308.37 × 100% = 6%.

This satisfies the carrier limit (i.e. not more than 15% oversized). If one machines fails, then the percentage of demand satisfied at design will be 2 × 109/308.37 × 100% = 71%, which is good.

• 2 chillers: The margin of design-oversizing is (2 × 180−308.37)/308.37 × 100% = 16.7%.

This is just over the carrier limit. Hence this solution is not recommended.

Therefore, 3 Carrier 30RH-120 chillers are the only possible selection from this chiller range (Figure 5-106).

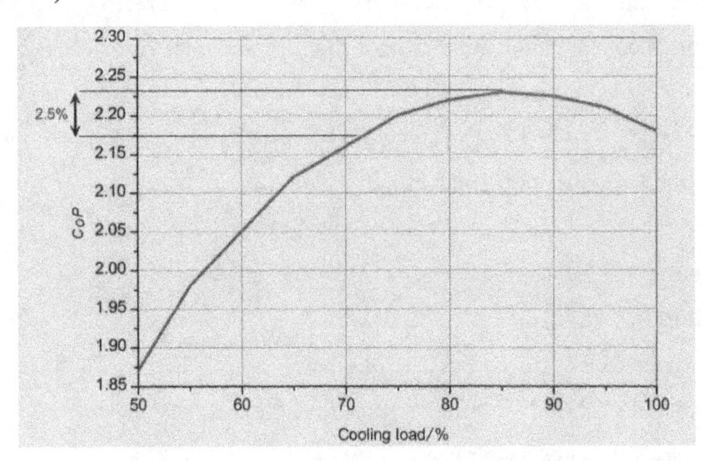

Figure 5-106 Typical chiller part load performance

2. Boiler Selection

Select a boiler plant and refrigeration plant from the following manufacturers data:

- Boilers—Stokvis & Sons "Econoflame" range (from Opus'97) (Figure 5-107).
- Chillers—Carrier air cooled compression chillers (Carrier catalogue).

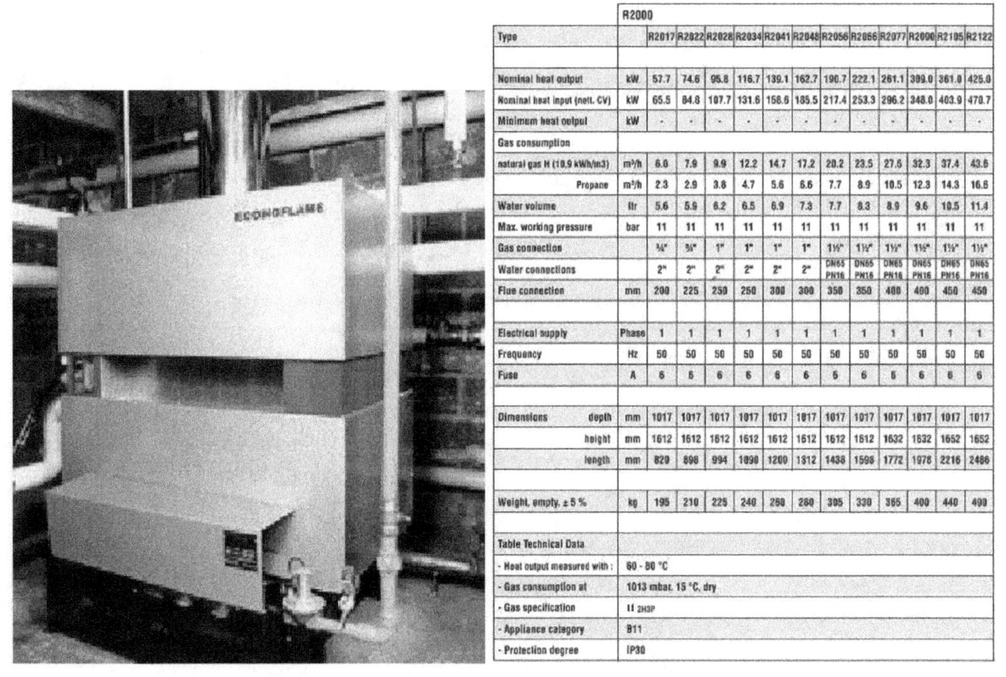

Type		R2017	R2022	R2028	R2034	R2041	R2048	R2056	R2066	R2077	R2090	R2105	R2122
Nominal heat output	kW	57.7	74.6	95.8	116.7	139.1	162.7	190.7	222.1	261.1	309.0	361.0	425.0
Nominal heat input (nett. CV)	kW	65.5	84.8	107.7	131.6	158.6	185.5	217.4	253.3	296.2	348.0	403.9	470.7
Minimum heat output	kW	-	-	-	-	-	-	-	-	-	-	-	-
Gas consumption													
natural gas H (10,9 kWh/m3)	m^3/h	6.0	7.9	9.9	12.2	14.7	17.2	20.2	23.5	27.5	32.3	37.4	43.6
Propane	m^3/h	2.3	2.9	3.8	4.7	5.6	6.6	7.7	8.9	10.5	12.3	14.3	16.6
Water volume	ltr	5.6	5.9	6.2	6.5	6.9	7.3	7.7	8.3	8.9	9.6	10.5	11.4
Max. working pressure	bar	11	11	11	11	11	11	11	11	11	11	11	11
Gas connection		¾"	¾"	1"	1"	1"	1"	1½"	1½"	1½"	1½"	1½"	1½"
Water connections		2"	2"	2"	2"	2"	2"	DN65 PN16	DN65 PN16	DN65 PN16	DN65 PN16	DN65 PN16	DN65 PN16
Flue connection	mm	200	225	250	250	300	300	350	350	400	400	450	450
Electrical supply	Phase	1	1	1	1	1	1	1	1	1	1	1	1
Frequency	Hz	50	50	50	50	50	50	50	50	50	50	50	50
Fuse	A	6	6	6	6	6	6	5	6	6	6	6	6
Dimensions depth	mm	1017	1017	1017	1017	1017	1017	1017	1017	1017	1017	1017	1017
height	mm	1612	1612	1612	1612	1612	1612	1612	1612	1632	1632	1652	1652
length	mm	820	898	994	1090	1200	1312	1438	1598	1772	1978	2216	2486
Weight, empty, ± 5 %	kg	195	210	225	240	260	280	305	330	365	400	440	490

Table Technical Data	
- Heat output measured with :	60 - 80 °C
- Gas consumption at	1013 mbar, 15 °C, dry
- Gas specification	II 2H3P
- Appliance category	B11
- Protection degree	IP30

Figure 5-107 Stokvis & Sons "Econoflame" range

1) Required capacity

Option A—Preheat period:

- Design boiler capacity = 175 kW (fan coils) × 1.25 (25%) = 218.75 kW (preheat capacity).
- Primary air discounted since it would not run during the preheat period.

Option B—Occupied period:

- Design boiler capacity = 175 kW (fan coils) + primary air vent load 40.4 kW = 215.4 kW.

Option A is the larger; therefore select for Q=219 kW.

2) Solution

Boilers (require 219 kW).

3) Possibilities

- 3 units (Stokvis R2022), each rated at 74.6 kW (Table 5-28):

  3 × 74.6 kW = 223.8 kW (effectively a 28% preheat margin)

- 2 units (Stokvis R2034), each rated at 116.7 kW (Table 5-28):

  2 × 116.7 kW = 233.4 kW (effectively a 33.5% preheat margin)

• Note that the turndown of these boilers is 20%.

Table 5-28 Nominal heat output corresponding to different types

Type	R2000												
	Unit	R2017	R2022	R2028	R2034	R2041	R2048	R2056	R2066	R2077	R2090	R2105	R2122
Nominal heat output	kW	57.7	74.6	95.8	116.7	139.1	162.7	190.7	222.1	261.1	309.0	361.0	425.0

• 3 boilers: Min. controllable load will be 20% × 74.6 kW = 14.9 kW, which is 14.9/215.4 × 100% = 6.9% of design.

If one boiler fails, there will be:

2 × 74.6/215.4 × 100% = 69% design load cover

• 2 boilers: Min. controllable load will be 20% × 116.7 kW = 23.3 kW, which is 23.3/215.4 × 100% = 11.8% of design.

If one boiler fails, then the remaining boiler will satisfy

1 × 116.7/215.4 × 100% = 54% design load cover

• Therefore, 3 boilers would be a better choice since they would give better light load control and more security (Figure 5-108).

3. AHU Selection

What is volume flow rate dependant upon?

• The sensible cooling load;

• The Δt (room air temperature−supply temperature);

• Recommended range of Δt = 11 (max) to 8 (min);

• Equates to 12 ℃ to 15 ℃ (23 ℃ room temperature).

$$Q_{\mathrm{S}} = mC_{\mathrm{p}}\Delta t$$

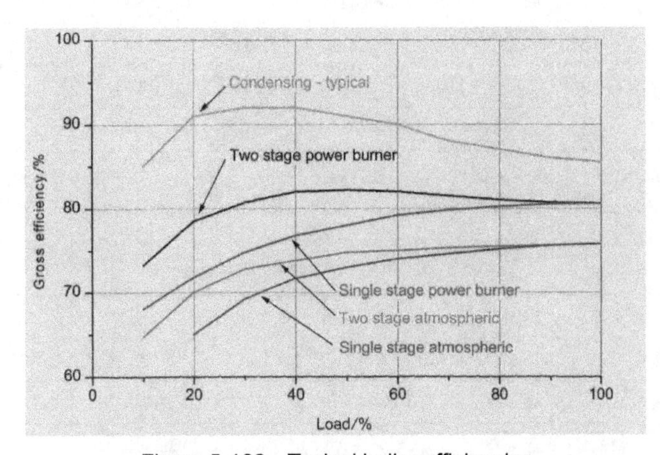

Figure 5-108 Typical boiler efficiencies

How to determine the volume flow rate?

Step 1: Calculate the air flow rate required.

• You should consider adding a margin of 5%-10% onto the volume flow rate to allow for duct leakage.

• Read HVDW144 Spec for Sheet Metal Ductwork 2016 of CIBSE.

Step 2: Determine the components required in the AHU.

• Frost coil;

• Mixing box or some other forms of heat recovery (recirculation OK or full fresh air);

• Humidifier;

• Access sections;

• Think too about the configuration (Figure 5-109).

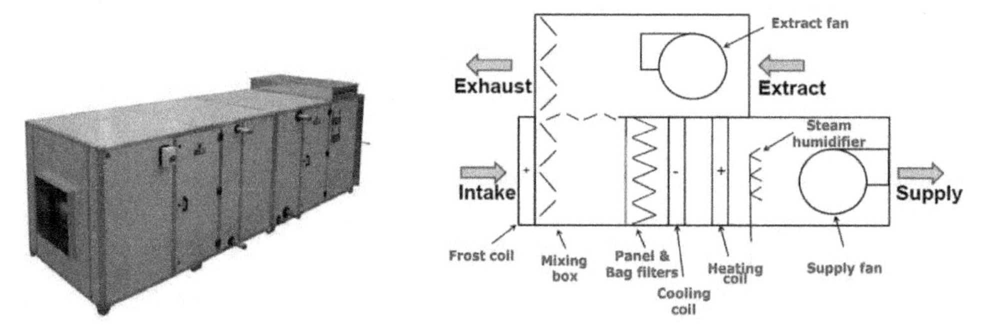

(http://www.eatonwilliams.com/colmanmoducel/moducel_products/moducel_lkp_lks.php)

Figure 5-109 Constant volume AHU with mixing box

Step 3: Determine the AHU frame size required (Table 5-29, Table 5-30).

Table 5-29 LKP air volumes (m^3/s)

Model	LKP-1	LKP-2	LKP-3	LKP-4	LKP-5	LKP-6	LKP-7	LKP-8	LKP-9	LKP-10	LKP-11	LKP-12	LKP-13	LKP-14
2.5 m/s	0.43	0.875	1.315	1.972 5	1.75	2.625	3.505	3.285	4.38	5.475	3.945	5.25	6.575	7.885
3.5 m/s	0.602	1.225	1.841	2.761 5	2.45	3.675	4.907	4.599	6.132	7.665	5.523	7.35	9.205	10
4 m/s*	0.688	1.4	2.104	3.156	2.8	4.2	5.608	5.256	7.008	8.76	6.312	8.4	10	10

Notes: ① LKP is a flush-mounted square diffuser with a square face plate for installation in ceiling systems. LKP is suitable for the horizontal supply of cooled air and has a large dynamic range. ② Air volumes are based on filter velocities. *Extract only.

Table 5-30 LKP nominal dimensions

Model	LKP-1	LKP-2	LKP-3	LKP-4	LKP-5	LKP-6	LKP-7
30 mm frame	672 W × 437 h	672 W × 742 h	972 W × 742 h	972 W × 1 047 h	1272W × 742 h	1272W × 1047 h	1272W × 1352 h
40 mm frame	673 W × 438 h	673 W × 743 h	973 W × 743 h	973 W × 1 078 h	1273W × 743 h	1273W × 1048 h	1273W × 1353 h
50 mm frame	715 W × 480 h	715 W × 785 h	1 015 W × 785 h	1015W × 1090 h	1315W × 785 h	1315W × 1090 h	1315W × 1395 h
70 mm frame	—	—	—	—	—	—	—

Continued

Model	LKP-8	LKP-9	LKP-10	LKP-11	LKP-12	LKP-13	LKP-14
30 mm frame	—	—	—	—	—	—	—
40 mm frame	1 573 W × 1 048 h	1 573 W × 1 353 h	1 573 W × 1 658 h	1 873 W × 1 048 h	1 873 W × 1 353 h	1 873 W × 1 658 h	1 873 W × 1 963 h
50 mm frame	1 615 W × 1 090 h	1 615 W × 1 395 h	1 615 W × 1 700 h	1 915 W × 1 090 h	1 915 W × 1 395 h	1 915 W × 1 700 h	1 915 W × 2 005 h
70 mm frame	1 623 W × 1 098 h	1 623 W × 1 403 h	1 623 W × 1 708 h	1 923 W × 1 098 h	1 923 W × 1 403 h	1 923 W × 1 708 h	1 923 W × 2 013 h

Step 4: Determine the length of each component.

• Write the lengths of each section on your AHU diagram.

• Add the relevant lengths to find the total AHU length: this along with the frame size gives you the overall dimensions of the AHU (Table 5-31).

Table 5-31 Approximate section lengths (mm)

LKP Model	LKP-1	LKP-2	LKP-3	LKP-4	LKP-5	LKP-6	LKP-7
Fan	700	1 000	1 000	1 300	1 300	1 300	1 600
Diffuser	400	700	700	700	700	700	700
Mixing box	300	300	300	300	600	600	600
Damper	165	165	165	165	165	165	165
Pad filter	200	200	200	200	200	200	200
Pad/bag filter	600	600	600	600	600	600	600
Bag filter	500	500	500	500	500	500	500
Steam humidifier	1 000	1 000	1 000	1 000	1 000	1 000	1 000
Coil 1 row	200	200	200	200	300	300	300
Coil 2 row	200	200	200	200	300	300	300
Coil 4 row	500	500	500	500	500	500	500
Coil 5 row	500	500	500	500	500	500	500
Coil 6 row	600	600	600	600	600	600	600
Coil 7 row	700	700	700	700	700	700	700
Coil 8 row	700	700	700	700	700	700	700
Electric coil	800	800	800	800	800	800	800
Silencer	900	900	900	900	900	900	900
Recuperator	On application						
Heat wheel	1 300	1 300	1 300	1 300	1 300	1 300	1 300
indirect heater	On application						

Example using an alternative manufacturer: Barkell Ltd. (http://www.barkell.co.uk)

• Select a primary air AHU (full fresh air) to supply 2 000 m^2 of floor area (assume 1 person per 10 m^2) (Figure 5-110).

Figure 5-110 A primary air AHU

Determine the volume flow rate:

2000 m^2 divided by m^2 per person gives an occupancy of 200 people.

Each person should receive 10 l/s of primary air, so we need to deliver $10 \times 200 = 2\ 000$ l/s or 2 m^3/s.

Select the AHU based on the volume flow rate (Table 5-32):

Table 5-32 AHU sizing guide

Unit size	External width /mm	External height /mm	Standard base height/mm	High volume flow rate (3 m/s)/(m^3/s)	Mid volume flow rate (2.5 m/s)/(m^3/s)	Optimum volume flow rate (2 m/s)/(m^3/s)
1	750	590	74	0.51	0.43	0.34
2	750	830	74	0.84	0.70	0.56
3	1 050	830	74	1.38	1.15	0.92
4	1 250	950	74	1.85	1.72	1.38
5	1 350	950	74	2.28	1.90	1.52
6	1 350	1 110	74	2.66	2.32	1.86
7	1 350	1 190	74	3.02	2.52	2.02
8	1 350	1 350	74	3.40	2.94	2.35
9	1 450	1 350	90	3.89	3.22	2.58
10	1 550	1 350	90	4.16	3.50	2.80
11	1 650	1 350	90	4.62	3.78	3.02
12	1 750	1 350	90	4.75	4.06	3.25
13	1 950	1 350	90	5.35	4.62	3.70
14	1 950	1 430	90	5.94	4.95	3.96
15	1 950	1 670	90	7.06	5.89	4.71

• Unit size 7 is optimal.

• Unit size 6 is OK if we are happy to go to 2.5 m/s (higher pressure drop and therefore higher energy consumption).

• Note that Unit size 5 should not be selected for the design of the cooling airflow of the air-conditioning units because the figure of 3.0 m/s is too small. This is because at this velocity some of the condensate can be blown down the AHU and cause corrosion and other problems.

Determine the total length according to the components (Table 5-33):

Table 5-33 Approximate length of air handling unit components for unit size (mm)

Description	1	2	3	4	5	6	7	8	9	10	11	12	13	14	15
Fan section	1 200	1 200	1 285	1 395	1 395	1 530	1 605	1 605	1 675	1 675	1 870	1 870	1 870	1 870	1 975
Diffuser	500	500	500	500	600	600	600	600	700	700	700	700	700	700	700
Attenuator-1200 mm long	1 300	1 300	1 300	1 300	1 300	1 300	1 300	1 300	1 300	1 300	1 300	1 300	1 300	1 300	1 300
Panel filter	500	500	500	500	600	600	600	600	600	600	600	600	600	600	600
Bag filter	500	500	500	500	600	600	600	600	600	600	600	600	600	600	600
Panel and bag filter	500	500	500	500	600	600	600	600	600	600	600	600	600	600	600
Rigid filter	400	400	400	400	400	400	400	400	400	400	400	400	400	400	400
Panel and rigid filter	400	400	400	400	400	400	400	400	400	400	400	400	400	400	400
Carbon filter	750	750	750	750	750	750	750	750	750	750	750	750	750	750	750
HEPA filter	400	400	400	400	400	400	400	400	400	400	400	400	400	400	400
Water heating coil	280	280	280	280	310	310	310	310	310	310	310	310	310	310	310
Steam heating coil	350	350	350	350	350	350	350	350	350	400	400	400	400	400	400
Water cooling coil	900	900	900	900	900	900	900	900	900	900	900	900	900	900	900
DX cooling coil	900	900	900	900	900	900	900	900	900	900	900	900	900	900	900
Runaround coil	600	600	600	600	600	600	600	600	600	600	600	600	600	600	600
Access	500	500	500	500	625	625	625	625	625	625	625	625	625	625	625
Electric heater	400	400	400	400	400	400	400	400	400	400	400	400	400	400	400
Indirect gas heater	860	860	860	860	860	860	860	860	1 570	1 570	1 570	1 570	1 570	1 570	1 865

5.2.3.8 Functionality of Different Devices

1. CHP (Combined Heat & Power) (Figure 5-111)

• Heating only.

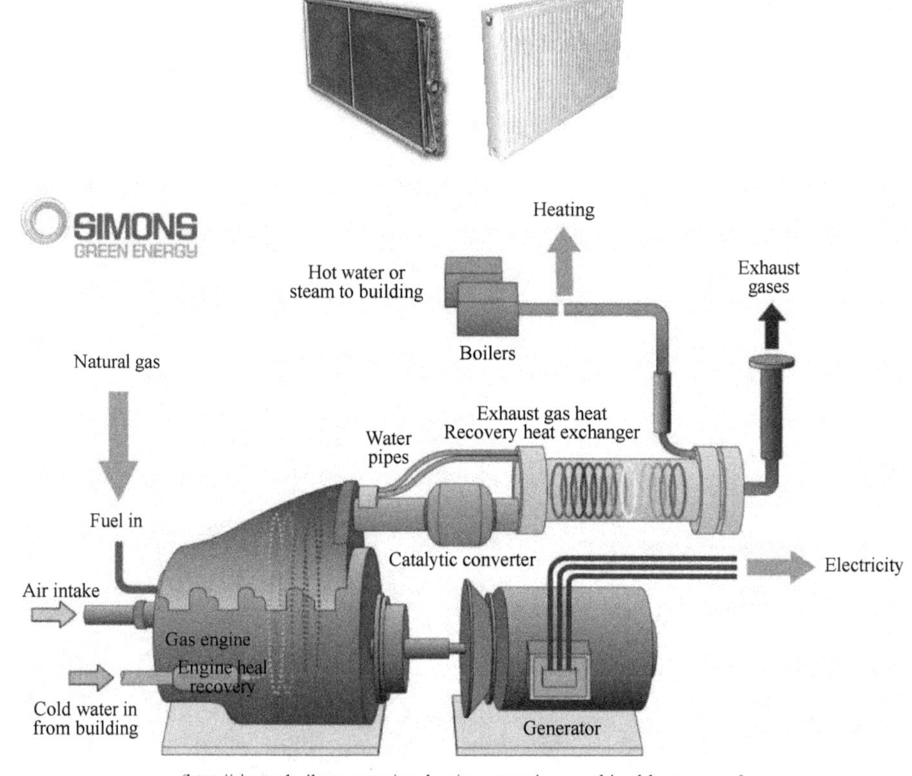

(http://simonsboiler.com.au/product/cogeneration-combined-heat-power/)

Figure 5-111 CHP

2. Heat Pump (Heating + Cooling) (Figure 5-112)

- Ground source HP;
- Water source HP;
- Air source HP.

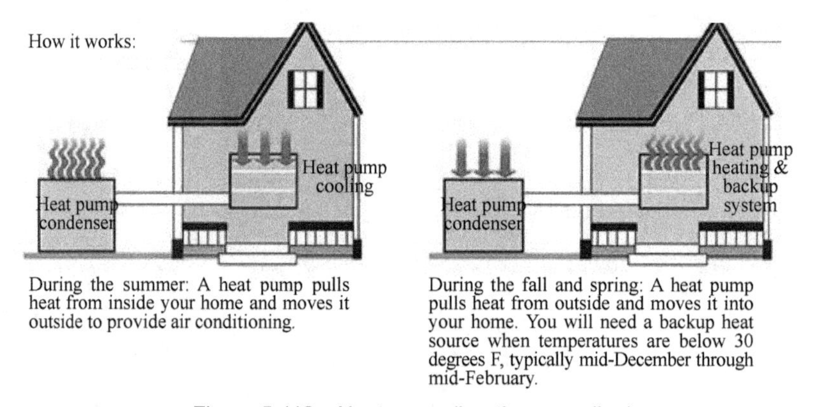

Figure 5-112 Heat pump (heating + cooling)

(https://www.mnpower.com/ProgramsRebates/AirSourceHeatPumps)

3. Chillers (Figure 5-113)

• Cooling only.

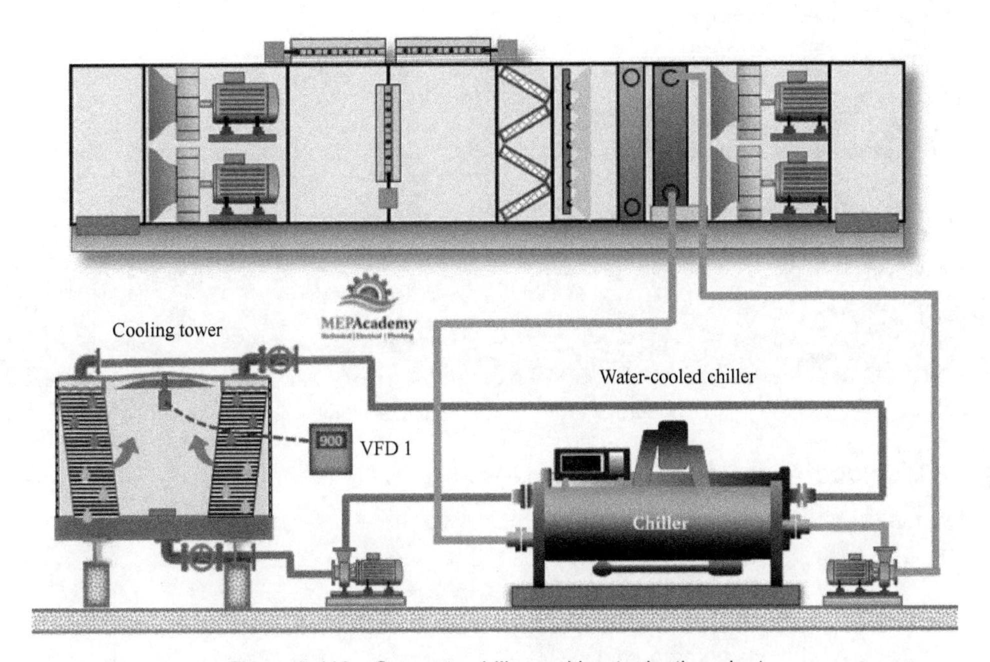

Figure 5-113 Separate chiller and heat rejection plant

4. Ice Storage (Figure 5-114)

• Cooling only.

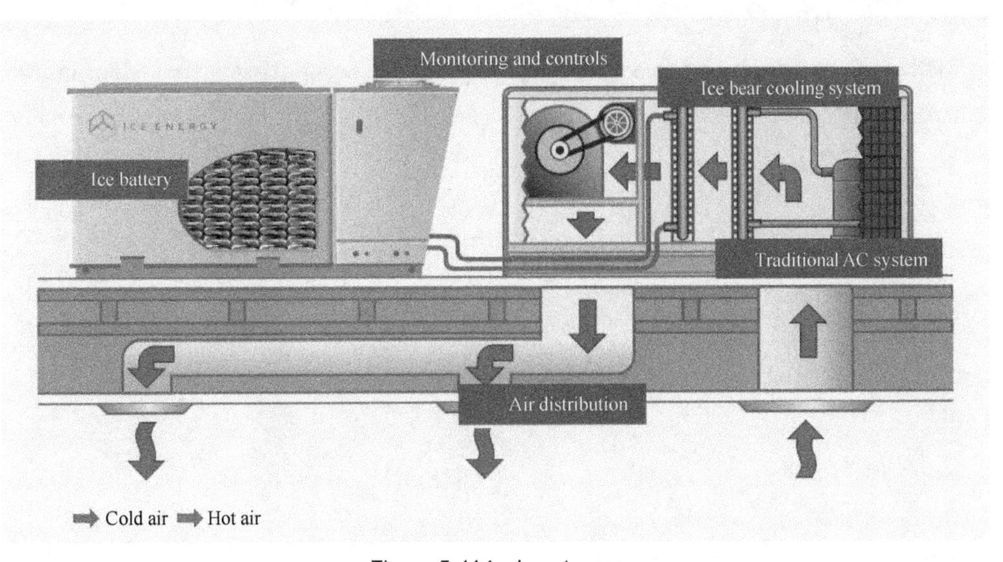

Figure 5-114 Ice storage

(https://www.ice-energy.com/technology/)

5.2.3.9 Diffusers

Types of diffusers/grills (Figure 5-115):

• Nozzles, jet diffusers;

• Square grilles, linear louvres, slot diffusers;

• Swirl diffusers, displacement diffusers;

• Perforated face diffusers, integrated diffusers.

Figure 5-115 Types of diffusers/grills

1. Displacement Flow (Figure 5-116)

Piston (or plug) flow:

• Air moves slowly across a room;

• Good choice for fixed pollutant source;

• Prevents pollutants being mixed.

Displacement ventilation:

• Air introduced at low level;

• Convection currents carry air to high level;

• Air extracted at high level.

Both systems provide excellent air quality within the occupied zone and result in a temperature gradient in the space.

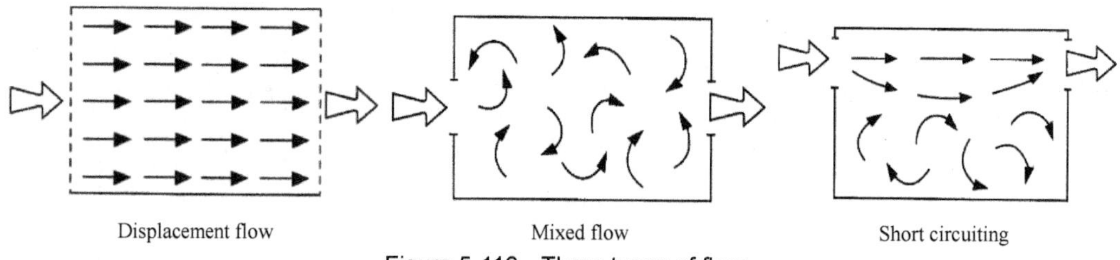

Figure 5-116 Three types of flow

2. Mixed Flow

Characteristics:

• Mixes pollutants with room air;

• Suitable where the pollution source is mobile;

• Enables high cooling loads to be met;

• Produces a more uniform temperature distribution.

Design issues:

• Perfect mixing is difficult to achieve;

• Short circuiting must be avoided (Supply air extracted before mixing thoroughly with the room air is termed as short circuiting);

• Air discharged at significant velocity—noise issues.

Important terminology (Figure 5-117):

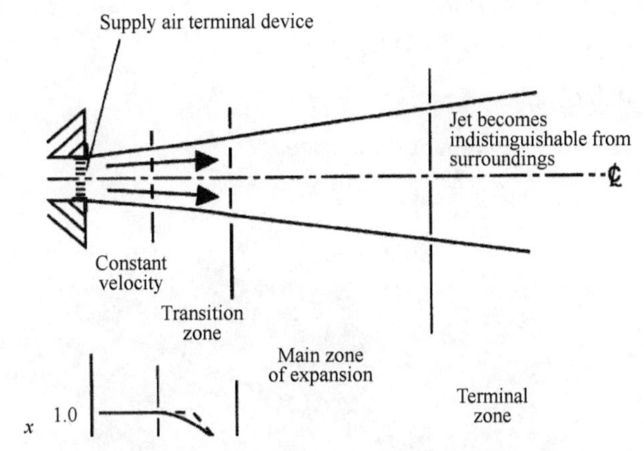

Figure 5-117 Air stream characteristics & terminology

Throw:

• The distance to which the jet has slowed to the "terminal velocity";

• The standard terminal velocity is 0.5 m/s.

Buoyancy effects:

• Cause jet to drop or rise;

• Avoid cold air dumping into the occupied zone.

Surface effects:

• Air discharged at a speed greater than 1.5 m/s along a ceiling clings to the ceiling due to the "Coanda effect" (Figure 5-118).

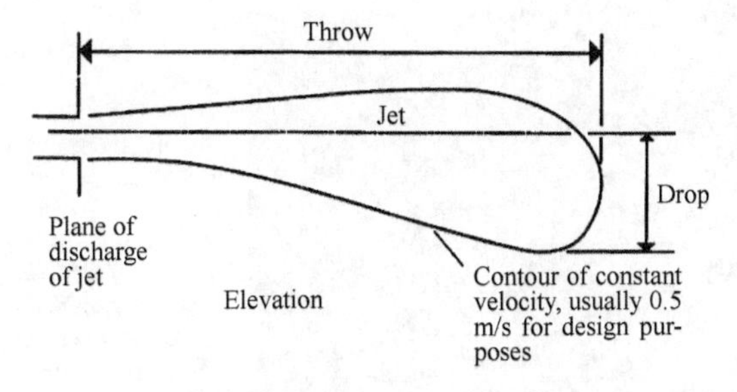

Figure 5-118 Throw and drop

Positioning extract grilles (Figure 5-119):

Velocity:

- Depends upon the volume flow rate and cross sectional area;
- Soon drops off with distance;
- Position is therefore not critical.

It is advantageous if you can position the extract grilles so that:

- Pollutants can be extracted as directly as possible;
- Warmed air can be removed rather than mixed.

Installation location:

- Position above photocopier/printer;
- Position above the water chamber of the heater;
- Position of the ceiling void.

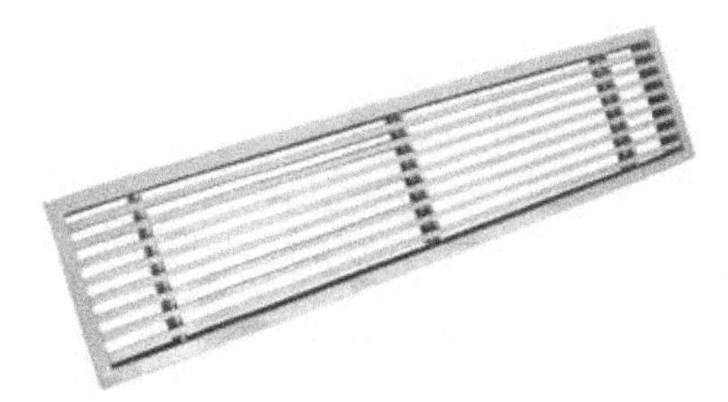

Figure 5-119 Extract grilles

5.2.3.10 Diffuser Selection

Rules for diffuser selection (Figure 5-120):

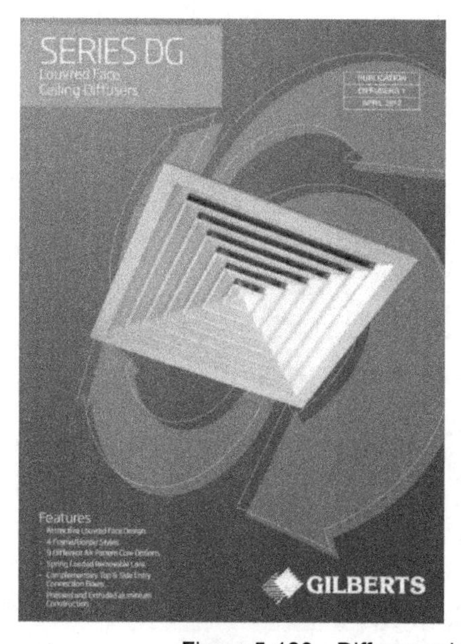

Plenty of other manufactures out there

If you have never selected diffusers before then stick with simple data such as that presented by Gilberts

Figure 5-120 Diffuser selection

• The zone width must not be more than 3 times the floor-to-ceiling height;

• Zone aspect ratio ≤ 1.5 ∶ 1.0 (length ∶ width), square zones are best;

• Throw distance (0.5 m/s) should be 75% of the distance to the closest zone boundary i.e. the width rather than the length, and it is impossible to achieve exactly 75% (65% to 85% is an acceptable range) (Figure 5-121).

Having established the position where terminals can be sited, refer to data showing core pattern details and select the suitable direction pattern required. Knowing the volume and throw for each diffuser in question then check:

a) Recommended limit of volume per direction according to ceiling height (Table 1) with throw of air required lying between the max and min values.

b) Note sound level from performance data and check this recommendation shown on table.

c) Determine the total pressure drop from performance data.

Table 1

Ceiling height	Max vol per direction (each diffuser)	Cooling differential maximum
2.5 m	0.090 m³/s	12 ℃
3.0 m	0.200 m³/s	12 ℃
3.5 m	0.350 m³/s	12 ℃

Figure 5-121 Selection procedure

• Throw: The first figure in the choice box is the throw that corresponds to 0.5 m/s, and the second figure is the throw that corresponds to 0.25 m/s (Figure 5-122).

Example:

Throw =1 m, corresponding to 0.5 m/s;

Throw =2 m, corresponding to 0.25 m/s.

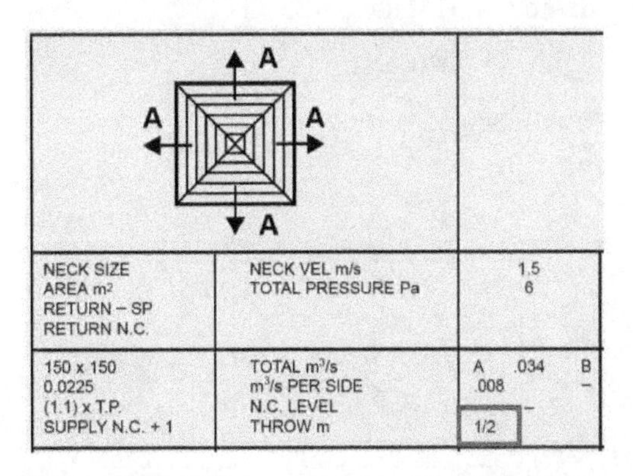

Figure 5-122 Figures corresponding to the throw

5.2.3.11 Diffuser Positioning

Diffuser positioning demo (1) (Figure 5-123):

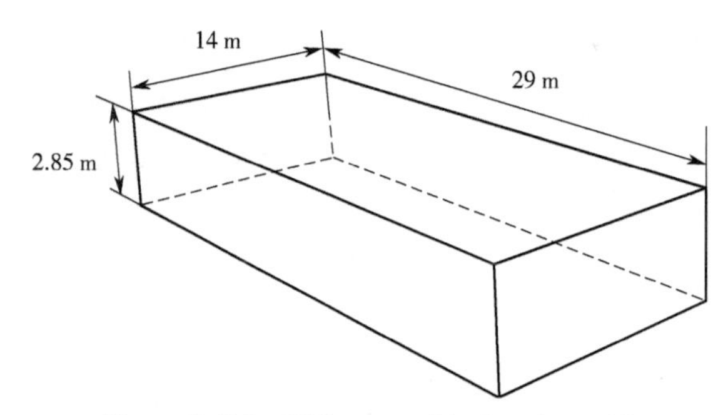

Figure 5-123 Diffuser positioning demo (1)

Diffuser positioning demo (2) (Figure 5-124):

• 3 times the floor-to-ceiling height is 8.55 m;

• This is less than the room width (14 m)—so best to make 2 rows.

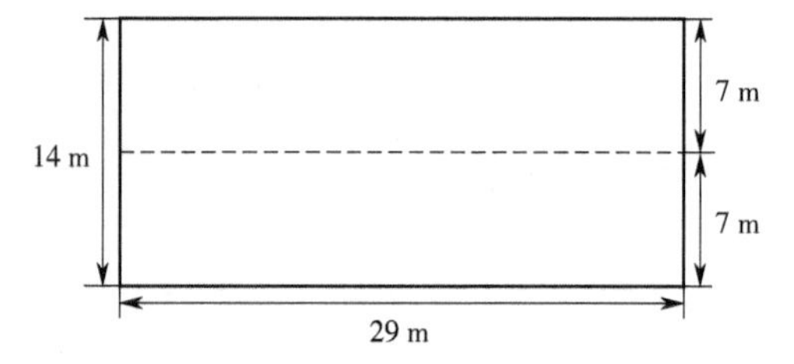

Figure 5-124 Diffuser positioning demo (2)

Diffuser positioning demo (3) (Figure 5-125):

• Max zone length = 1.5 × 7 = 10.5 m;

• The room could be divided into three down its length;

• Remember that we would like to have square zones if possible.

Diffuser positioning demo (3a) (Figure 5-125):

• Option 1—three columns (6 diffusers);

• Offers the minimum number of diffusers;

• 9.67 m ∶ 7.0 m = 1.38 ∶ 1.0.

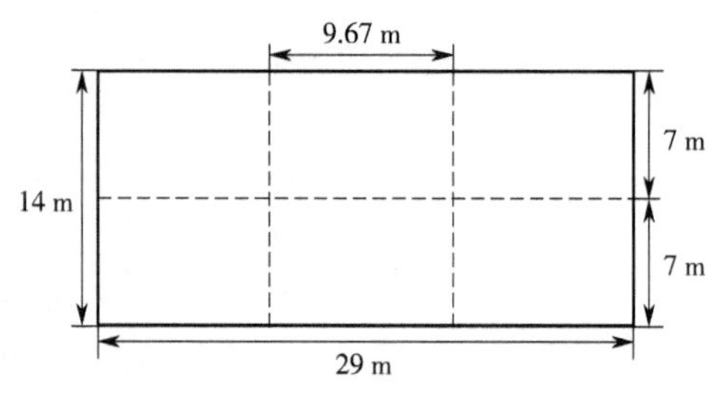

Figure 5-125 Diffuser positioning demo (3) or (3a)

Diffuser positioning demo (3b) (Figure 5-126):

• Option 2—four columns (8 diffusers);

• More square zones;

• 7.25 m ∶ 7.0 m= 1.04 ∶ 1.0.

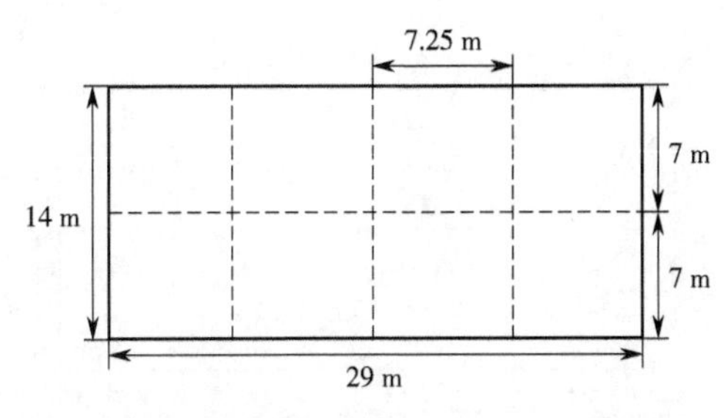

Figure 5-126 Diffuser positioning demo (3b)

Diffuser positioning demo (3c) (Figure 5-127):

• Option 3—five columns (10 diffusers);

• More square zones;

• 5.8 m ∶ 7.0 m= 0.83 ∶ 1.0.

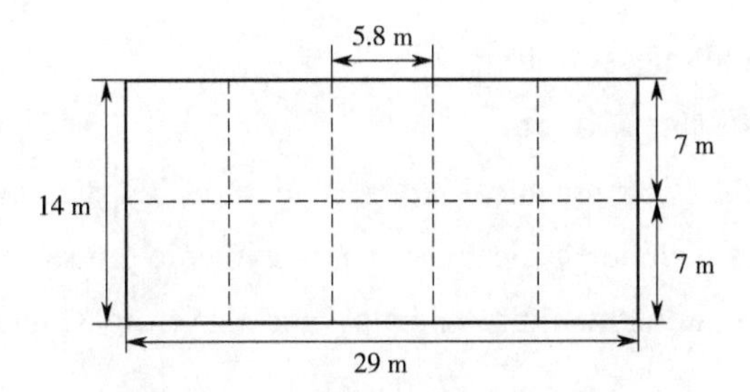

Figure 5-127 Diffuser positioning demo (3c)

Diffuser positioning demo (4) (Figure 5-128):

• Optimal solution;

• Option 2—four columns (8 diffusers), 7.25 m ∶ 7.0 m= 1.04 ∶ 1.0;

• The ideal throw = 0.75 × 7/2 = 2.625 m.

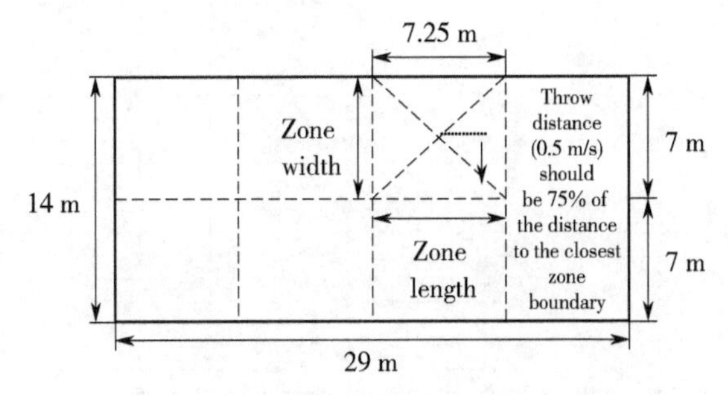

Figure 5-128 Diffuser positioning demo (4)

Diffuser positioning demo (5) (Figure 5-129):

• The throw can be in the range of 65% to 85%;

• Minimum throw = 0.65 × 7/2 = 2.275 m≈2.3 m;

• Maximum throw = 0.85 × 7/2 = 2.975 m≈3.0 m.

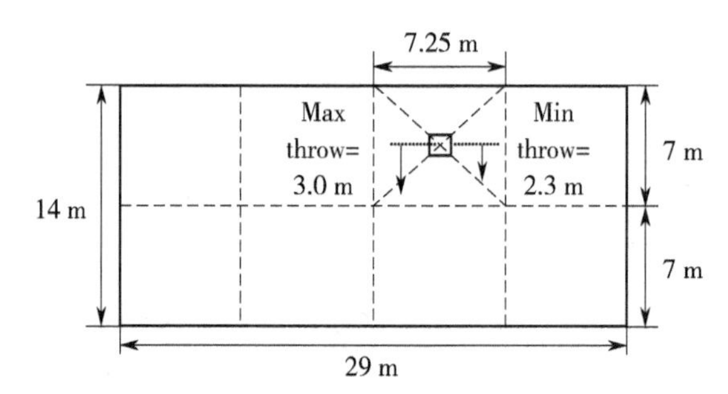

Figure 5-129 Diffuser positioning demo (5)

Diffuser positioning summary:

• 8 diffusers are optimal;

• The diffuser throw (0.5 m/s and at the design air flow rate) must be between 2.3 m and 3.0 m;

• We can now start to select diffusers.

5.2.3.12 Psychrometric Chart

A psychrometric chart is a graphical representation of the psychrometric processes of air. Psychrometric processes include physical and thermodynamic properties such as dry bulb temperature, wet bulb temperature, humidity, enthalpy, and air density (Figure 5-130).

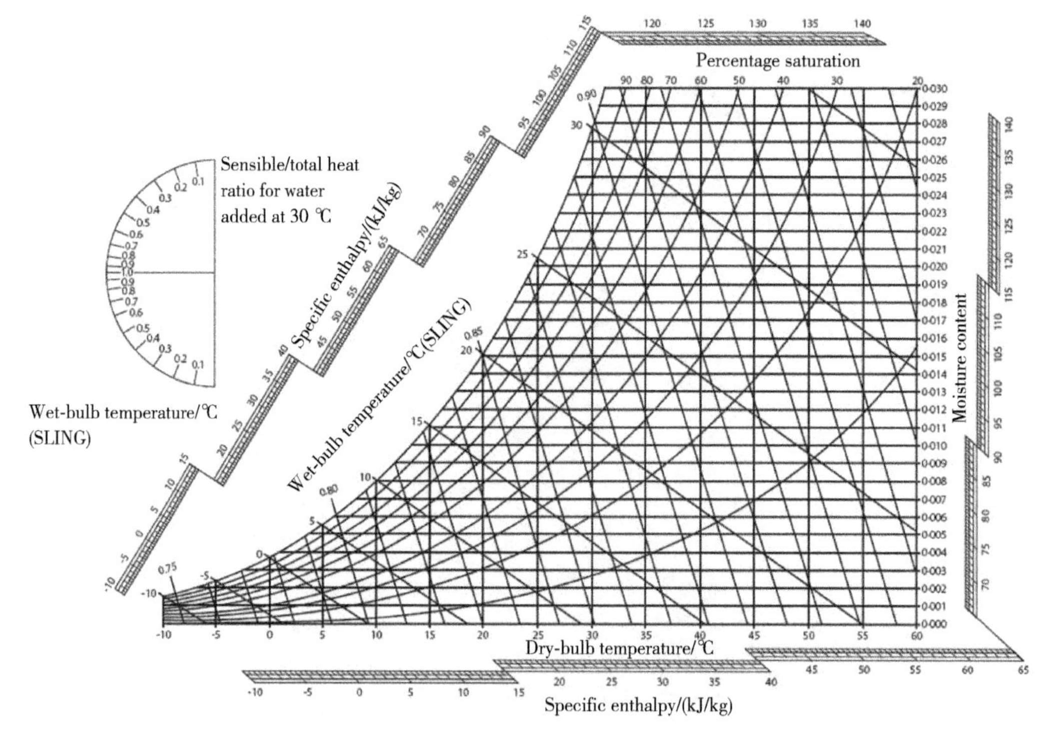

Figure 5-130 Standard CIBSE psychrometric chart

(https://www.cibsejournal.com/cpd/modules/2009-04/)

The chart is also often used by building service engineers to dynamically plot points that represent the exterior air conditions and understand the process the air must go through to reach comfortable conditions for the occupants inside a building. When using the psychrometric chart (Figure 5-131) for this purpose, the data points move around the chart.

• Enthalpy (H): A measurement of the total heat energy in a thermodynamic system. It is equal to the internal energy of the system plus the product of pressure and volume. The unit of measurement for enthalpy in the International System of Units (SI) is the joule.

$$H = U + pV$$

where H is the enthalpy of the system; U is the internal energy of the system; p is the pressure of the system; V is the volume of the system.

• Room ratio line (RRL) (Figure 5-131):

$$\mathrm{RRL} = \frac{Q_{\mathrm{sensible}}}{Q_{\mathrm{sensible}} + Q_{\mathrm{latent}}}$$

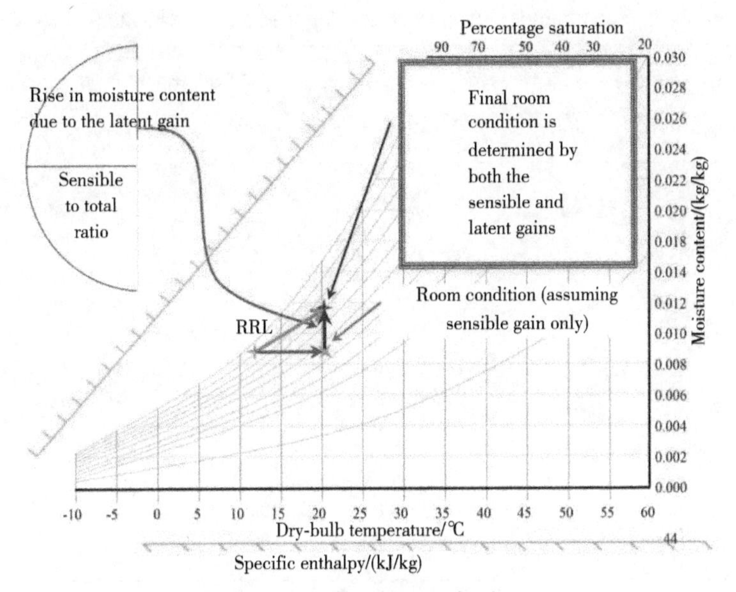

Figure 5-131 Psychrometric chart

Examples (Figure 5-132):

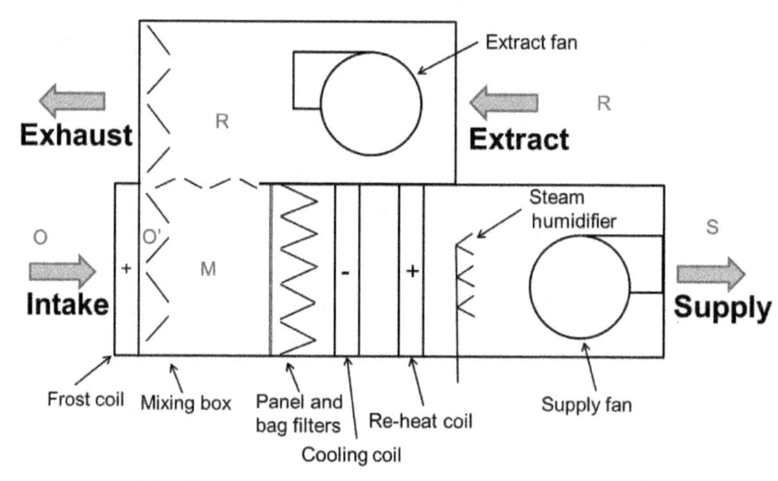

Figure 5-132 Air treatment process in a constant air volume (CAV) system

• Determination of indoor condition (R) (Table 5-34).

Table 5-34 Recommended comfort criteria for specific applications (CIBSE Guide A)

Building/room type	Customary winter operative temperatures for stated activity and clothing levels			Customary summer operative temperatures (air conditioned buildings†) for stated activity and clothing levels		
	Temp. /°C	Activity / met	Clothing / clo	Temp. /°C	Activity / met	Clothing / clo
Airport terminals:						
— baggage reclaim	12-19	1.8	1.2	21-25	1.3	0.6
— check in areas	18-20	1.4	1.2	21-25	1.3	0.6
— concourse (no seats)	19-24	1.8	1.2	21-25	1.3	0.6
— customs area	18-20	1.4	1.2	21-25	1.3	0.6
— departure lounge	19-21	1.3	1.2	22-25	1.2	0.6
Art galleries — see Museums and art galleries						
Banks, building societies, post offices:						
— counters	19-21	1.4	1.0	21-25	1.3	0.6
— public areas	19-21	1.4	1.0	21-25	1.3	0.6
Bars/lounges	20-22	1.3	1.0	22-25	1.3	0.6
Bus/coach stations — see Railway/coach stations						
Churches	19-21	1.3	1.2	22-25	1.3	0.6
Computer rooms	19-21	1.4	1.0	21-25	1.3	0.6
Conference/board rooms	22-23	1.1	1.0	23-25	1.1	0.6
Drawing offices	19-21	1.4	1.0	21-25	1.3	0.6

A constant volume air conditioning system should be designed to provide cooling and ventilation to the computer rooms on the ground floor to maintain 50%RH +/- 5% year round.

RH from design requirement or standard

• Determination of outdoor condition (O) (Table 5-35).

Table 5-35 Design conditions (CIBSE Guide A)

Station Information

Station name	WMO#	Lat	Long	Elev	StdP	Hours +/- UTC	Time zone code	Period
1a	1b	1c	1d	1e	1f	1g	1h	1i
AUGHTON	033220	53.55N	2.92W	56	100.65	0.00	GMT	8296

Annual Heating and Humidification Design Conditions

Coldest month	Heating DB		Humidification DP/MCDB and HR						Coldest month WS/MCDB				MCWS/PCWD to 99.6% DB	
			99.6%			99%			0.4%		1%			
	99.6%	99%	DP	HR	MCDB	DP	HR	MCDB	WS	MCDB	WS	MCDB	MCWS	PCWD
2	3a	3b	4a	4b	4c	4d	4e	4f	5a	5b	5c	5d	6a	6b
2	-3.4	-1.9	-6.9	2.1	-1.9	-5.2	2.4	-0.6	14.3	6.7	12.7	6.7	3.4	140

Annual Cooling, Dehumidification, and Enthalpy Design Conditions

Hottest month	Hottest month DB range	Cooling DB/MCWB						Evaporation WB/MCDB						MCWS/PCWD to 0.4% DB	
		0.4%		1%		2%		0.4%		1%		2%			
		DB	MCWB	DB	MCWB	DB	MCWB	WB	MCDB	WB	MCDB	WB	MCDB	MCWS	PCWD
7	8	9a	9b	9c	9d	9e	9f	10a	10b	10c	10d	10e	10f	11a	11b
7	6.2	24.4	17.6	22.4	16.8	20.6	16.0	18.5	23.1	17.5	21.3	16.7	19.7	4.0	90

Dehumidification DP/MCDB and HR									Enthalpy/MCDB					
0.4%			1%			2%			0.4%		1%		2%	
DP	HR	MCDB	DP	HR	MCDB	DP	HR	MCDB	Enth	MCDB	Enth	MCDB	Enth	MCDB
12a	12b	12c	12d	12e	12f	12g	12h	12i	13a	13b	13c	13d	13e	13f
16.8	12.1	19.8	16.0	11.5	19.0	15.3	10.9	18.1	52.3	23.2	49.5	21.4	46.9	19.9

Note:Annual: 99% ≈ 88 Hours; 99.6% ≈ 35 Hours.

- Determination of mixing point (M).
- Mixing point should be on the line between R and O.
- Enthalpy is determined by

$$m_{\mathrm{O}} \cdot h_{\mathrm{O}} + m_{\mathrm{R}'} \cdot h_{\mathrm{R}'} = m_{\mathrm{M}} \cdot h_{\mathrm{M}}$$

- Determination of supply condition (S).

Winter: RRL from R + Sensible heating line from M.

Summer: RRL + 8-11 ℃ temperature difference between R and S.

- Determination of Frost Temperature (O′).
- Sensible heating line from O + Dry-bulb temperature of 5 ℃ .
- Heating load calculation (Figure 5-133):

$$Q_{\mathrm{heat}} = m_{\mathrm{M}} C_p (t_{\mathrm{s}} - t_{\mathrm{M}})$$

$$Q_{\mathrm{frost}} = m_{\mathrm{O}} C_p (t_{\mathrm{O}'} - t_{\mathrm{O}})$$

where Q is heating load (kW); m is mass flow rate (kg/s); C_p is specific heat (J/(kg · K)); t_{S} is supply temperature (℃); t_{M} is mixing temperature (℃); $t_{\mathrm{O}'}$ is preheat temperature (℃); t_{O} is outdoor temperature.

- Cooling load calculation (Figure 5-134):

$$Q_{\mathrm{cool}} = m_{\mathrm{M}} (H_{\mathrm{M}} - H_{\mathrm{S}})$$

where Q is cooling load (kW); m is mass flow rate (kg/s); H_{M} is mixing enthalpy (kJ/kg); H_{S} is supply enthalpy (kJ/kg).

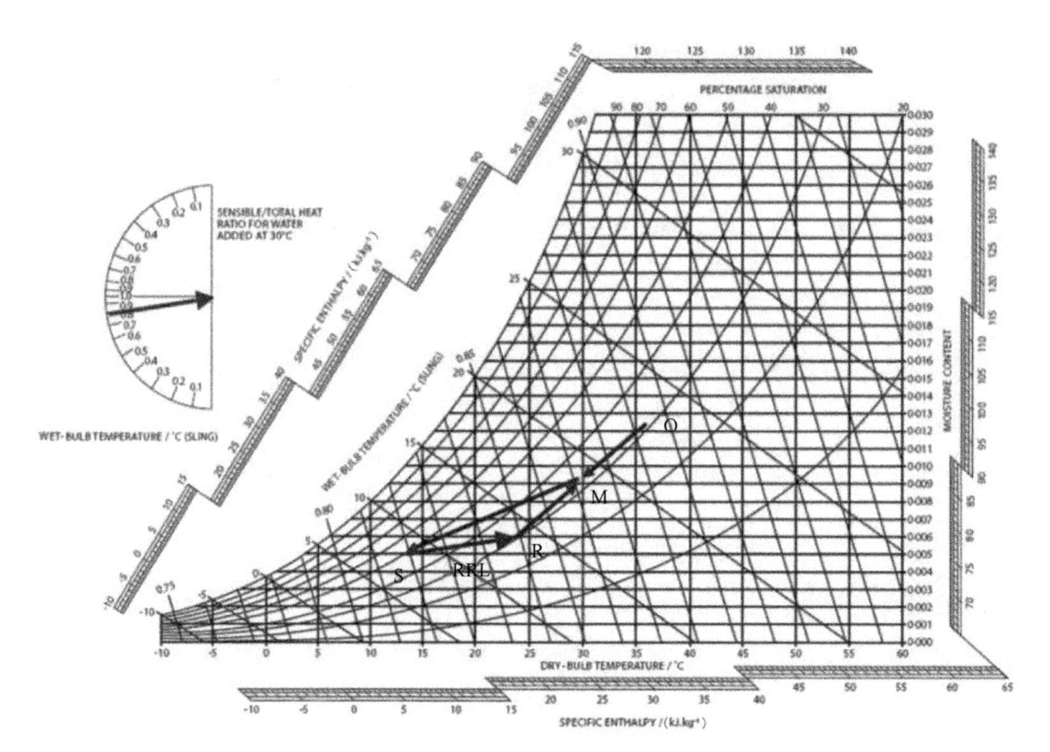

Figure 5-133 Heating load calculation in winter

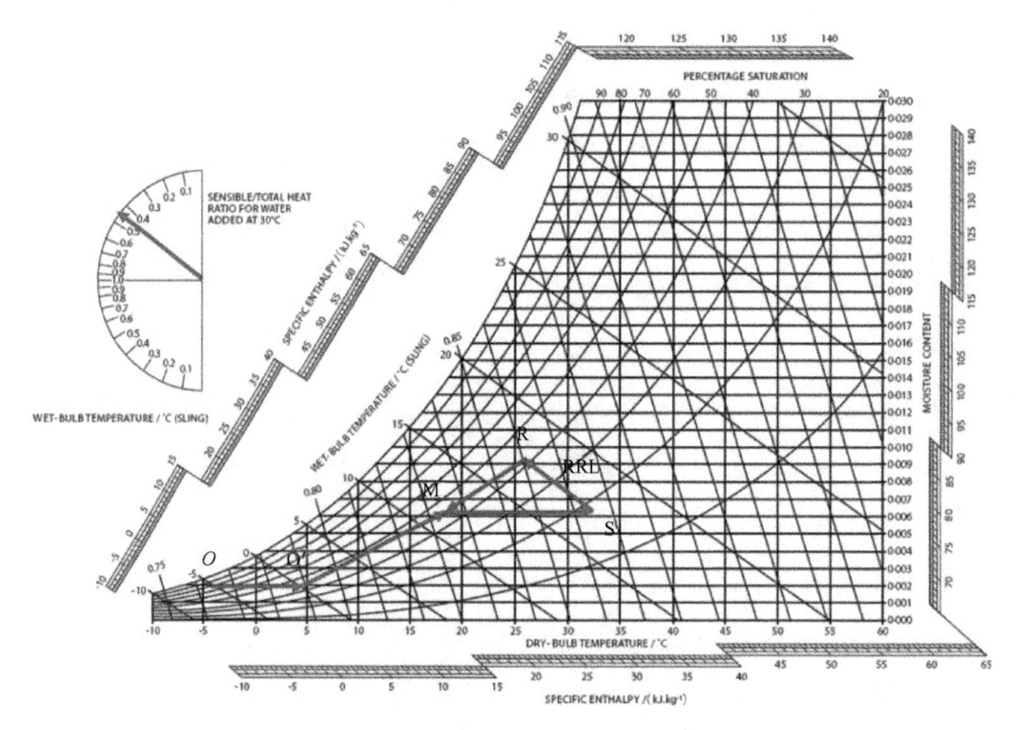

Figure 5-134 Cooling load calculation in summer

5.2.4 Lighting System

5.2.4.1 Definition of Lighting

Light can be considered as a human sensation in a similar manner to smell, taste, sound etc. Like other sensations it is necessary to stimulate the senses, in the case of light sensation the stimulant is electromagnetic radiation falling onto the retina of the eye. The sensation of light can be considered as a combination of electromagnetic radiation and the eyes response to it (Figure 5-135).

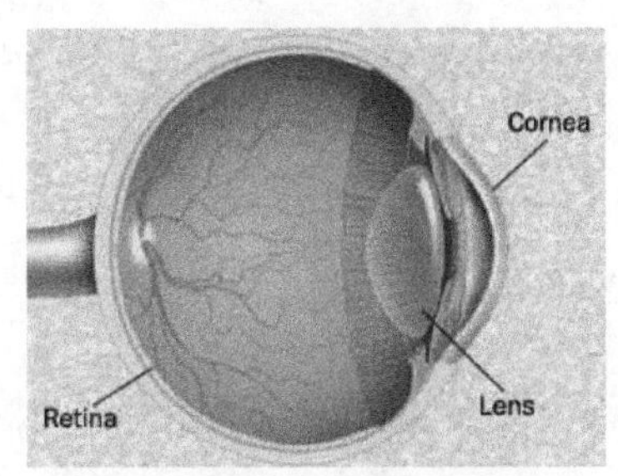

Figure 5-135 Eyeball structure

5.2.4.2 Electromagnetic Radiation

Electromagnetic radiation is a form of energy that can pass from one place to another without the need for any material substance in the intervening space.

Light sensation is caused by a relatively narrow band of the electromagnetic spectrum with wavelengths from 380 nm to 780 nm (1 nm = 10^{-9} m), and this band can be further sub-divided into the range of color sensations that humans can experience (Figure 5-136). The eye discriminates within this range by the sensation of color.

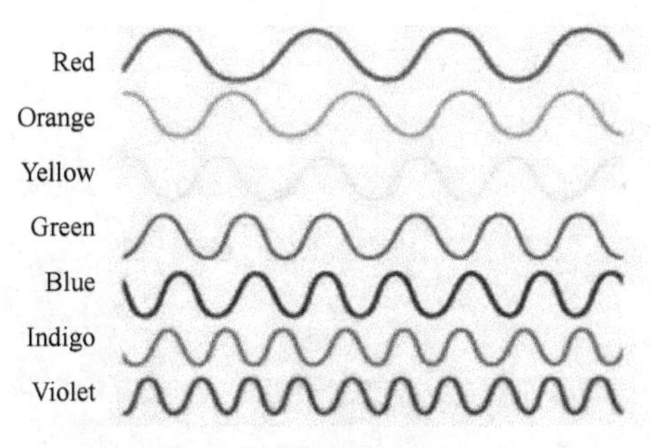

Figure 5-136 Color band

It is usual to refer to electromagnetic radiation in terms of wavelength: Electromagnetic radiation (photons) travels at a speed of approximately 300 000 km/s (3×10^8 m/s) in a vacuum.

Frequency = Velocity/Wavelength

Wavelength = Velocity/Frequency

Hence wavelength is proportional to 1/frequency.

In order to have a comprehensive understanding of various electromagnetic waves, people arrange these electromagnetic waves in order of their wavelength or frequency, wave number and energy, which is the electromagnetic spectrum (Figure 5-137). According to the wavelength, frequency and wave source, the electromagnetic spectrum can be roughly divided into radio waves, microwaves, infrared, visible light, ultraviolet, X-rays and gamma rays (Figure 5-138).

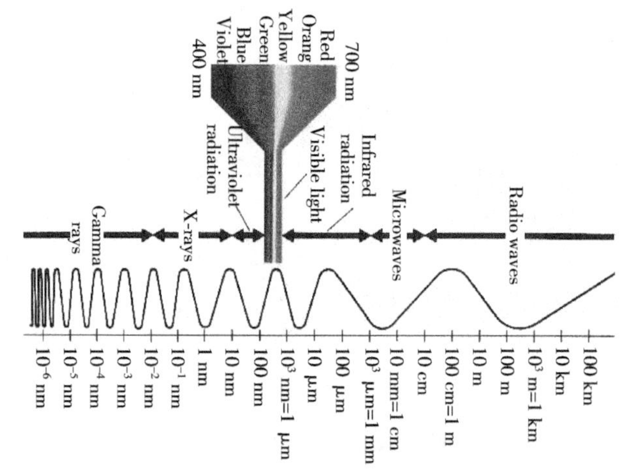

Figure 5-137 Various electromagnetic waves

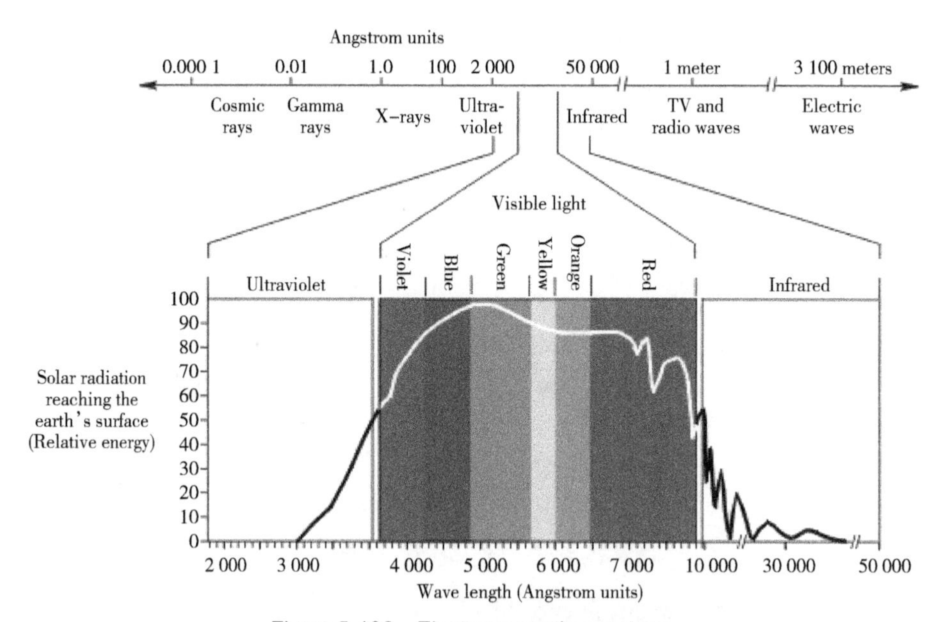

Figure 5-138 Electromagnetic spectrum

5.2.4.3 Visual Response

1. Visual Response: Seeing and Understanding

• Light brings human being into contact with its surroundings via sight.

• Approximately 80% of our sensory input is visual.

• In the process of designing a lighting system in the built environment, it is important that the designer has an understanding of the visual processes and characteristics of vision if he/she is to provide acceptable lighting for the visual process to function properly (Figure 5-139).

Figure 5-139 Seeing and understanding

The process of seeing and understanding comprises of three functions:

• The production of an optical image on the light sensitive receptors of the eye;

• The production of a signal in the nerve system from the light receptors to the brain;

• The interpretation of these signals.

2. Visual Response: Eye Sensitivity

• The eye responds over the visible part of the spectrum 400 nm to 700 nm, however their quantitative response to each wavelength (color) varies i.e. the sensitivity of the eye is not uniform but varies with wavelength (Figure 5-140).

• The maximum sensitivity of the eye is in the yellow/green part of the spectrum at 555 nm.

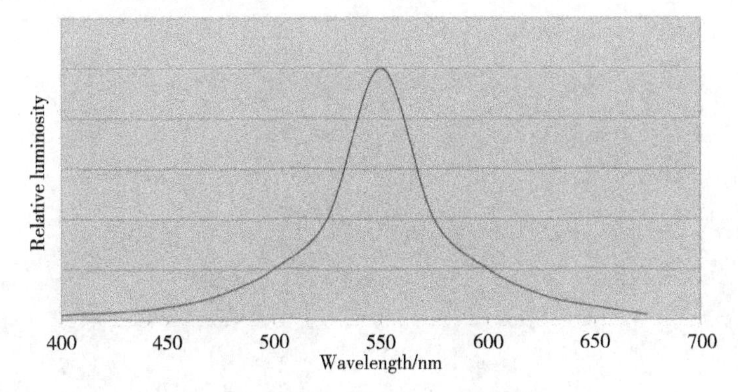

Figure 5-140 Eye sensitivity

5.2.4.4 Relative Physical Parameters

1. Luminous Flux—*f*, Lumen (lm)

Luminous flux (f) is the total light emitted by a source or received by a surface in all directions, and is measured in lumens (Figure 5-141).

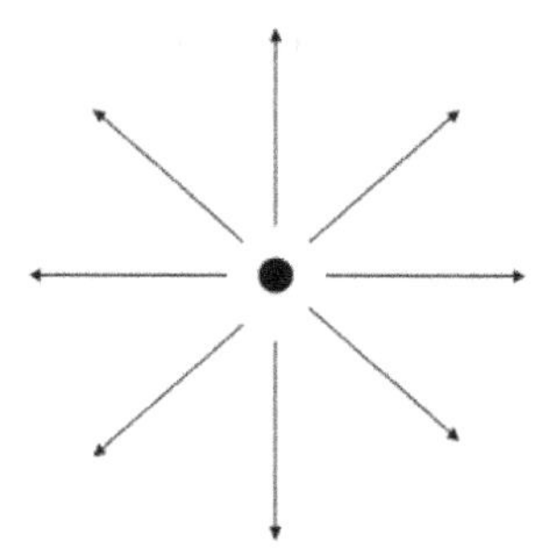

Figure 5-141 Luminous flux (f)

2. Luminous Intensity—I, Candela (cd)

Luminous intensity (I) is the power of a source to emit light in a given direction, and is measured in candelas. For a point source the intensity is the luminous flux per unit solid angle in the direction in question. One candela is equal to one lumen per steradian (Figure 5-142).

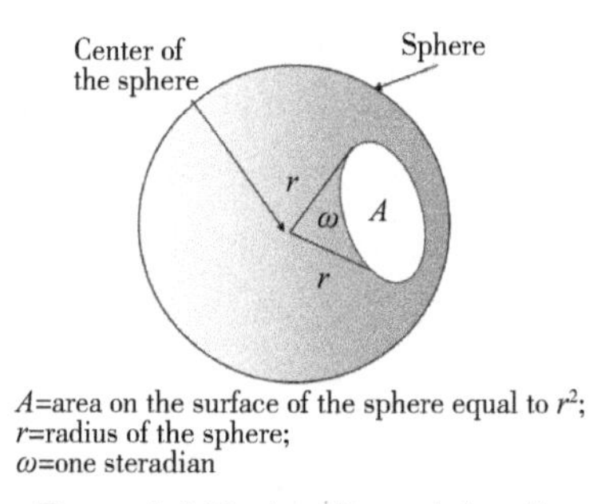

Figure 5-142 Luminous intensity

3. Steradian (Sr)

One steradian is a unit solid angle subtending an area on the surface of a sphere equal to the square of the sphere radius (Figure 5-143).

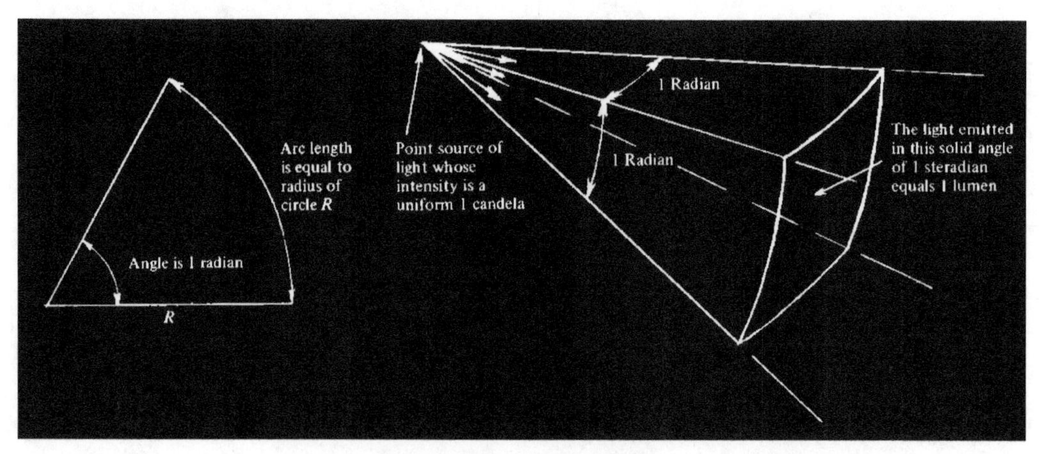

Figure 5-143 Steradian (Sr)

4. Surface Properties

Surface properties are shown in Figure 5-144.

Reflectance (ρ): the ratio of luminous flux reflected from a surface to the luminous flux incident on it.

Transmittance (τ): the ratio of luminous flux transmitted by a surface to the luminous flux incident on it.

Absorptance (α): the ratio of luminous flux absorbed by a surface to the luminous flux incident on it.

$$\rho + \tau + \alpha = 1$$

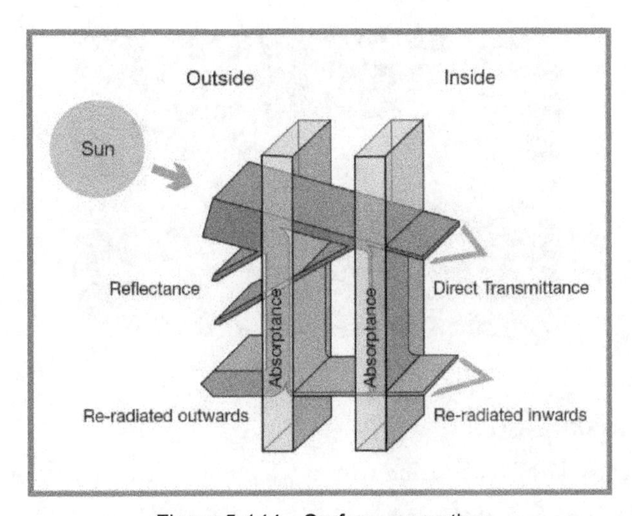

Figure 5-144 Surface properties

(https://www.pilkington.com/en-gb/uk/architects/types-of-glass/solar-control-glass/how-it-works)

Reflectance is dimensionless, always less than unity and is expressed as a percentage or as a decimal. Reflectance can be perfectly diffused (i.e. off a matt surface such as a painted wall) or perfectly specular (i.e. off a mirror)—in reality it is always somewhere in between.

Transmittance is dimensionless, always less than unity and is expressed as a percentage or as a decimal. It can also be diffused or specular.

5.2.4.5 Visual Comfort

1. Visual Comfort: Brightness

The physical measure for describing the brightness of an object is luminance (in the unit of candela per square meter or cd/m^2). The eye does not behave like a linear receptor—an object that has twice the luminance of another one does not necessarily appear twice as bright to our eyes. There is no linear relationship between the luminance and the perceived brightness (Figure 5-145).

How we perceive the brightness of an object depends on the following factors:

• The luminance (measured brightness) of the object;

• The state of the adaptation of the eyes;

• The luminance of the surrounding of the object, for example, the same object can appear very bright when looked at in a dark environment or rather dark when put against a well lit background.

A typical example of how the perceived brightness depends on the environment is a TV screen. When we watch television during the day, the display is not much brighter than the surrounding. Since the eyes are adapted to the surrounding, the screen appears rather dim. On the contrary, given the same TV set put into a dark room, the picture will look much brighter.

Figure 5-145 Brightness

2. Visual Comfort: Adaption

The eyes are capable of functioning over a wide luminance range from 10^{-6} cd/m^2 to 10^6 cd/m^2 (candela per square meter). The eyes cannot handle the whole of this range simultaneously. At any one operating level the eye self-optimises or adapts to operate efficiently. Following a sudden change in field luminance, vision is impaired whilst adaptation to the new conditions takes place.

3. Visual Comfort: Contrast

Contrast is a measure of the difference of luminance levels between two areas, or between an object and its background (Figure 5-146). An object can only be seen because of its contrast with its surroundings. Maximum contrast, for a given illumination, is obtained by putting matt black on a matt white surrounding and vice versa.

Figure 5-146 Contrast

Contrast is usually defined as below:

$$C=\frac{L_{\text{object}}-L_{\text{background}}}{L_{\text{background}}}$$

Although different definitions are also widely used, unfortunately, there is no agreement in the literature on the method of expressing contrast.

The human visual system can adapt to a very wide range of luminance for any given scene. However, this range will be much reduced if both, very bright and very dark objects, are in the field of view (high contrasts); the dark ones will appear black, while the bright ones look completely washed out. It is then impossible to distinguish any details (Figure 5-147).

Figure 5-47 High contrasts

Contrast can also be achieved by color. The eye is most sensitive at about 555 nm, which is yellow or green, the brightest colors to the eye. At the other end of the sensitivity curve, the eye is least sensitive to violet 400 nm or red 750 nm, the darkest colors to the eye. Therefore there is not only a color difference but also a considerable contrast difference between yellow and violet.

4. Visual Comfort: Luminance and Glare

Luminance is a photometric measure of the luminous intensity per unit area of light travelling in a given direction. It is the physical measure for describing the brightness of an object.

Glare is caused by a significant ratio of luminance between the task (the object that is being looked at) and the glare source. Factors such as the angle between the task and the glare source and eye adaptation have significant impacts on the experience of glare (Figure 5-148).

(http://sunposition-ralphb.blogspot.co.uk/2015/07/sun-glare-while-driving-june-2015.html)

Figure 5-148 Luminance and glare

5. Visual Comfort: Color Rendering

The color rendering of a light source is an indicator of its ability to realistically reproduce the color of an object.

• Good color rendering: If a lamp has good color rendering properties, then different colors of the illuminated surface will be easy to distinguish. An example would be surfaces illuminated by a filament lamp. The filament lamp emits light at all wavelengths of the visible spectrum (400-700 nm) with an emphasis on the red part of the spectrum. The color rendering will therefore be good (Figure 5-149).

Figure 5-149 Poor color rendering

• Poor color rendering: If a lamp has poor color rendering properties, then different colors of the illuminated surface will be difficult to distinguish. An example would be a low-pressure sodium lamp. This lamp emits light at a single wavelength (600 nm) and will therefore only reveal that color (yellow/orange). It has no ability to distinguish color but is satisfactory for applications such as street lighting.

The CIE (International Lighting Commission) produce an index for color rendering—"R_a" between 0 and 100. Where lower values indicate poor color rendering (Figure 5-150) and higher ones good color rendering.

Figure 5-150 Good and poor color rendering

To make a comparison of the color rendering qualities of light sources easier, color rendering groups have been defined (Table 5-36):

Table 5-36 The CIE color rendering groups

Group	R_a	Importance	Typical application
1A	90 to 100	Accurate color matching	Galleries, medical examinations, color mixing
1B	80 to 90	Accurate color judgement	Home, hotels, offices, schools
2	60 to 80	Moderate color rendering	Industry, offices, schools
3	40 to 60	Accurate color rendering is of little importance	Industry, sports halls
4	20 to 40	Accurate color rendering is of no importance	Traffic lighting

(https://www.new-learn.info/packages/clear/visual/people/comfort/colour_rendering.html)

Some tasks such as color matching in the printing industry have high demands in accurate color rendering and require special attention from the lighting designer. For normal offices, however, the color rendering group will be 1B or 2, which is easily achieved with normal fluorescent lamps (Table 5-37).

Table 5-37 Color rendering indices for different light sources

Light source	Color rendering group
Incandescent lamp	1A
Metal halide lamp	1A to 2
Fluorescent lamp	1A to 3
High-pressure sodium lamp	1B to 4
Low-pressure sodium lamp	4

(https://www.new-learn.info/packages/clear/visual/people/comfort/colour_rendering.html)

The reason for lamps with a poor color rendering such as high- and low-pressure sodium lamps being used is their high efficacy. They output more light per watt of electrical power than lamps that provide a good color rendering.

6. Visual Comfort: Efficacy (ε)

Efficacy (ε) is a measure of the efficiency of a light source—it is the ratio of the luminous flux emitted by a light source to the input power consumed by the light source, expressed in lumens/W or lm/W.

Examples: tungsten lamp, 10 lm/W; halogen lamps, 15-18 lm/W; fluorescent lamps, 60-70 lm/W; LED lamp, 80-90 lm/W (Figure 5-151).

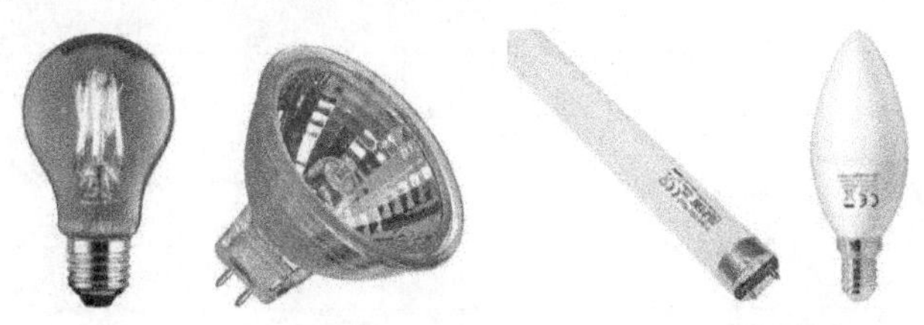

Figure 5-151 Tungsten, halogen, fluorescent and LED lamps

7. Visual Comfort: Color Appearance

Most lamps produce some form of white light that can be described as being from cool to warm in appearance, and this is their color appearance (Figure 5-152). A cooler appearance is generally associated with high levels of lighting and working conditions, whereas, a warmer appearance tends to be associated with relaxing environments. Lamps are provided in a range of appearances and the designer is required to make the most suitable choice.

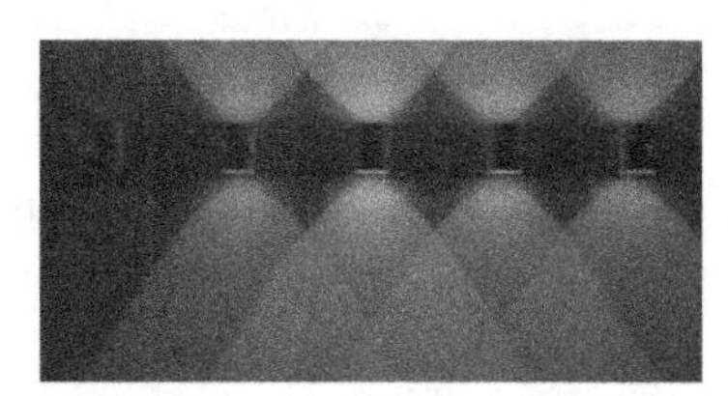

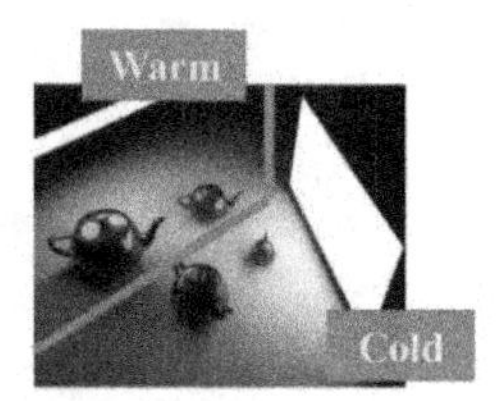

Figure 5-152 Color appearance

A higher color temperature describes a source that is a "cooler" color (i.e. more blue), and a lower color temperature describes a source that is a "warmer" color (i.e. more red). The color appearance of an incandescent lamp is described by its color temperature; this is the temperature of the filament. A tungsten filament lamp has a filament temperature of approximately 2 800 K and can be described as warm in appearance. The color appearance of a discharge lamp is described by it correlated color temperature (CCT); this is the temperature of the blackbody radiation that appears closest to the color appearance of the lamp. A tubular fluorescent lamp can have a CCT between 2 700 K and 6 500 K (Table 5-38).

"White" lamps in general use have CCTs ranging between around 2 500K and 6 500K—the values are published by lamp manufacturers. Most common lamps are "warm white" (2 700-2 900 K).

Table 5-38 CCTs of different light sources

CCT	Light source
1 500 K	Candlelight
2 680 K	40 W incandescent lamp
3 000 K	200 W incandescent lamp
3 200 K	Sunrise/sunset
3 400 K	Tungsten lamp
3 400 K	Dusk/dawn light
4 500-5 000 K	Xenon lamp/light arc
5 500 K	Sunny daylight around noon
5 500-5 600 K	Electronic photo flash
6 500-7 500 K	Overcast sky
9 000-12 000 K	Blue sky

8. Visual Comfort: Indoor Lighting Requirement

Illuminance (*E*) is the total luminous flux incident on a surface, per unit area, in a unit of lux (Table 5-39).

Table 5-39 Recommended comfort criteria for specific applications

Building/room type	Customary winter operative temperatures for stated activity and clothing levels*			Customary summer operative temperatures (air conditioned buildings†) for stated activity and clothing levels*			Suggested air supply rate / ($L \cdot s^{-1}$ per person unless stated otherwise)	Filtration grade‡	Maintained illuminance¶ / lux	Noise criterion§		
	Temp. / ℃	Activity / met	Clothing / clo	Temp. / ℃	Activity / met	Clothing / clo				NR	dBA§	dBC§
Airport terminals:												
— baggage reclaim	12–19[1]	1.8	1.2	21–25	1.3	0.6	10[2]	F6–F7	200	45	50	75
— check-in areas[3]	18–20	1.4	1.2	21–25	1.3	0.6	10[2]	F6–F7	500[4]	45	50	75
— concourse (no seats)	19–24[1]	1.8	1.2	21–25	1.3	0.6	10[2]	F6–F7	200	45	50	75
— customs area	18–20	1.4	1.2	21–25	1.3	0.6	10[2]	F6–F7	500	45	50	75
— departure lounge	19–21	1.3	1.2	22–25	1.2	0.6	10[2]	F6–F7	200	40	45	70
Art galleries — see Museums and art galleries												
Banks, building societies, post offices:												
— counters	19–21	1.4	1.0	21–25	1.3	0.6	10[2]	F6–F7	500	35–40	40–45	65–70
— public areas	19–21	1.4	1.0	21–25	1.3	0.6	10[2]	F5–F7	300	35–45	40–45	65–70

5.2.4.6 Artificial Lighting

1. Light Distribution

• Direct lighting: Direct lighting is where the majority of the luminous flux from the light source(s) reaches the surface being lit directly, without reflection off surrounding surfaces.

• Indirect lighting: Indirect lighting is where the majority of the luminous flux from the light source(s) reaches the surface being lit only after reflection off other surfaces.

• Direct-indirect lighting: Direct-indirect lighting is where near equal proportions of the luminous flux from the light source(s) reach the surface being lit with and without reflection off other surfaces (Figure 5-153).

Figure 5-153 Direct and indirect lighting

2. Lighting Strategies

1) Ambient/general lighting

It refers to the main source or general illumination in any room. Windows, skylights, and centrally located ceiling fixtures can provide the main source of light (Figure 5-154). You can also achieve this by using recessed fixtures and decorative lighting such as chandeliers and larger pendants.

Figure 5-154 Ambient/general lighting

2) Accent lighting

It is used to focus light on a particular area (Figure 5-155). An example of this would be using an adjustable track light to highlight wall art or decorative items on display. Track light, picture lights, small adjustable recessed lights and sconces are all considered to be accent lighting.

Figure 5-155 Accent lighting

3) Task lighting

It is used to perform certain tasks (Figure 5-156), such as preparing food, working, and applying makeup. There are many products on the market that you can use for the task at hand, such as under cabinet lights, puck lights, and portable lamps. This will allow you to have light directly where you need it.

Figure 5-156 Task lighting

4) Wall washing

It is a technique used to wash vertical spaces in a room with light (Figure 5-157). This allows the light to add ambiance or create a dramatic look in a space. Wall washing is a great way to highlight architecture or textures that have been applied to the walls. The wall wash technique looks better when the light source is concealed.

Figure 5-157 Wall washing

5.2.4.7 Types of Lamps

1. Incandescent Lamp

Tungsten lamp: Oldest available form of incandescent lamp, in which electricity is used to heat up a coiled (tungsten) filament to emit light (Figure 5-158).

- 10 lm/W for 1 000 hours.

Tungsten-halogen lamp: Incandescent light sources utilizing the halogen regenerative cycle to prevent blackening of the lamp envelope during life. Usually more compact and longer life than comparable standard incandescent sources, often low-voltage (12 V or 24 V).

- 15-20 lm/W for 2 000-3 000 hours.

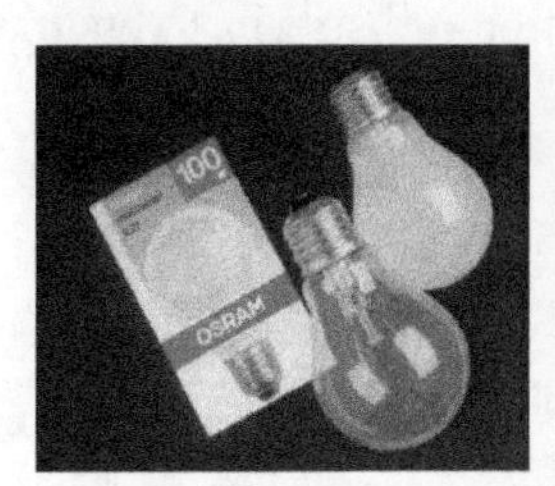

Figure 5-158 Incandescent lamp

2. Fluorescent Lamp/CFL (Compact Fluorescent Lamp)

It is an energy-efficient type of lamp that produces light through the activation of the phosphor coating on the inside surface of a glass envelope by mercury vapor that has been ionized by an electric arc (Figure 5-159).

- 70-80 lm/W for 8 000-10 000 hours.

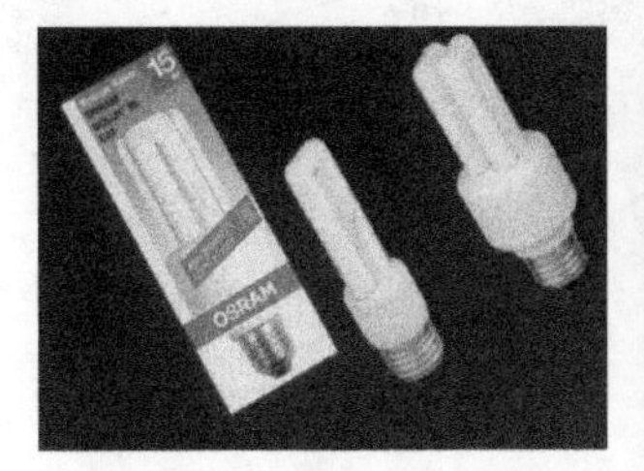

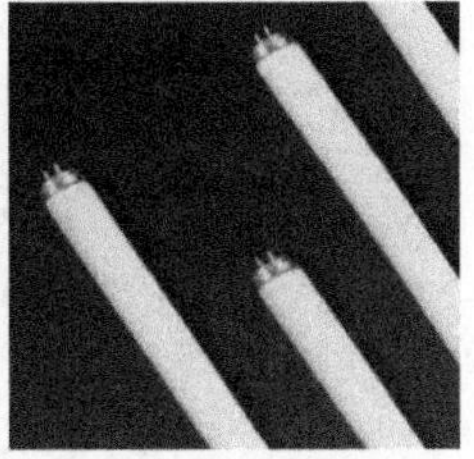

Figure 5-159 Fluorescent lamp/CFL

3. Dimmable CFL (Figure 5-160)

- 1 360 lumens, 20 W;
- 68 lm/W;
- 16 000 hours;
- Standard replacement.

Figure 5-160　Dimmable CFL

4. High-intensity Discharge (HID) Lamp (Figure 5-161)

HID lamp passes a high-pressure electron arc stream through a gas vapor, and examples are:

- Mercury lamp—65 lm/W;
- Metal halide lamps—75 lm/W;
- High-pressure sodium lamps—100 lm/W;
- Low-pressure sodium lamps—200 lm/W.

Up to 25 000 hours.

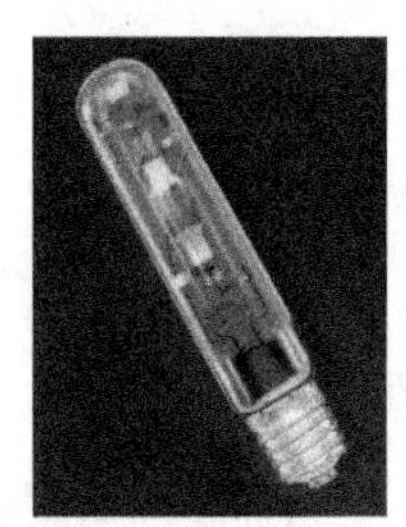

Figure 5-161　HID lamp

Summary of the advantages and disadvantages of different lamps is shown in Table 5-40.

Table 5-40 Advantages and disadvantages of different lamps

Type	Incandescent lamp		Fluorescent lamp		High intensity discharge lamp	
	Tungsten	Halogen	Linear	Compact	Metal halide	Sodium
Advantages	Point source Low intial cost No control gear required Not affected by ambient temperature Simple to dim Good color rendering	Point source Improved efficacy No control gear required if mains voltage Not affected by ambient temperature Simple to dim if mains voltage Good color rendering	Low initial cost High efficacy Very long life Low brightness Low operating temperature Good color rendering can be achieved (with reduced efficacy)	Higher output with smaller dimensions High efficacy Very long life Low brightness Low operating temperature Good color rendering can be achieved (with reduced efficacy)	Point source Very good efficacy Long life High output for compact size Light output not affected by ambient temperature	Point source Very good efficacy Long life High output for compact size Light output not affected by ambient temperature
Disadvantages	Very low efficacy High operating temperature Very sensitive to voltage variations Very short life	Low efficacy High operating temperature Very sensitive to voltage variations Requires transformer if low voltage Short life	Not a point source Requires control gear Affected by ambient temperature Requires special control gear to dim	Not a point source Requires control gear Affected by ambient temperature Requires special control gear to dim	Relatively poor color rendering High initial cost Requires control gear Cannot dim Long warm-up & restrike times Problems starting in cold weather	Poor color rendering Color appearance is orange Requires control gear Cannot dim Long warm-up & restrike times Problems starting in cold weather

5. Other Lamp Types

1) Cold cathode lamp

Cold cathode lamp is a neon-like electric-discharge light source primarily used for illumination (neon is often used for signage or as an art form) (Figure 5-162). Cold cathode can sometimes be used where fluorescent tubes would be too large or too hard to re-lamp, and can be colored or curved.

Figure 5-162 Cold cathode lamp

2) Induction/electrodeless lamp

It adopts similar technology to fluorescent lamp, but it uses microwaves rather than an electric arc to ionize the mercury, resulting in an extremely long lamp life (60 000 hour vs. 15 000 hours) (Figure 5-163).

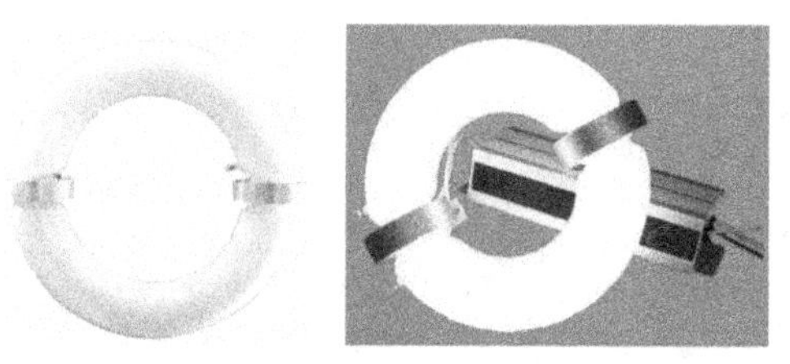

Figure 5-163 Induction/electrodeless lamp

3) Light emitting diode (LED) lamp

Recent advances in LED technology in both quantity of luminous flux output and color (white LEDs are increasingly available) have made LED lamps become a viable choice for architectural lighting (Figure 5-164), with the advantages of a very low power consumption and an extremely long life. Still very low lumen levels are available.

Figure 5-164 LED lamp

- 810 lumens, 10 W;
- 81 lm/W;
- Warm white color;
- Standard UK fitting;
- Highest output standard, lamp replacement available;
- Specialist units approach 100 lm/W.

5.2.4.8 Control of Artificial Lighting

- Occupancy-based control by PIR sensors;
- Dimmable control, automatic or manual;
- Zonal control (Figure 5-165).

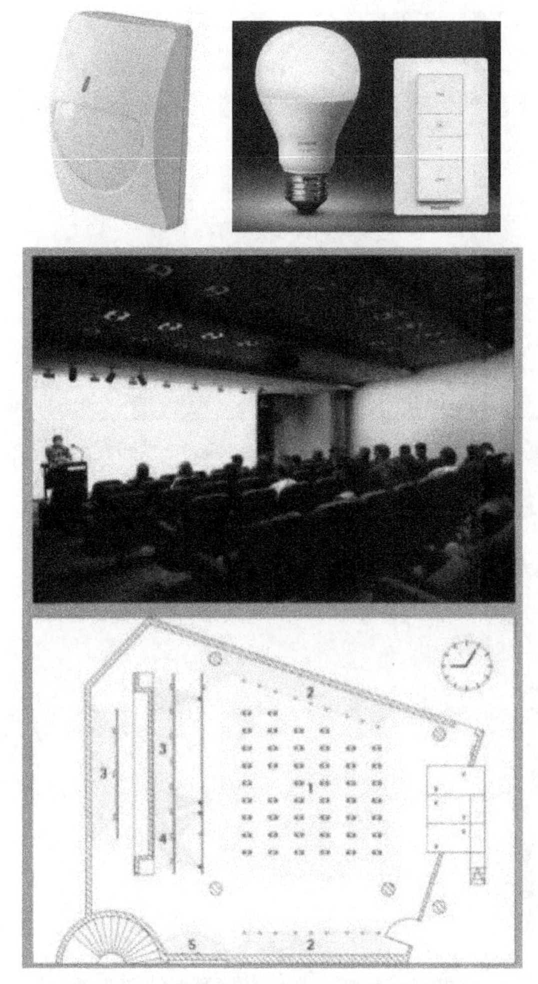

Figure 5-165 Control of artificial lighting

5.2.4.9 Office Lighting

• Considers balancing luminance in the visual field for comfort;

• Design guidance to ensure a comfortable environment in spaces with display screens;

• Considers glare on screens from luminaires, daylight and high contrast;

• Includes guidance on luminance limits for luminaires at various angles to avoid reflections on computer screens;

• Considers screen technology and software standard (Figure 5-166).

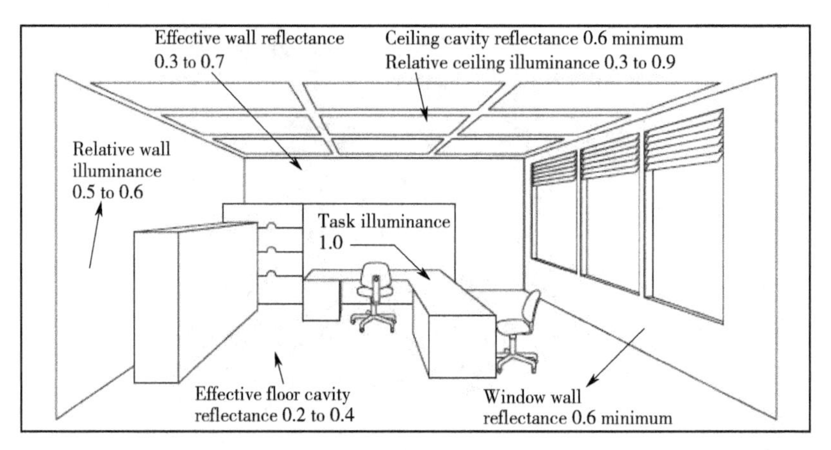

Figure 5-166 Considerations of office lighting (CIBSE LG7)

5.2.4.10 Lighting Software

1. Lighting Analysis Software (Figure 5-167)

- LIGHT;
- DIALux.

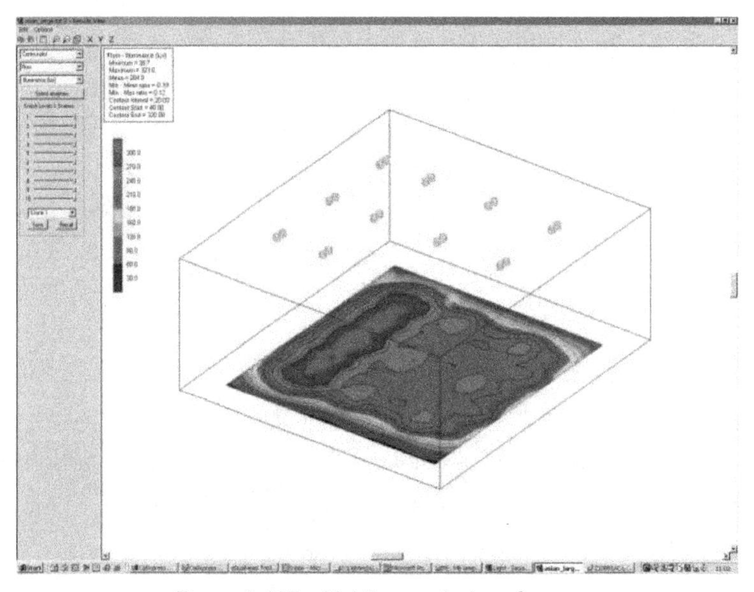

Figure 5-167 Lighting analysis software

2. Visualization Software

Visualization software includes:

- Radiance;
- Lightscape;
- Lighting Visualization.

Usage characteristics of visualization software are listed below:

• Computer modeling of spaces to obtain very accurate results and/or photo-realistic images;

• Much more complex to use;

• These programs accurately simulate the properties of light and materials using a technique called ray-tracing;

• Required inputs include complex geometry, accurate surface colors and material properties, luminaire data, daylight data etc.;

• Radiance is extremely accurate and flexible visualization software, but is difficult to use.

5.3 Modern Intelligent System

5.3.1 Building Management System

5.3.1.1 Definition of Building Management System

Building management system is digitally based control and management systems, it can take all control and operation systems for all systems. Major systems include operator console at building manager's office, smaller systems operated via touch pads in control panels. It can include wireless remote access via laptop or smart phone or tablet or remote Internet access.

5.3.1.2 Origins of Building Management Systems

Origins of building management system are listed in Table 5-41.

Table 5-41 Origins of building management systems

Controls were once pneumatic	
Mid-1970s: Single-chip controllers appeared	
1980s: Direct digital control (DDC) systems appeared	
1990s: BMS systems utilized the power of the personal computer	

5.3.1.3 Direct Digital Control (DDC)

It is a method of monitoring and controlling HVAC systems, by collecting, processing and sending information using sensors, actuators, and microprocessors.

A DDC system encompasses all the sensors, inputs, process algorithms, and output actuators, which can implement any control scheme (Figure 5-168).

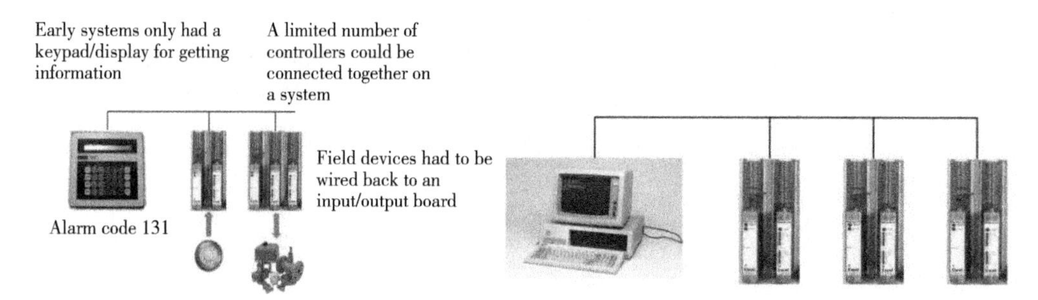

Figure 5-168 DDC system from the early 1980s

The development of the personal computer allowed BMS to show text information in an easily understood way with no codes to fathom out e.g. Alarm code 131 (fan tripped)!

5.3.1.4 Components of Modern BMS

There is an intrinsic interconnection between many devices, so there is a need for perfect automation management. The equipment management system is set up for integrated management, scheduling, monitoring, operation and control of the equipment (Figure 5-169).

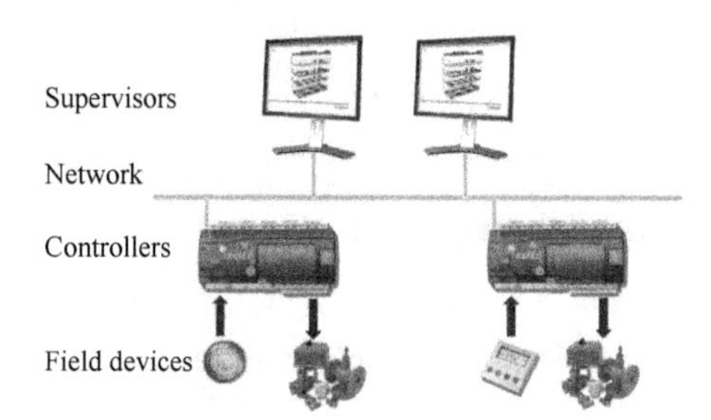

Figure 5-169 Components of modern BMS

Field devices (Figure 5-170):

Figure 5-170 Field devices

1. Analog Inputs (Figure 5-171)

• Temperature;

• Thermistors;

- Resistance temperature detectors (RTDs);
- Transmitters;
- Pressure;
- Humidity;
- Voltage;
- Current;
- CO_2;
- Flow (CFM, GPM).

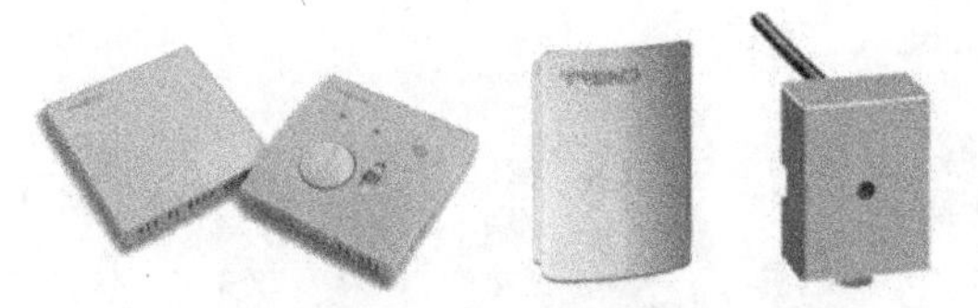

Figure 5-171 Analog inputs

2. Digital Inputs (Figure 5-172)

- Switch dry contact (open or closed);
- Airflow;
- Water;
- Differential pressure;
- High/low limit switch (alarm or normal);
- Freeze alarm;
- Smoke detectors;
- Meter pulses (pulse initiator).

Figure 5-172 Digital inputs

3. Analog Outputs (AO: 0-10 volts DC) (Figure 5-173)

- Damper actuators;
- Modulating valves;
- Variable speed drives;
- Plant and equipment utilizing modulating signal.

Figure 5-173 Analog outputs

4. Digital Outputs (Figure 5-174)

- Relays and contactors;
- Lighting;
- Plant and equipment.

Figure 5-174 Digital outputs

5.3.1.5 Network

Within the whole building, through the whole set of building automatic control system and its built-in optimal control program and preset time program, all electromechanical equipment is centrally managed and monitored (Figure 5-175). On the premise of meeting the control requirements, it realizes comprehensive energy saving and replaces the daily operation and maintenance work with the control function of the controller, which greatly reduces the daily workload and reduces the loss of control of equipment or equipment damage caused by the work errors of maintenance personnel.

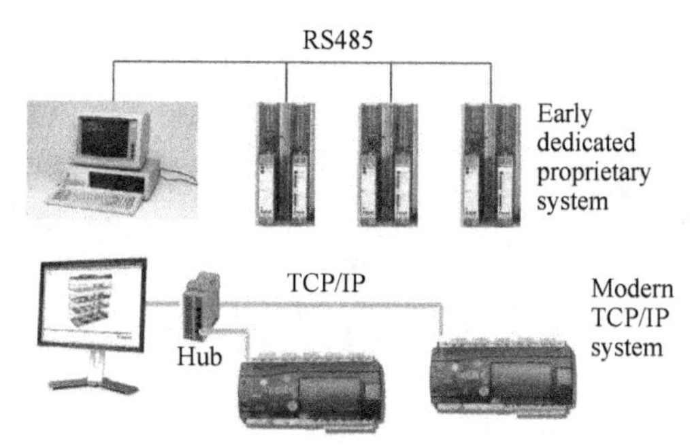

Figure 5-175 Network control

5.3.1.6 Importance of Sensor Location

The best way to take full advantage of a BMS is to deploy sensors, which add another dimension to data collection in a building (Figure 5-176). IoT sensors can collect a vast amount of data, including asset temperatures, occupancy rates, and even the number of open doors and windows. Because each of these sensors is designed for a specific purpose, they can be deployed in a targeted manner to collect data that the building management system might otherwise miss. For example, occupancy sensors can determine when a public room is unoccupied and can be further integrated into the BMS to automatically turn off lights when unoccupied. Data collected from sensors over time can also reveal occupancy trends, which can lead to data-driven decision making to optimize energy consumption. Using a BMS has proven to be an effective way for facility managers to reduce energy consumption and a building's carbon footprint. Sensors are an excellent complement to any BMS, as they can collect more comprehensive data and provide deeper insight into energy and resource consumption. By utilizing both technologies together, facility managers can truly maximize energy efficiency and sustainability gains.

Figure 5-176 Sensor location

5.3.1.7 Supervisors (Figure 5-177) or "Front Ends"

- Operators access into the building energy management system (BEMS);
- Graphical real-time view of system parameters;
- Temperature, pressure, plant status etc.;
- Animated and dynamic graphics giving real-time visibility;
- Adjustment of parameters;
- Time zones, setpoints, switches;
- Diary function for pre-programming of future events, bank holidays etc;
- Data logging;
- Alarm management;
- System faults instantly reported;
- Alarms can be routed easily;
- User friendly;
- Flexible & intuitive.

Figure 5-177 Supervisors

5.3.1.8 Modern Day BMS Options

Modern day BMS options are shown in Figure 5-178.

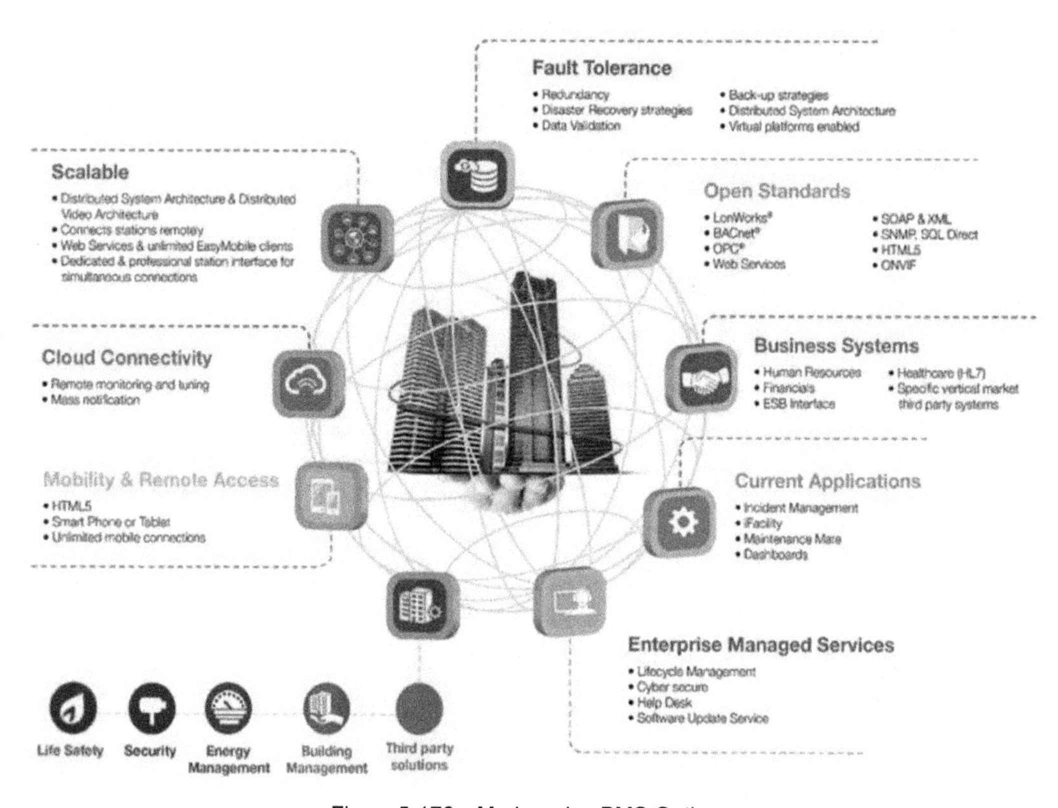

Figure 5-178 Modern day BMS Options

5.3.1.9 BMS Functions (Figure 5-179)

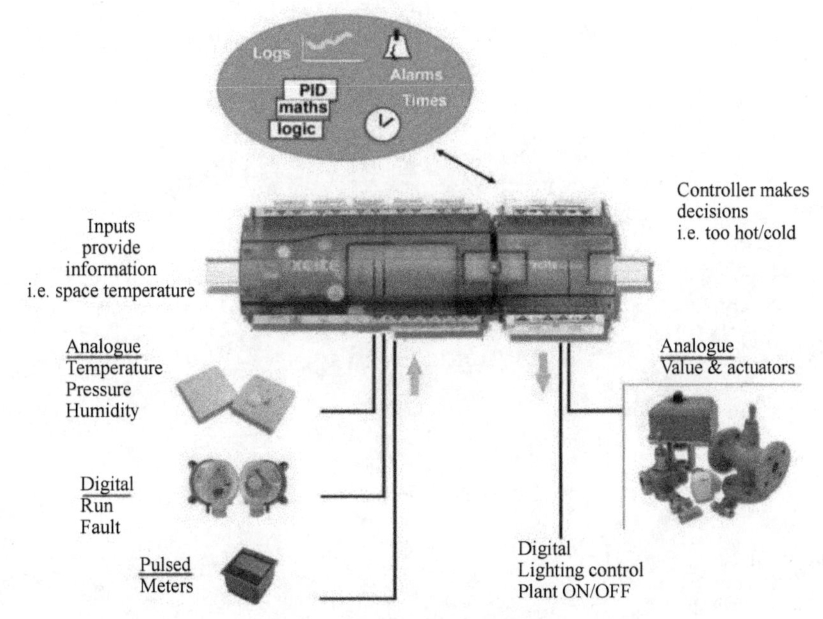

Figure 5-179 BMS functions

1. System Control

System control is divided into two main aspects: temperature control and humidity control.

1) Temperature control

• Variable temperature & constant flow, e.g. mixing circuit,variable plant output or compensated circuit;

• Variable flow & constant temperature, e.g. VAV or diverting circuit.

2) Humidity control

• Spray Humidifiers;

• Steam Humidifiers;

• De-humidification (mechanical or chemical).

2. Time Operations (Start/Stop)—Time Clock (Figure 5-180)

• Multiple time-zones;

• 7 days/5 days;

• GMT/BST;

• Annual calendar;

• Sun angle for any time and date.

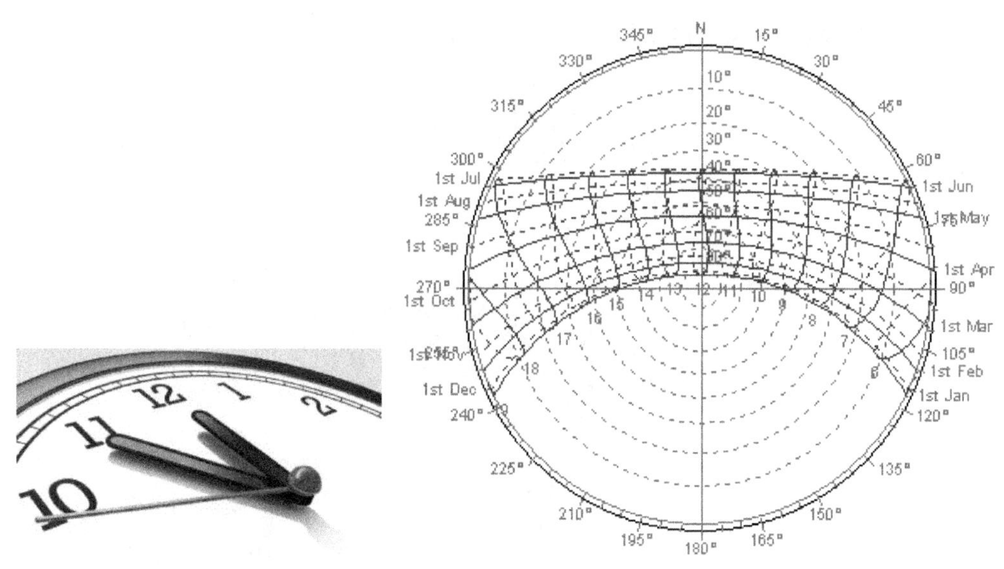

Figure 5-180 Time operations

Manual override:

• Out of hours working;

• Special events;

• Different departments' operating hours.

Optimization:

• Optimum start;

• Start heating earlier on cold days;

• Start the system at the latest responsible time;

• Optimum Off;

• Turn the plant off at earliest opportunity.

3. System Monitoring (Figure 5-181)

• Observe/review any point on system;

• Check critical values, e.g. temperature and humidity;

• Check plant operation, e.g. fault detection and maintenance programming;

• Metering, e.g. maximum demand control and billing (whole building or zones).

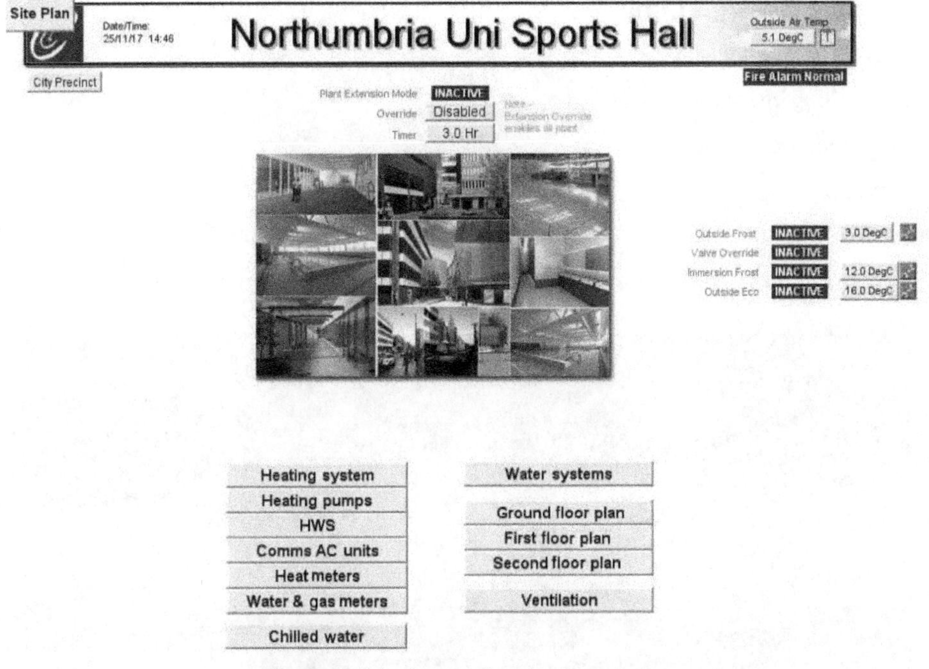

Figure 5-181 Items of system monitoring

4. Trend Logging

- Logging history of any point on system;
- Historical data, graph or chart;
- Management reports, including operation, performance and energy;
- Records, e.g. critical values.

5. Fault Alarms (Figure 5-182)

- Alert appropriate staff to any fault on system;
- Warning signal to operator console;
- Advise criticality;
- Divert to alternative console at night;
- Modem to remote location, i.e. not via Internet-secure communication;
- Maintenance alerts, e.g. filter change.

Figure 5-182 Fault alarms

6. Energy Management

- Recording energy use;
- Controlling energy saving regimes;
- Maximum demand control;
- Preventive;
- Predictive;
- Demand control (Figure 5-183).

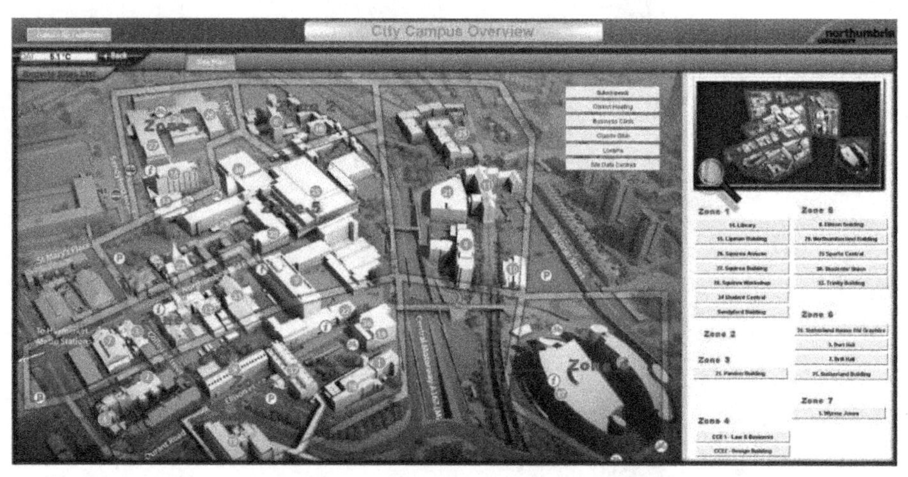

Figure 5-183 Demand allocation of the zone

7. System Interlocks

- Duplicate plant changeover;
- Equal plant use;
- Changeover on plant failure;
- Related plant control, e.g. laboratory ventilation plant;
- Safety interlocks, e.g. fire alarm overrides.

8. Planned Maintenance

- Daily maintenance schedule;
- Varied by hours run;
- Varied by trend log;
- Plant & equipment data;
- Spares listings;
- Spares locations;
- O & M manuals;
- Indexing;
- Manufacturers' data;

• Repair instructions.

5.3.1.10 Instant Monitoring

Instant monitoring—graphic interface (plant) (Figure 5-184):

• Allows easy access to a complex system with real-time data and status feedback.

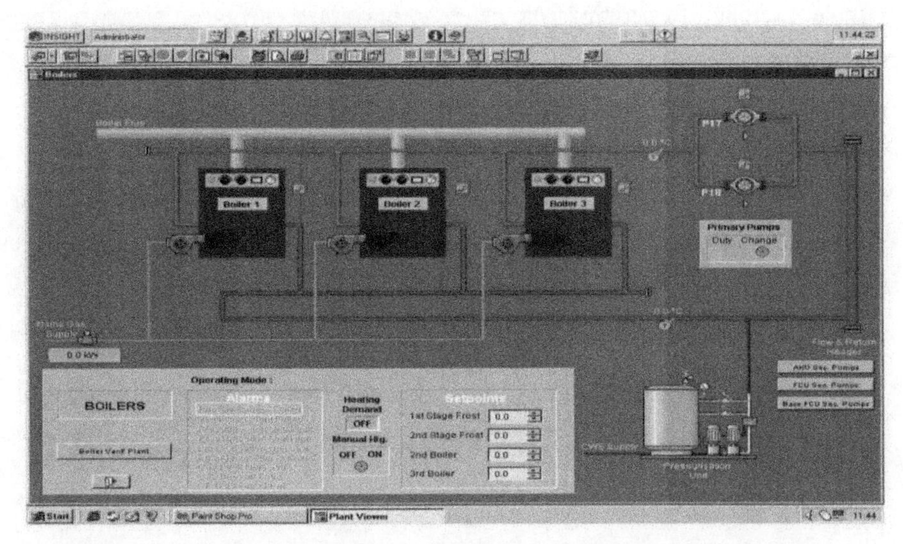

Figure 5-184 Instant monitoring—graphic interface (plant)

Instant monitoring—graphic interface (AHU) (Figure 5-185):

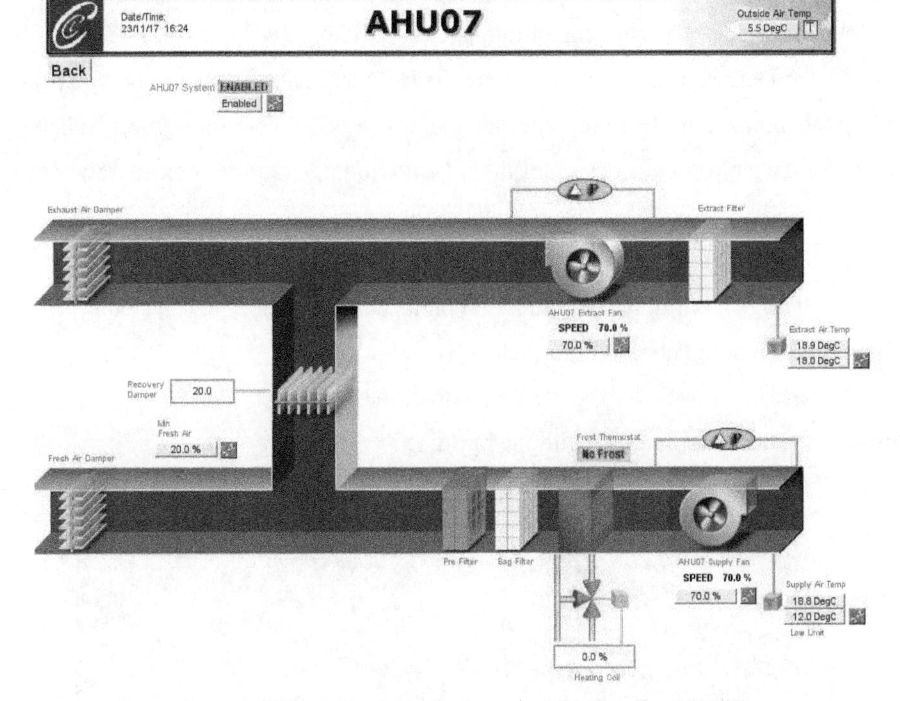

Figure 5-185 Instant monitoring—graphic interface (AHU)

• Allows easy access to a complex system with real-time data and status feedback.

5.3.1.11 Major Contributions of the BMS to Energy Efficient Buildings

• Optimizing latest start & earliest stop times based on weather data and deep learning network, shut down plant based on weekends, holidays, special events, etc.;

• Remote resetting of operating parameters to maximize energy efficiency, e.g. temperatures, pressures, speed, etc.;

• Automatic adjustment of internal &external shading to minimize solar gain;

• Auto dimming of internal lighting to maximize daylight use;

• Monitors load demand & tariffs to get lowest cost & avoid max. demand charges;

• Provides operators with real-time data on equipment performance for informed resetting of operating parameters ;

• Presents performance data to allow planned maintenance to keep plant operating at peak efficiency.

5.3.2 Intelligent Buildings

5.3.2.1 Definition of Intelligent Buildings

Intelligent building (Figure 5-186) refers to the optimal combination of building structure, systems, services and management according to the needs of users, so as to provide users with an efficient, comfortable and convenient humanized building environment. Intelligent building is a product of modern science and technology. Its technical basis is mainly composed of modern building technology, modern computer technology, modern communication technology and modern control technology. The design of intelligent buildings requires consideration of the following issues:

Can buildings be intelligent or are they purely reactive?

Does an intelligent building succeed based on its control devices and systems?

What level of intelligence is appropriate?

How do these controls influence the aesthetic of the building?

How can the occupant interact with the building?

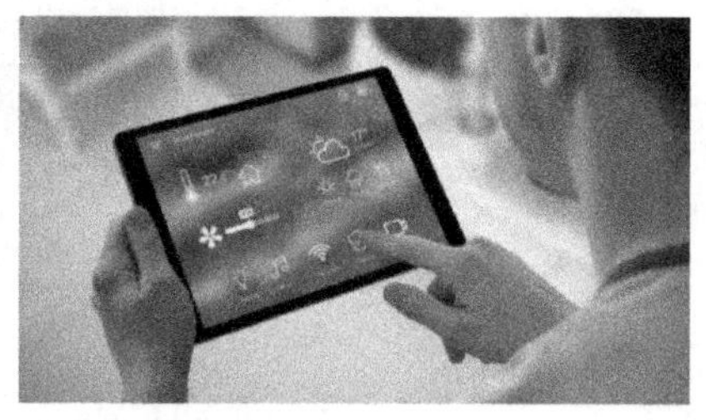

Figure 5-186 Intelligent building

5.3.2.2 Control Programs

1. Psychology of Control

• Adjustment versus full range control.

• Dummy controls: The dummy controls do nothing except to give the occupants the impression that they have control of the HVAC system and the psychological effect of having control of their work environment (Figure 5-187).

Figure 5-187 Dummy controls

2. Motorized Windows (Figure 5-188)

• Motors can be bulky;

• Complex coordination issues;

• Design responsibility;

• User override.

Figure 5-188 Motorized Windows

3. Ventilation (Figure 5-189)

• Hot air vent.

• Visual issues obviously.

• Keep out the rain!

• Openings need to be accessible or motorized.

• High-level maintenance.

Figure 5-189 Ventilation

4. Smoke Ventilation (Figure 5-190)

- During and after fires;
- Very robust construction;
- Critical operation with fail safe modes.

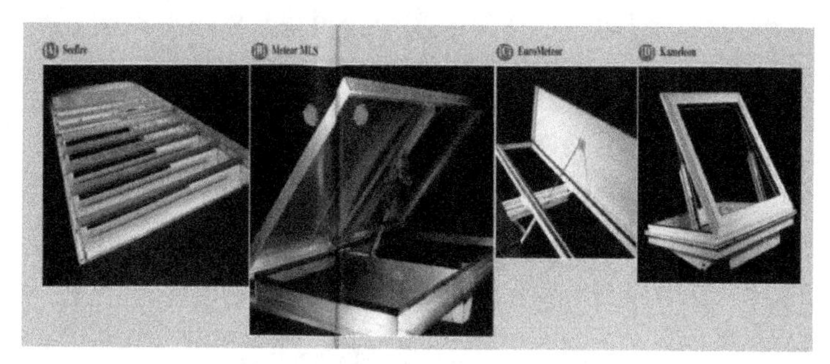

Figure 5-190 Smoke ventilation

5. Motorized Louvres (Figure 5-191)

- Motor is difficult to hide;
- Cabling needs resolving;
- Louvres not airtight;
- Louvres not watertight.

Figure 5-191 Motorized louvres

6. External Motorized Louvres (Figure 5-192, Figure 5-193)

- Control is crucial to comfort;
- Controls are expensive;
- Expensive external maintenance;
- Resist high wind forces;
- Highly visible;
- Expensive access for cleaning and maintenance;
- Ditto for window cleaning;
- Multiple motors/linkages & lubrication.

Figure 5-192 External motorized louvres

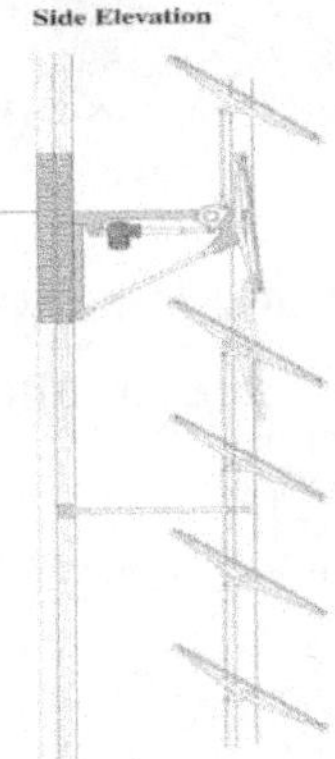

Figure 5-193 Motorized louvres for curtain walls

7. Sun Shade Control (Figure 5-194)

• Daylight sensitive;

• Arab Institute in Paris with light sensitive wall using motorized iris for daylight control, which is an intelligent facade (time/solar exposure/temperature);

• Air pollution generated corrosion and "stiction" ;

• Too complex external elements—they never worked;

Figure 5-194 Sun shade control

• Active shading to maximize views and provide solar shading (Figure 5-195);

• Overrides provided for building occupant use and closure in high wind;

• Irritating auto open/close and noise in wind.

Figure 5-195 Active shading

8. Internal Lighting Control (Figure 5-196)

• Often achieved by PIR sensors to detect occupancy of the room.

• It is good for improving building energy efficiency.

• Do occupants really love this solution?

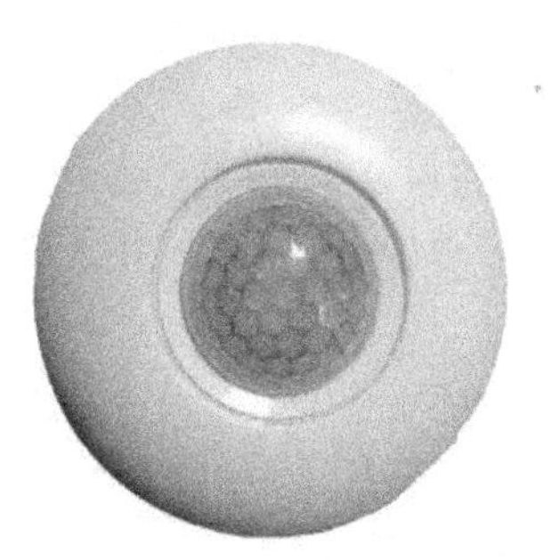

Figure 5-196 Internal lighting control

9. Weather Station and Controls (Figure 5-197)

(1) Allow us to react to conditions of:

• Temperature;

• Humidity;

• Wind;

• Rain;

• Sun exposure.

(2) Allow closed loop control.

Figure 5-197 Weather station and controls

10. Lift Control (Figure 5-198)

• No matter the age or style of the lift, modern adaptive & predictive control can be applied;

• Lift can be monitored and managed through PLC, BMS or remotely through an Internet link;

• Users still press up & down call buttons together—the lift comes faster.

Figure 5-198 Lift control

5.3.2.3 Relevant Systems

1. Building Automation Systems (Figure 5-199)

• Digital control;

• BMS (slow) versus PLC (fast);

• Control of service systems;

• Integrated systems;

- External access (Figure 5-200);
- Telephone interfaces;
- Wi-Fi access;
- Person recognition;
- Security interfaces;
- Energy billing;
- Fire & emergency;

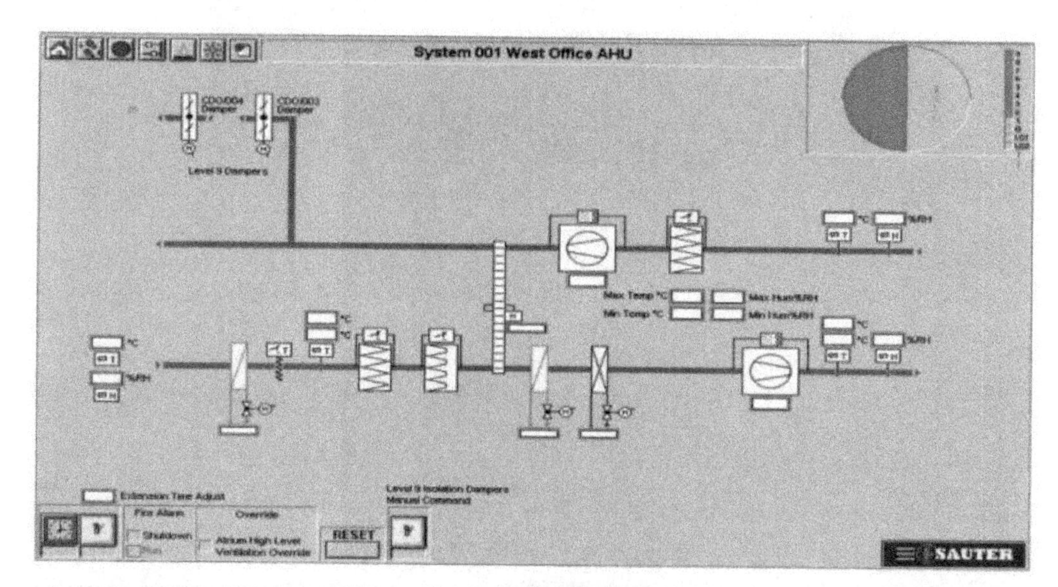

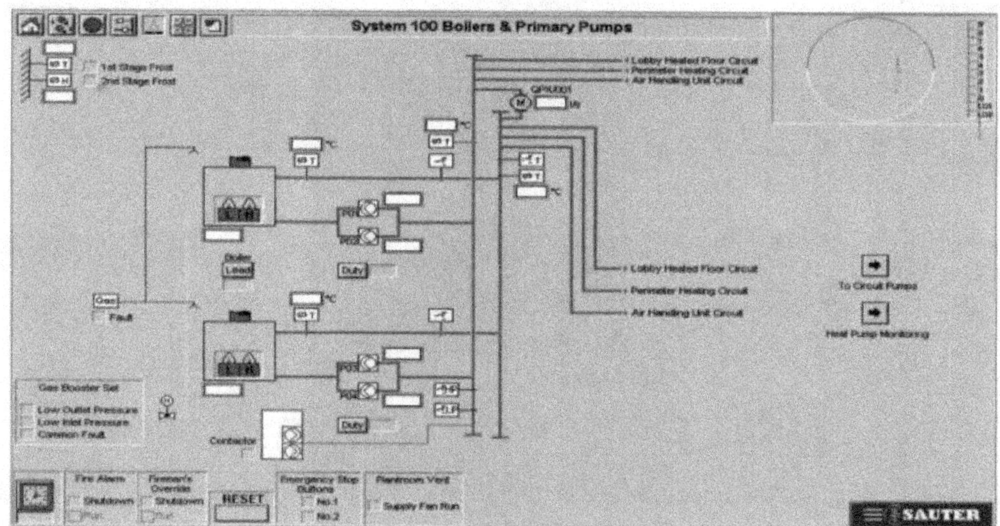

Figure 5-199 Building automation system

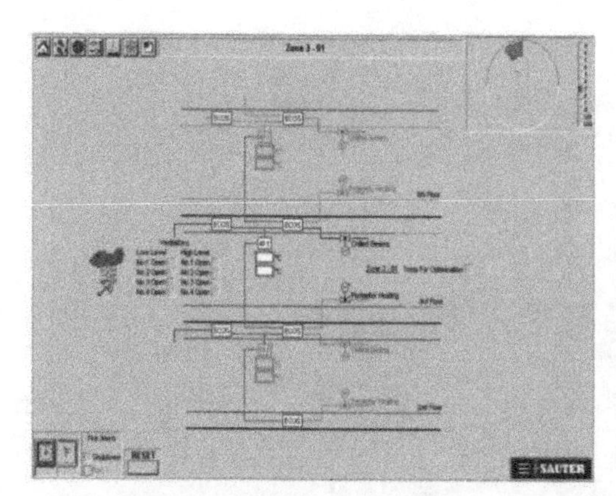

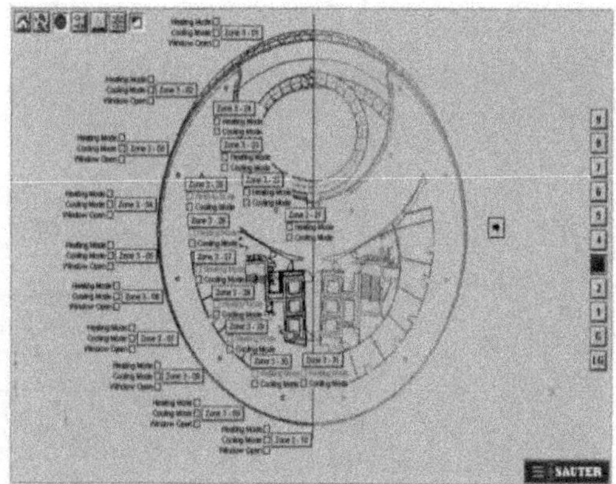

Figure 5-200 Allowing easy access to a complex system with real-time data and status feedback

- Individual space control;
- Web based graphics for all building users.

2. Building Energy Monitoring Systems (BEMS)

- Energy metering;
- Load shedding;
- Load management;
- Trend logging;
- Maintenance scheduling;
- Automatic lighting level control;
- Zoning control;
- Individual lighting control for flexibility;
- User interface & override;
- Automatic maintenance and checking schedules;
- Scene setting;
- Global vs. local control.
- System design—Programmable Logic Controller (PLC) (Figure 5-201);
- Similar to BMS but dedicated to one function;
- Can report to BMS and be overridden;
- Individually addressable light fittings and switches;
- Software coordinates all the links;
- Programming via PC.

3. Audio Visual Systems (Table 5-42)

- Integrated systems;
- Video conferencing;
- Links to lighting for best camera environment;

• Highlighting speaker;

• Loudspeaker locations programmable to suit layout;

• Microphone layout ditto;

• Speech interference management, e.g. rustling papers.

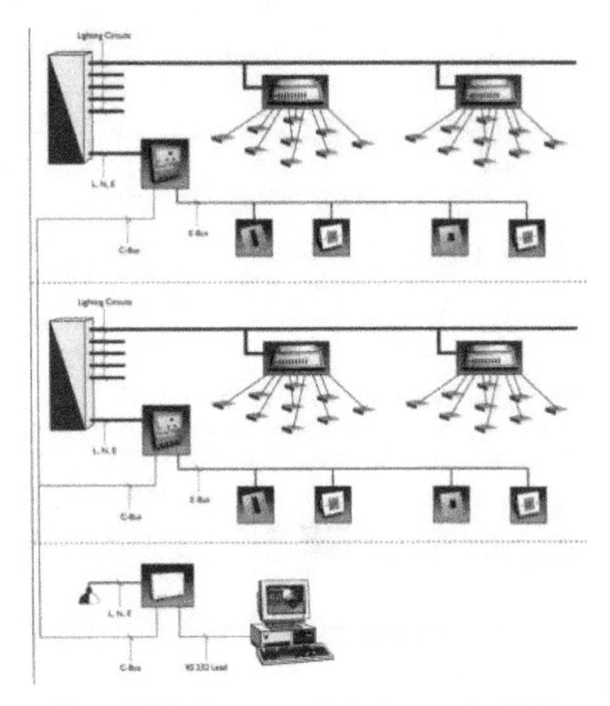

Figure 5-201 Programmable logic controller (PLC)

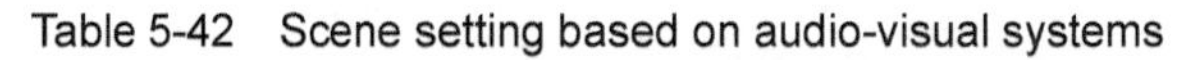

Table 5-42 Scene setting based on audio-visual systems

Auditorium	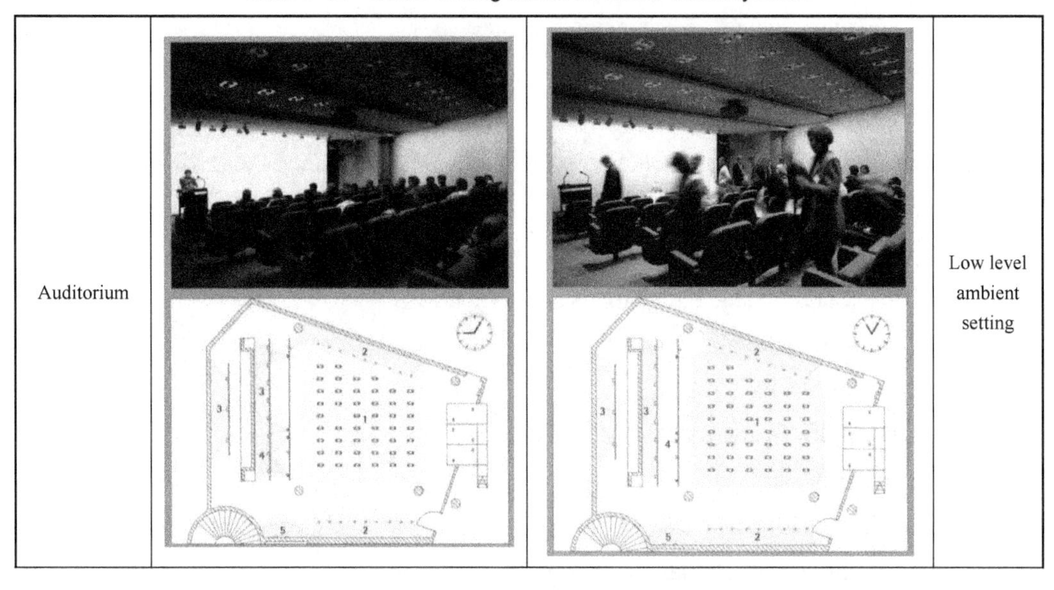	Low level ambient setting

Continued

Presentation	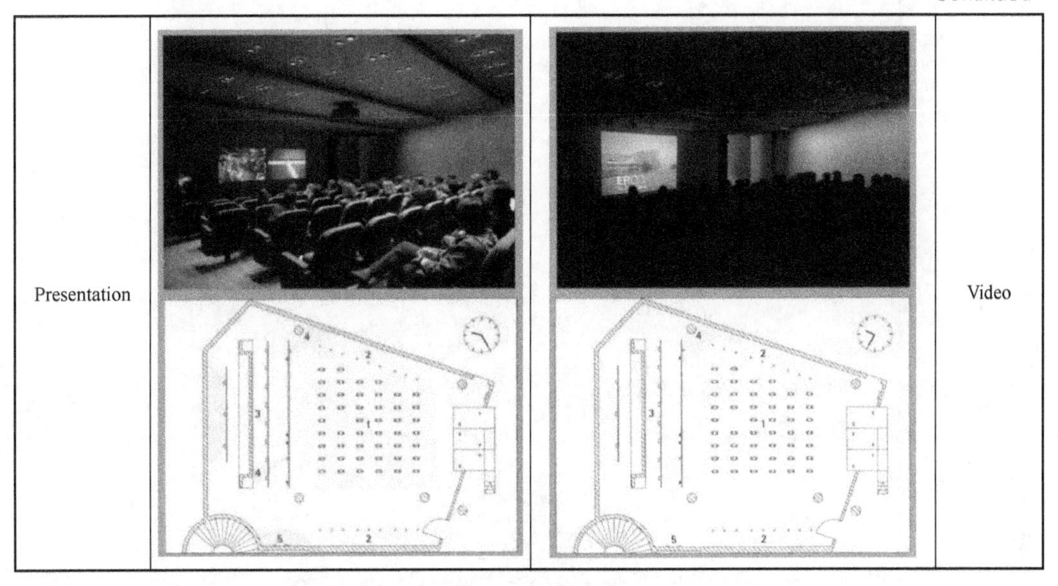	Video

5.3.2.4 Summary (Key Issues to Contemplate)

- Intelligence is provided by automation.
- Automation is required if the users are unable or unwilling to control the environment appropriately.
- Controls need to be reliable, robust and accessible.
- Manual controls need to be intuitive.
- Automatic controls can only react based on history and the current conditions. They are not very good at being predictive.
- Therefore buildings are not intelligent. Neither are the users!

5.3.3 Dynamic Building Performance Simulation

5.3.3.1 Definition of Building Performance Simulation

Building performance simulation (BPS) is the replication of aspects of building performance using a computer-based mathematical model created on the basis of fundamental physical principles and sound engineering practice. It is a complicated model with a number of energy flowpaths (Figure 5-202).

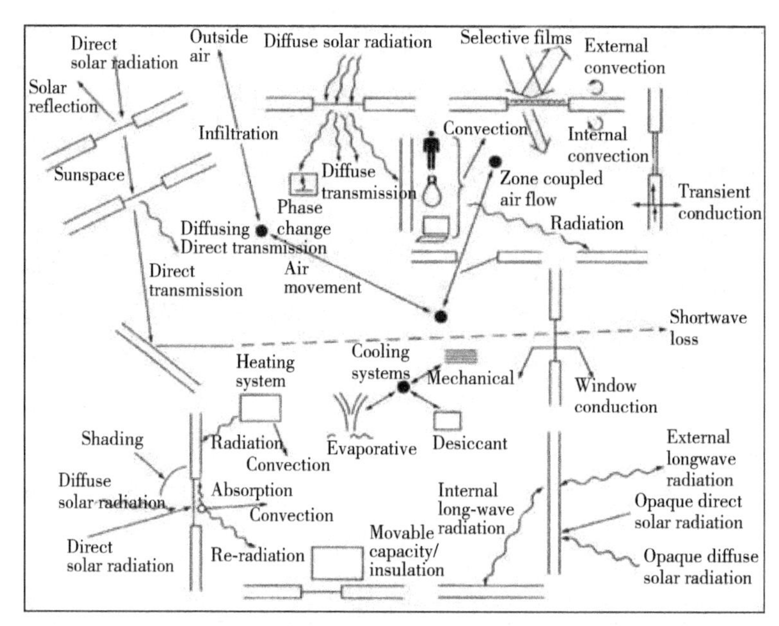

Figure 5-202 Building energy flowpaths (Clark, 2001)

5.3.3.2 Importance of Building Performance Simulation

Buildings consume approximate 40% of the society energy in developed countries, such as USA, Japan and UK to maintain the thermal conditions and lighting inside the building. Prediction-based decision making is important at the design stage of buildings.

Building performance simulation is a useful, reliable, user-friendly approach of predicting the performance of the building under different design scenarios (e.g. different systems, constructions and weather conditions).

5.3.3.3 Available Calculation Methods

- Empirical, such as Rules of Thumb and Benchmarking;
- Steady State Method;
- Steady Cyclic Method;
- Simplified Building Energy Model (SBEM);
- Dynamic Thermal Models (DTM);
- Computational Fluid Dynamics (CFD);
- Machine Learning Models, e.g. MLR, SVM, Decision Tree.

5.3.3.4 Building Energy Software Tools Dictionary

- Hosted by U. S. Department of Energy (DoE).
- A comprehensive list of building energy softwares in the world, with different capabilities.
- Website: https://www.buildingenergysoftwaretools.com/.
- Select suitable tools based on capabilities: whole building energy simulation, load calcu-

lations, HVAC system selection and sizing, parametrics and optimization, energy conservation measures, code compliance, ratings and certificates, etc.

• Select right tools based on building type: commercial, residential, multifamily, industrial, portfolio scale, district scale, urban scale, subsystem level.

5.3.3.5 Commonly Used Building Performance Simulation Tools

• EnergyPlus (USA), DesignBuilder (UK), IES VE(USA), ESP-r(UK), TRNSYS(USA), eQuest (USA), OpenStudio (USA) and DeST (China) (Figure 5-203).

• EnergyPlus and Apache are two major energy simulation engines.

• DesignBuilder, OpenStudio are developed to provide easy user interfaces for EnergyPlus; and IES VE uses Apache.

Figure 5-203 Building performance simulation tools

5.3.3.6 Goals of Building Performance Simulation

For design and retrofit:

• Reliability estimation for natural ventilation and mixed mode strategies;

• Load calculations for sizing of equipment such as fans, chillers and boilers;

• Energy and environment estimation for selected building envelops, system and control strategies, under various climatic conditions.

For operation:

• Prediction-based building control enabling better building performance.

For policy making:

• Providing policy makers with useful advices in terms of building energy efficiency (policy, incentive), such as for the UK Green Deal Scheme.

5.3.3.7 Reasons Why Simulation Can Save Energy

Building energy simulation allows one to model a building before it is built or before renovations start.

Simulation allows various energy alternatives and design options to be investigated and compared to one another.

Simulation can provide valuable evidence on design solutions and lead to an energy-optimized buildings. Simulation is much more cost effective and time efficient than experimentations (every building is different).

5.3.3.8 Main Features That Influence Building Performance

• Climate: ambient air temperature, relative humidity, direct and diffuse solar radiation, wind speed and direction.

• Site: location and orientation of the building, shading by topography and surrounding buildings, ground properties.

• Geometry: building shape and zone geometry.

• Envelope: materials and constructions, windows and shading, thermal bridges, infiltration and openings.

• Internal gains: lighting, equipment and occupants including schedules for operation/occupancy.

• Ventilation system: transport and conditioning (heating, cooling, humidification) of air.

• Room units: local units for heating, cooling and ventilation.

• Plant: central units for transformation, storage and delivery of energy to the building.

• Controls: for window opening, shading devices, ventilation systems, room units, plant components.

Dynamic interactions of sub-systems in buildings are shown in Figure 5-204.

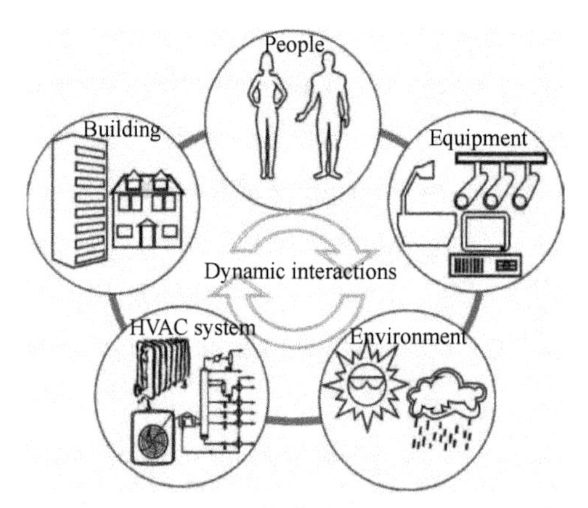

Figure 5-204 Dynamic interactions of sub-systems in buildings (Hensen and Lamberts, 2012)

Climate:

• Specifying the outdoor environment of a building;

• Simulation weather data;

• Winter design weather data;

• Summer design weather data (Figure 5-205).

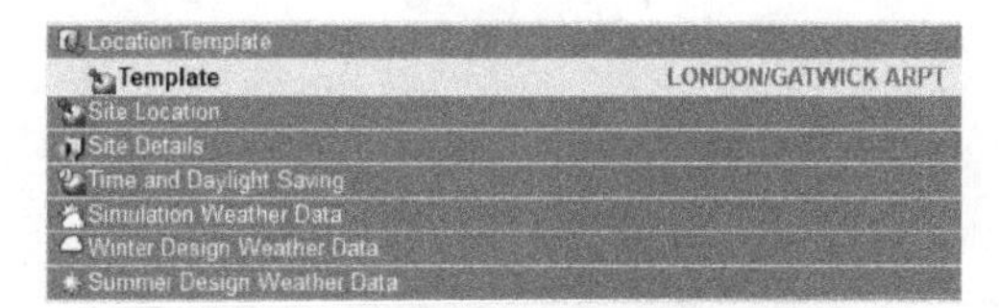

Figure 5-205 Weather data

Weather characteristics (Figure 5-206):

• Some parameters change slowly in a day, such as temperatures.

• Many parameters vary within minutes, e.g. wind speed and direction, solar radiation.

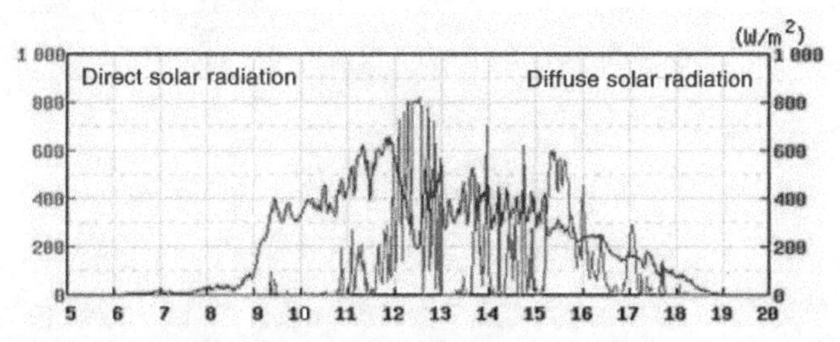

Figure 5-206 Direct/Diffuse solar radiation in 20/05/2010 in Hong Kong

5.3.3.9 Building Performance Prediction

1. Simulation Weather Data

• Contain hourly data for 8 760 h (a whole year) energy simulation.

• Important parameters: dry-bulb temperature, dew-point temperature, relative humidity, solar direct/diffusive radiation, solar illuminance, sky temperature, etc.

• File format: EPW, TMY3.

2. Design Day Data

• For load calculation and HVAC system sizing and selection.

• The building heating/cooling load is defined as the amount of heat that must be removed (cooling load) or added (heating load) to maintain a constant room air temperature.

• The recommended approach to the determination of the space design cooling load is to assume steady cyclic conditions where a single design day is repeated until the load does not vary from day to day.

• Two types of design day: summer design day and winter design day.

3. 3D Geometry Structure Hierarchy

3D geometry structure hierarchy is shown in Figure 5-207.

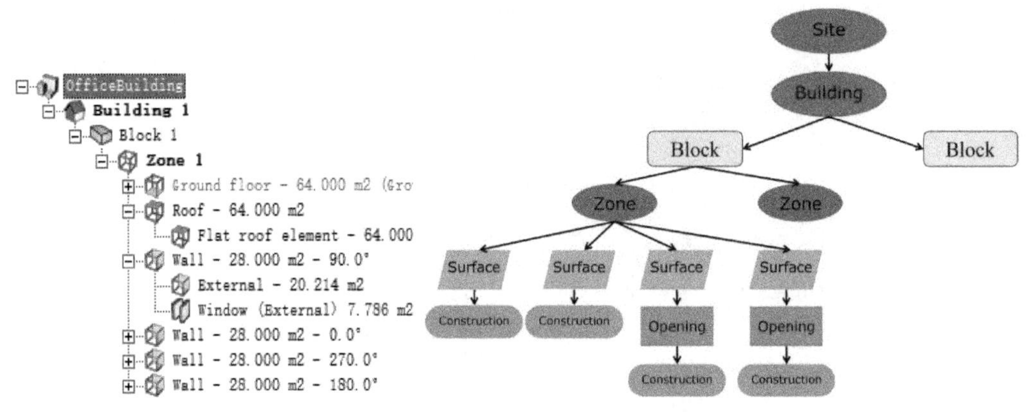

Figure 5-207 3D geometry structure hierarchy

4. Data Inheritance

Default data (setting) for all levels is inherited from the level above in the hierarchy, so block data is inherited from building level, zone data is inherited from block data and surface data from zone data.

For example, settings made at the block level can change data for all zones/surfaces in the block.

Users can set "user data" for a certain level to influence that level and all levels below, and will be highlighted in red (Figure 5-208).

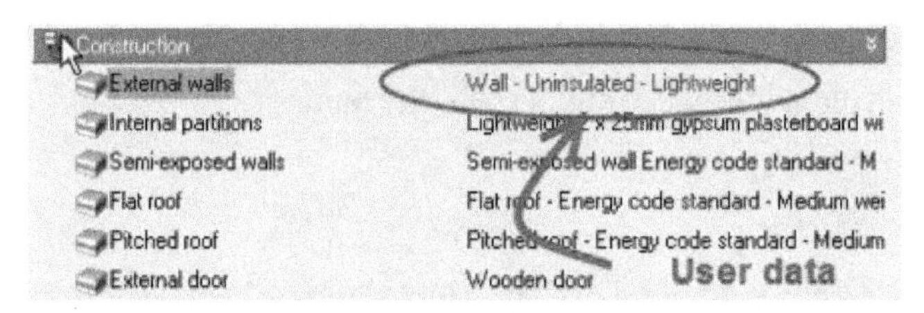

Figure 5-208 User data

5. Default Data

• Construction and opening data is used at the surface level.

• Activity, lighting, HVAC, equipment data is used at the zone level.

• The above data can be either default data or user data.

• "Clear to default" command can be used to delete user settings for the corresponding level and all other levels below.

• Partitions separate zones within a single block. These partitions within a block inherit their data from the block level.

6. Site Level (Figure 5-209)

• Location of the building: The location defines the geographical location and weather data for all buildings on this site.

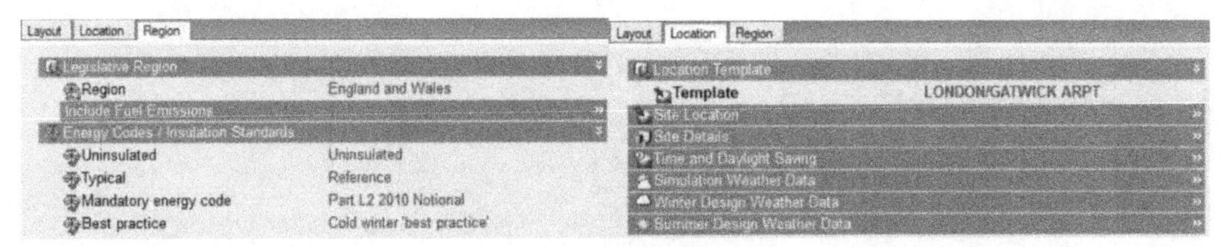

Figure 5-209 Site level

7. Building Level (Figure 5-210)

- Building type;
- LEED/ASHRAE 90.1 Building Settings;
- Project details;
- Assessor details;
- Owner details.

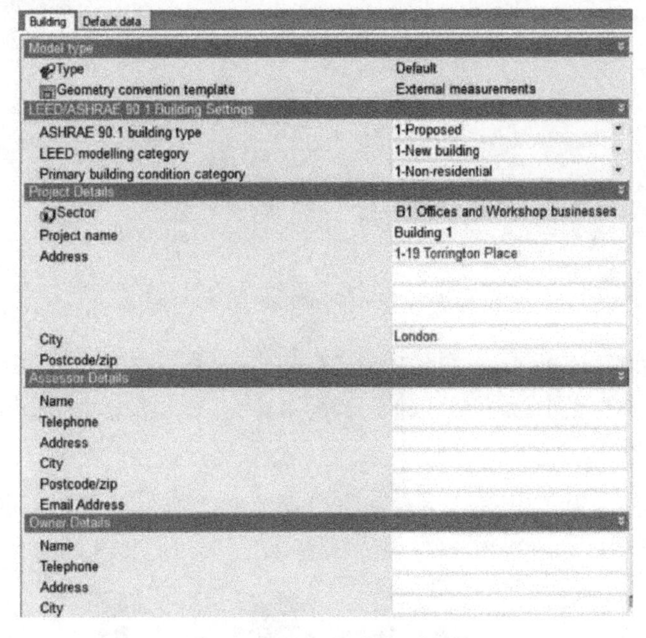

Figure 5-210 Building level

8. Block Level

Blocks are the basic elements used to create 3D DesignBuilder models.

- Building block: used to draw the model or a section of the model.
- Outline block: to assist in the creation of building block geometry and do not affect the model.
- Component block: to create visual and shading structures which do not contain zones.
- Standard: for shading, reflection and visualization.

• Ground: for setting the adjacency of any touching building block surfaces as being "adjacent to ground".

• Adiabatic: for setting adiabatic adjacency for modeling adjacent buildings at similar conditions, shading, reflection and visualization.

9. (Thermal) Zone Level (Figure 5-211)

• The building is generally divided into "zones" —special enclosed spaces, each of which has a uniform air mass.

• The air in ONE zone has uniform physical properties. In other words, there is no vertical and horizontal temperature difference in a zone.

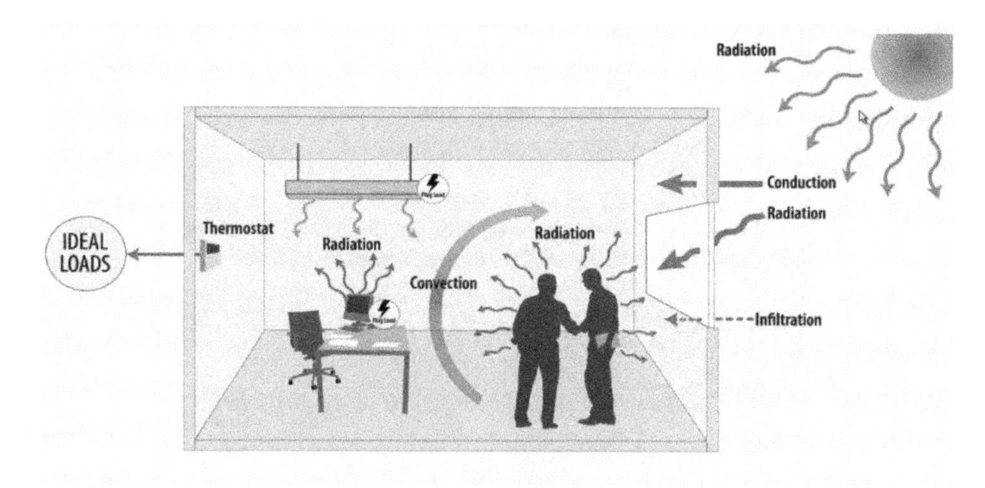

Figure 5-211 (Thermal) Zone level

A zoning example is shown in Figure 5-212.

• In Figure 5-215(a), 1, 2 refer to classroom; 3 refers to office; 4, 7 refer to reception; 5,6 refer to office;

• In Figure 5-215(b) and (c), 1, 2 refer to heat pump; 3 refers to Fan coil I; 4, 7 refer to VAV I; 5 refers VAV II; 6 refers Fan coil II.

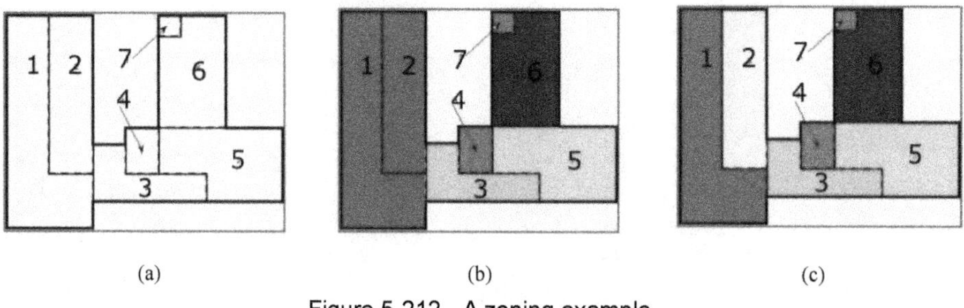

Figure 5-212 A zoning example

(a)Building plan (b)Zoning result 1 (c)Zoning result 2

10. Surface (Figure 5-213)

• Walls, roofs, ceilings, floors, partitions are represented by surfaces.

• Surfaces may contain one or more subsurfaces or openings.

• Surfaces are defined by "construction".

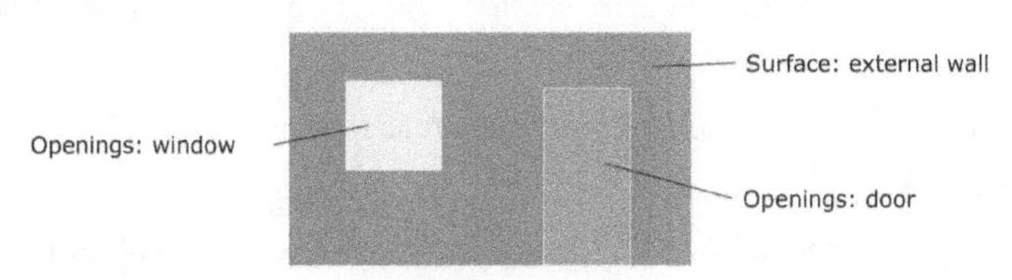

Figure 5-213 Surface

11. Layer-by-layer Construction (Figure 5-214)

• Building envelops are built layer by layer in dynamic building performance simulation, for both opaque and transparent envelops.

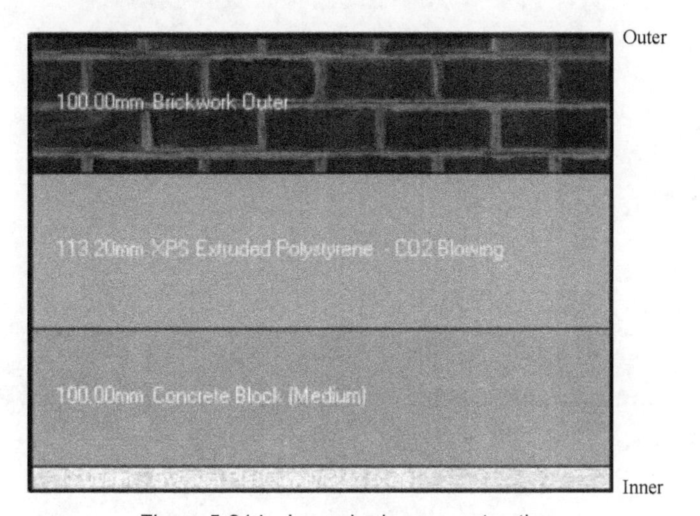

Figure 5-214 Layer-by-layer construction

12. Opening Level

• Window (yellow).

• Sub-surface (blue): opaque elements within the surface that have different properties from the main construction.

• Hole (green).

• Door (cyan).

• Vent (magenta).

13. Internal Heat Gains

• Internal heat gain is the sensible and latent heat emitted within an internal space from any source, which is to be removed by mechanical cooling, ventilation or other means.

• Three main types: people, equipment (computers, coffee machines, cooking appliances, etc.) and lighting.

14. People (Figure 5-215)

• Effects include internal heat gain, comfort, system operation, and even carbon dioxide simulation.

• Heat gains related to people include latent part and sensible part.

• Sensible heat warms air directly.

• Convection, thermal radiation (long wavelength), visible radiation (short wavelength).

• Latent heat gains are related to moisture/humidity changes in a zone.

Floor Areas and Volumes	
Occupied floor area (m2)	81.1
Occupied volume (m3)	265.0
Unoccupied floor area (m2)	0.0
Unoccupied volume (m3)	0.0
Occupancy	
Occupancy density (people/m2)	0.0188
Schedule	TM59_Living_Occ
Metabolic	
Activity	TM_59
Factor (Men=1.00, Women=0.85, Children=0.75)	1.00
CO2 generation rate (m3/s-W)	0.0000000382
Clothing	
Winter clothing (clo)	1.00
Summer clothing (clo)	0.50

Figure 5-215 Simulation about people

15. Equipment (Figure 5-216)

• Modeling computers, printers, projectors, coffee machines, etc.

• The internal heat gains for equipment are normally allocated as an allowance in watts per square meter (W/m^2) of net usable floor area.

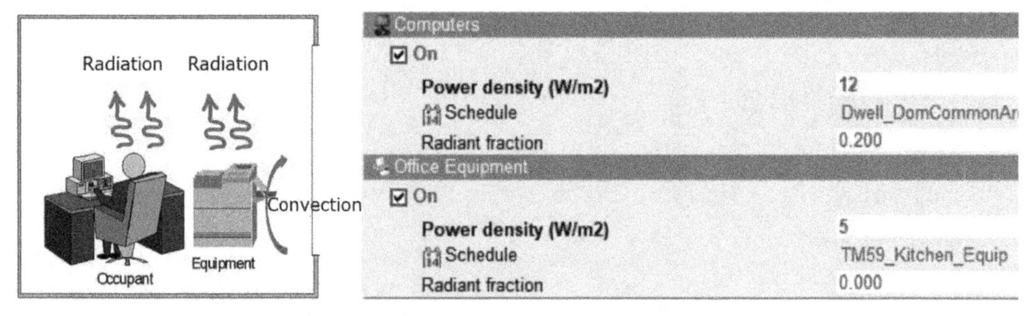

Figure 5-216 Simulation about equipment

16. Lighting (Figure 5-217)

• Modeling electric lighting system (heat gain and illuminance).

• The following must be known to determine the internal heat gains due to lighting: total electrical input power, fraction of heat emitted that enters the space, radiant, convective and conductive components (CIBSE Guide A, Section 6-4).

General Lighting	
☑ On	
Normalised power density (W/m2-100 lux)	15.0000
Schedule	TM59_Default_Light
Luminaire type	1-Suspended
Return air fraction	0.000
Radiant fraction	0.420
Visible fraction	0.180
Convective fraction	0.400

Figure 5-217 Simulation about lighting

17. Solar Heat Gains

• The sun provides not only heat gains but also daylighting.

• Solar gains through windows include energy transmitted directly through the glass and energy absorbed by the glass and frame and then re-radiated into the space (Figure 5-218).

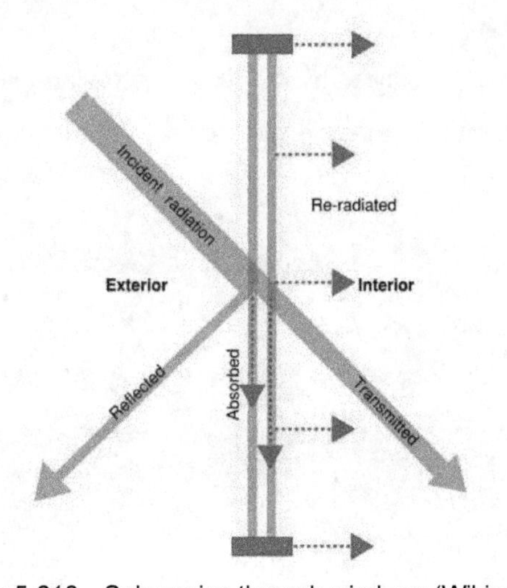

Figure 5-218 Solar gains through windows (Wikipedia)

Key parameters related to solar gains:

• SHGC (solar heat gain coefficient) is the ratio of transmitted solar radiation to incident solar radiation of an entire window assembly. This includes the transmitted solar radiation and the inward flowing heat from the solar radiation that is absorbed by the glazing.

• *g*-value (total solar energy transmittance) is the sum of the direct solar transmittance and the heat transferred by radiation and convection into the space (BSI, 2011) at normal incidence divided by the incident solar radiation normal to the surface. *g*-value is commonly used in EU, and SHGC in U.S.

18. Ventilation Related Heat Gains or Losses

• Infiltration refers to uncontrolled or unintended flow of outdoor air into a building due to cracks and other unintentional openings, normal use of exterior doors, infiltration of building materials.

• Airtightness is a measure of the air leakage rate through the building envelope. For the same intensity of driving force, the rate of infiltration reduces as the building becomes more air-tight (Figure 5-219).

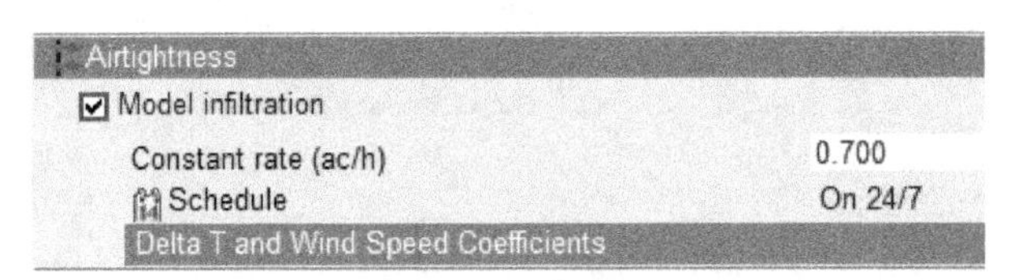

Figure 5-219 Airtightness

• Ventilation: purposeful opening of windows or doors to promote air exchange for natural ventilation and fresh air (Figure 5-220).

• This pressure difference is caused by the wind airflow around a building, which is called the wind effect, as well as the temperature difference between indoor and outdoor environments, which is called the stack effect.

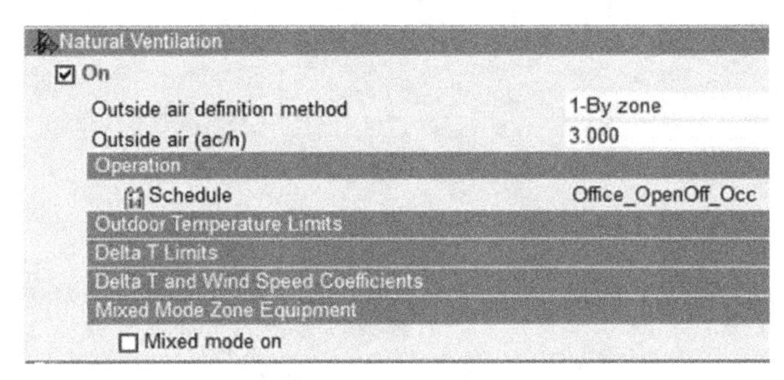

Figure 5-220 Natural ventilation

5.3.3.10 Dynamic Building Performance Simulation

HVAC systems (Figure 5-221):

• HVAC systems aims at creating a thermal comfort and acceptable air quality indoor environment;

• Loop based configurable HVAC systems (Figure 5-222);

• Pair-components are connected with common node.

Ideal load air systems:

• Used in situations where the user wishes to study the performance of a building without modeling a full HVAC system;

• Add or remove heat and moisture at 100% efficiency in order to produce a supply air stream at the specified conditions;

• In DesignBuilder, simple HVAC is modeled by the ideal load air system.

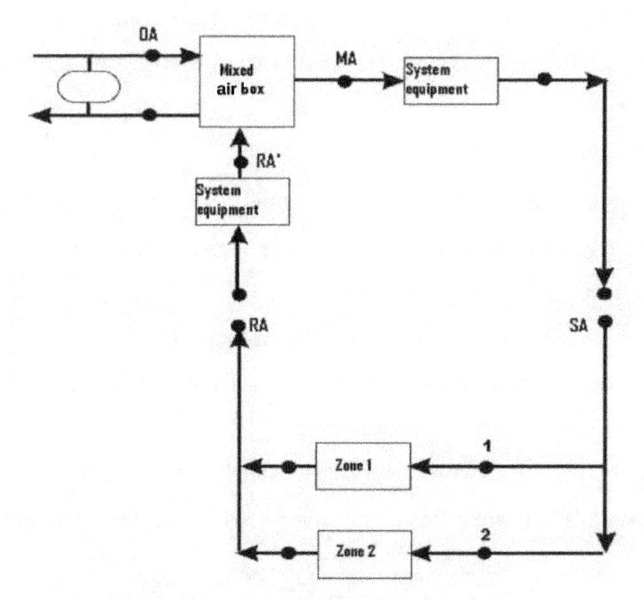

Figure 5-221 Air loop system connection

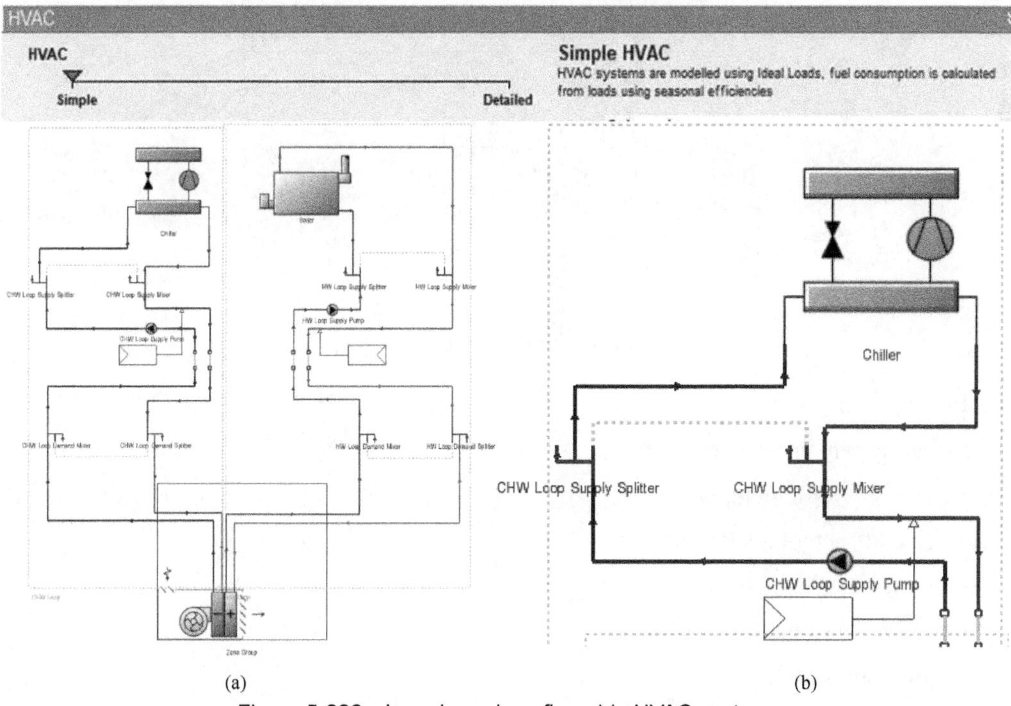

(a) (b)

Figure 5-222 Loop based configurable HVAC systems

(a) Fan coil unit with chilled water and hot water (b) Chilled water loop

Common HVAC systems:

• Zone equipment: packaged terminal heat pump (PTHP), fan coil, air terminals, variable refrigerant flow (VRF) , radiant heating system. etc.

• Air loop system: variable air volume (VAV), constant air volume (CAV), etc.

• Plant loop system: chilled water system, hot water system, cooling tower system, ground heat exchange system, etc.

5.3.3.11 Schedules

• Buildings are dynamic: people coming and leaving, systems on and off, etc.

• In general, schedules are a way of specifying how much or many of a particular quantity is present or at what level something should be set.

• Can be defined as either time-based or condition-based (Figure 5-223).

• Time-based schedules define time-varying parameters, such as occupancy density ratio with design level.

• Condition-based schedules work only when some conditions are satisfied, for example when room temperature is below heating setpoints to start the heating system.

	Time	Value
1	00:00	0.00
2	09:00	0.00
3	09:00	1.00
4	17:00	1.00
5	17:00	0.00
6	24:00	0.00

An example of time-based schedule

	Time	Value
1	00:00	0.00
2	09:00	0.00
3	09:00	(ta>23)
4	17:00	(ta>23)
5	17:00	0.00
6	24:00	0.00

An example of condition-based schedule

Figure 5-223 Schedules

Environmental control:

• Heating setpoint temperature;

• Cooling setpoint temperature;

• Humidity setpoint temperature;

• Comfort PMV setpoint;

• Ventilation setpoint temperature;

• Minimum fresh air;

• CO_2/contaminant setpoint.

• Lighting.

5.3.3.12 Major Steps Using DBPS

• Define the problem, goals and objectives;

• Choose right tools;

• Define the 3D geometry of the building;

• Select location and weather data;

• Define building constructions;

• Define internal heat gains and occupancy;

• Define building systems and their operations;

• Run simulation and analyse results.

5.3.3.13 Performance Gaps

The difference between actual and calculated energy is called the "energy performance gap" (Brom et al., 2018).

There are three groups of uncertainties:

• Workmanship and quality of building elements;

• Environment;

• Behavior.

Model calibration is the process of adjustment of the model parameters and forcing within the margins of the uncertainties (in model parameters and / or model forcing) to obtain a model representation of the processes of interest that satisfies pre-agreed criteria.

5.3.3.14 Related Organizations

1) International Building Performance Simulation Association (IBPSA)

• IBPSA is a non-profit international society of building performance simulation researchers, developers and practitioners, dedicated to improving the built environment.

• Regional affiliates: 28 regional IBPSA affiliates, spanning 5 continents.

• Free membership.

2) IBPSA-England (https://www.ibpsa-england.org)

• IBPSA-England was founded to advance and promote the science of building performance simulation in order to improve the design, construction, operation and maintenance of new and existing buildings, through the provision of a forum for the exchange of information between researchers, developers and practitioners operating in the area of building performance simulation and related issues.

5.3.3.15 Demonstration: Dynamic Building Performance Simulation

A case study (Figure 5-224):

Figure 5-224 Energy consumption of a vertical concrete house

Key information:

- Residential building.
- Square plan (4 m × 4 m), 3 stories (ground and 1st floor: 3 m, 2nd floor: 5 m).
- Bar window (1.5 m × 0.3 m, single glazing), wood door (thickness: 25 mm).
- Concrete wall (thickness: 200 mm) , glass roof.
- Human behavior: default setting.
- Location: London.
- Run period: 01/01-31/12; Design day: 1/21 (in winter), 7/21 (in summer).
- Interior gain: 2 persons; lights, 15 W/m^2; equipment: 10 W/m^2.
- Ventilation: 1 air change/hour.
- Air temperature setpoint: 22 ℃ (in winter), 24 ℃ (in summer).
- HVAC system: PTHP (CoP = 3).
- Annual energy consumption listed in Table 5-43.

Table 5-43 Renovation—energy performance of double glazing windows

Window construction	Annual energy consumption/(kW · h/m^2)
Clear 6 mm (base case)	302
Clear 6 mm+ air 13 mm+ clear 6 mm (double glazing)	—
Wall construction: 200 mm concrete (base case) + 200 mm insulation	170

5.3.3.16 CIBSE Cyclic (Admittance) Method for Peak Temperatures (Figure 5-225)

Figure 5-225 CIBSE Cyclic (Admittance) Method for Peak Temperatures

Peak temperature calculation:

• Maximum temperature guide (Table 5-44).

• Room nodes.

• Window solar heat gains: peak mean and swing.

• Fabric heat gains: peak mean and swing.

• Internal gains: peak mean and swing.

• Operative temperature (also known as dry resultant temperature): peak mean and swing.

Little UK guidance on benchmark summer peak temperatures or overheating criteria.

Table 5-44 Maximum temperatures (Category II expectation) for indoor environment in indoor spaces (clothing is assumed to be 1.0 clo in winter and 0.5 clo in summer)

Type and use of space	Assumed activity level/met	Maximum temperature for stated clothing level	
		Winter clo=1.0	Summer clo=0.5
Residential (sedentary) spaces	1.2	24.0	26.0
Residential (active) spaces	1.5	22.0	—
Offices	1.2	24.0	26.0
Public spaces (auditoria, café, etc.)	~1.2	24.0	26.0
Classrooms	1.2	24.0	26.0
Kindergarten	1.4	22.5	25.5
Shops	1.6	22.0	25.0

Which temperature best represents the room when determining the peak temperature?

It is the dry resultanttemperature, T_c, also called operative temperature.

T_c combines the air temperature and the mean radiant temperature into a single value to express their joint effect (Figure 5-226).

Correction of environmental node inputs is as follows.

• We calculate convective gains at the air node. We calculate radiative and convective gains at the environmental node.

• We calculate T_c with reference to total heat gain at the air node. This means that any heat gains realized at the environmental node require to be corrected using the following room admittance and transmittance factors:

Room transmittance factor (for mean gains): $F_{cu}=\dfrac{3(C_v+6\sum A)}{\sum AU+18\sum A}$

Room admittance factor (for cyclic gains): $F_{cy}=\dfrac{3(C_v+6\sum A)}{\sum AY+18\sum A}$

The response of a space to heat gains is a combination of two components:

• A daily mean value, which is characterized by the U-value (transmittance) of the space;

• A cyclic or "swing" value (which is the peak value minus the mean value), which is characterized by the Y-value of the space.

The Y-value is called the admittance, and is characteristic of the material surface. It represents the response to changes in environmental temperature and is a function of the surface emissivity, convective heat transfer coefficient and thermal properties of the structure (Table 5-45).

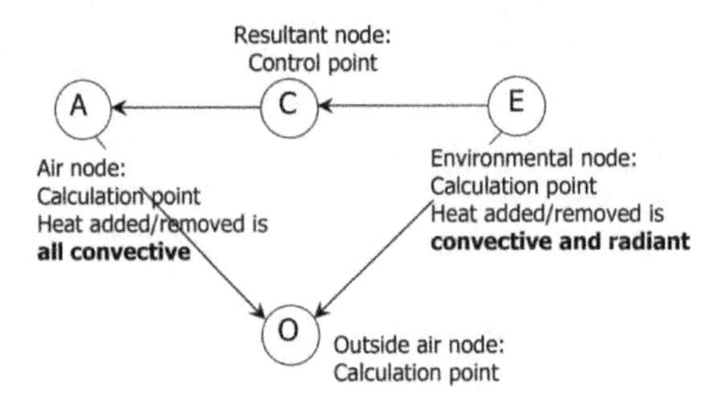

Figure 5-226 The link between nodes

Table 5-45 Heat transfer of the structure

Solar heat gains	Conduction heat gains	Internal gains
Transmission through glass Absorbed by surfaces: · Gains dampened and delayed by storage effect of surfaces; · Typically 1-2 hour delay depending on room thermal capacity	Through opaque fabric: · Effect relatively small due to decrement factor and time lag which reduces transmission of solar radiation, moderate internal/external temperature differences and presence of insulation	Equipment, lighting and people are the main sources: · A mix of convective and radiant sources (typical split 0.6/0.4)

Solar heat gains through glazing:

Due to the damping and delaying effect of the room surfaces, solar heat gains are calculated in two parts: 24-hour mean; instantaneous swing (deviation about the mean) at the particular time of interest.

• If there are no blinds, all solar gain through glazing is realized at the environmental node; if there are blinds, some of the energy is realized at the environmental node and some is realized at the air node.

Therefore, some design guides, for example, the American Society of Heating, ASHARE Handbook, provides solar heat gain factors (SHGFs) on cloudless days for daylight hours of the 21st day of each month for a given window orientation at a particular latitude and time of year.

Window solar gain factors:

• Solar gain factors based on transmittance values;

• In hand calculations using CIBSE method we would look up tables of values (Table 5.7 in CIBSE Guide A; Table 5.20 in new version);

• Usually we need to look up the mean and swing solar gain factor for the environmental node;

• Note that air node mean and swing solar gain factors are only used if there are blinds;

• Note also that the environmental node solar heat gain factors are dependent on whether the building is of low or high thermal capacity (denoted by additional subscripts "l" and "h", respectively).

Mean solar heat gains to a room:

• The mean solar heat gain to the room at the environmental node will be:

$$\bar{Q}_{\text{solar,e}} = \bar{S}_e \bar{I}_s A_g$$

where $\bar{S}_e$ is mean solar gain factor at the environmental node; $\bar{I}_e$ is mean total solar irradiance (W · m²); A_g is the area of glazing (m²).

$$\bar{S}_e = \frac{\text{Mean solar gain at the environmental node per m}^2\text{ of glazing}}{\text{Mean solar intensity incident on glazed facade}}$$

• If there are blinds, a mean solar heat gain component at the air node is also applicable:

$$\bar{Q}_{\text{solar,a}} = \bar{S}_a \bar{I}_s A_g$$

where $\bar{S}_a$ is mean solar gain factor at the air node.

$$\bar{S}_a = \frac{\text{Mean solar gain at air node per m}^2\text{ of glazing}}{\text{Mean solar intensity incident on glazed facade}}$$

• Refer to Table 2.27 in the CIBSE Guide A (Tables 2.13(g) in new version) for values of incident solar radiation, I_s at various orientations.

Solar gains through windows:

• A sample of solar gain factors is given in the following table, the value used is dependent on whether the building is lightweight or heavyweight for cyclic factors.

• Whether the building is of low ("lightweight") or high ("heavyweight") capacity determines the thermal weight ratio f_r as follows:

$$f_r = \frac{\sum AY + C_v}{\sum AU + C_v}$$

where $\sum AY$= sum of all room areas multiplied by their admittance values (other symbols have their usual meanings).

• Interpretation: if $f_r < 4$, the building is classified as "lightweight" and of "fast" response, otherwise it is classified as "heavyweight" and of "slow" response.

The diagram shows a room on the intermediate floor of an office block in London which has 4 occupants. The window is of single clear glass with no blind and the external wall is dark colored. For a mean effective ventilation rate of 3 air changes per hour, calculate the peak operative temperature in July.

Data (Figure 5-227):

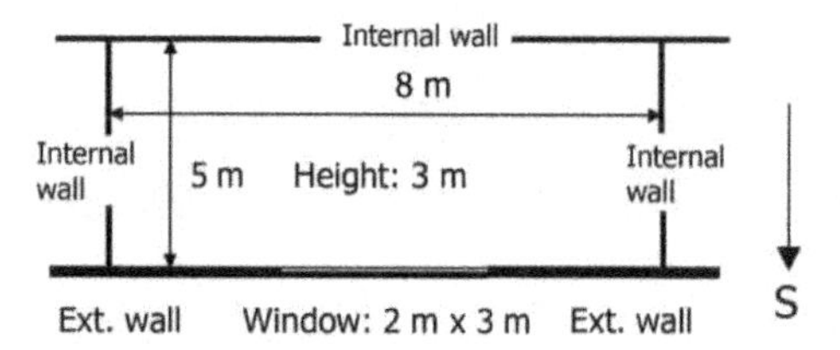

Figure 5-227 A room on the intermediate floor of an office block

External wall: U= 0.5 W/(m² · K); Y= 3.7 W/(m² · K); f= 0.4; φ= 6 h.

Internal wall: Y= 2.2 W/(m² · K).

Floor: Y= 3.7 W/(m² · K).

Ceiling: Y= 2.3 W/(m² · K).

Window: Y= 5.7 W/(m² · K); U= 5.7 W/(m² · K).

Occupant sensible gain: 90 W/person (08: 00-17: 00 h).

Stage 1: Is the room thermally lightweight or heavyweight?

Need to find f_r (thermal weight ratio) (Table 5-46).

Table 5-46 Thermal weight ratio

	A	U	Y	f	φ	AU	AY
Wall (external)	18	0.5	3.7	0.4	6	9	66.6
Wall (internal)	54		2.2				118.8
Floor	40		3.7				148
Ceiling	40		2.3				92

Continued

	A	U	Y	f	φ	AU	AY
Window	6	5.7	5.7			34.2	34.2
TOTALS	158					43.2	459.6

Note:Excluding the window area.

- Room volume = 8 m × 5 m × 3 m = 120 m^3
- C_v (ventilation conductance)=$NV/3$=3 × 120/3=120 W/K

C_v = 120 W/K ΣAY = 459.6 W/K ΣAU = 43.2 W/K

$$f_r = \frac{\Sigma AY + C_v}{\Sigma AU + C_v} = \frac{459.6+120}{43.2+120} = \frac{579.6}{163.2} = 3.55 < 4$$

The room is lightweight, so I need to use the query value for $\tilde{S}_{el}$.

Stage 2: Determine the mean solar heat gain to the room through the glazing at the environmental node (Table 5-47).

Table 5-47 Environmental node

Description	$\bar{S}_e$	$\bar{S}_{el}$	$\tilde{S}_{eh}$	$\bar{S}_a$	$\tilde{S}_a$
Single: clear	0.76	0.66	0.50	—	—
Single: clear with generic blind	0.34	0.33	0.29	0.11	0.11
Double: clear/clear	0.62	0.56	0.44	—	—
Double: clear/reflecting	0.36	0.32	0.26	—	—
Double: low emissivity/absorbing	0.43	0.38	0.32	—	—
Double: generic blind/low emissivity/clear	0.29	0.29	0.27	0.17	0.18

$$\bar{Q}_{solar,\,e} = \bar{S}_e \bar{I}_s A_g$$

where A_g is area of glass.

- A_g = 6 m^2
- I_s= 89 (Beam irradiance)+ 70 (Diffuse irradiance)= 159 W/m^2

Mean July irradiance on a south facing vertical surface (Table 2.30 in CIBSE Guide A (Table 2.13 (g) in new version)):

$$\bar{Q}_{solar,\,e} = 0.76 \times 159 \times 6 = 725.04 \text{ W}$$

Stage 2: Determine the mean solar heat gain to the room through the glazing at the air node.

$$\bar{Q}_{solar,\,a} = \bar{S}_e \bar{I}_s \bar{A}_g$$

The mean solar heat gain at the air node is zero, since in this example there are no blinds!

Mean internal heat gains to a room:

• Determine the mean internal heat gain (which is assumed to be entirely at the environmental node) as a 24-hour time-weighted average of all applicable sources.

• For typical rates of heat gain due to occupants and lighting, refer to Tables 6.1, 6.2 and 6.3 in CIBSE Guide A.

• To simplify the calculation it is assumed that all internal gains are received at the environmental node.

• For a total of N sources each acting for a daily duration of Δt hours:

$$\bar{Q}_{\text{internal, e}}=\frac{1}{24}\sum_{i}^{N}Q_i\Delta t_i$$

where Q_i is instantaneous heat gain from internal heat sources (W); Δt_i is the duration of internal heat sources (h).

Stage 3: Determine the mean internal heat gains to the room.

The room has 4 occupants, each occupant results in a heat gain to the room of 90 W over the period from 8: 00 to 17: 00

$$Q_1 = 90\text{ W},\ Q_2 = 90\text{ W},\ Q_3 = 90\text{ W},\ Q_4 = 90\text{ W}$$

$$t_1 = 9\text{ hours},\ t_2 = 9\text{ hours},\ t_3 = 9\text{ hours},\ t_4 = 9\text{ hours}$$

$$\bar{Q}_{\text{internal, e}}=\frac{1}{24}\sum_{i}^{N}Q_i\Delta t_i$$

$$\bar{Q}_{\text{internal, e}}=\frac{1}{24}\sum_{i}^{N}Q_i\Delta t_i=\frac{1}{24}\times 90\times 9\times 4=135\text{ W}$$

• Over a 24-hour period the mean operative temperature ( ) is obtained from a steady-state heat balance:

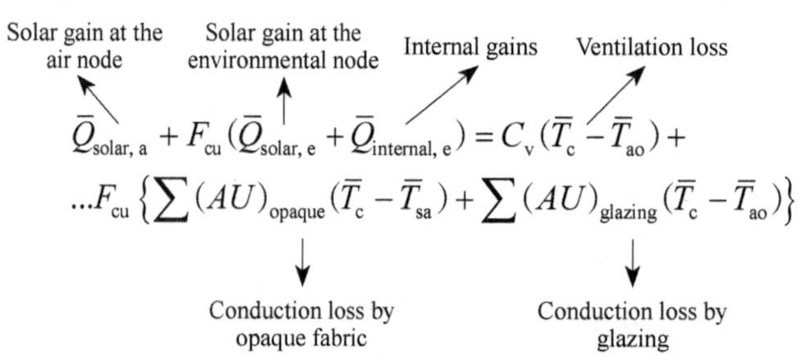

Stage 4: Determine the mean room operative temperature ($\bar{T}_{\text{c}}$).

Table 2.34 CIBSE Guide A (Table 2.14 in new version):

• July mean air temperature: $\bar{T}_{\text{ao}}$ = 19.6 ℃.

• July mean sol-air temperature on south facing black vertical surface: $\bar{T}_{\text{sa}}$ = 25.8 ℃.

Also, we need to know F_{cu} in order to calculate the mean room operative temperature:

$$F_{\text{cu}}=\frac{3(C_v+6\sum A)}{\sum AU+18\sum A}=\frac{3\times(120+6\times 158)}{43.2+18\times 158}=\frac{3\times 1\,068}{43.2+2\,844}=\frac{3\,204}{2\,887.2}=1.110$$

Calculate the mean room operative temperature $\bar{T}_{\text{c}}$.

Substitute those known parameters into the steady-state heat balance equation, then:

$$\bar{Q}_{\text{solar, a}} + F_{\text{cu}}\left(\bar{Q}_{\text{solar, e}} + \bar{Q}_{\text{internal, e}}\right) = C_{\text{v}}\left(\bar{T}_{\text{c}} - \bar{T}_{\text{ao}}\right) + \ldots F_{\text{cu}}\left\{\sum(AU)_{\text{opaque}}\left(\bar{T}_{\text{c}} - \bar{T}_{\text{sa}}\right) + \sum(AU)_{\text{glazing}}\left(\bar{T}_{\text{c}} - \bar{T}_{\text{ao}}\right)\right\}$$

$$0 + 1.110\times(725.04 + 135) = 120\times\left(\bar{T}_{\text{c}} - 19.6\right) + \ldots + 1.110\times\left\{\left(9\times\left(\bar{T}_{\text{c}} - 25.8\right) + 34.2\times\left(\bar{T}_{\text{c}} - 19.6\right)\right\}\right)$$

$$954.644 = 120\bar{T}_{\text{c}} - 2\,352 + \ldots + 1.110\times\left\{9\bar{T}_{\text{c}} - 232.2 + 34.2\bar{T}_{\text{c}} - 670.32\right\}$$

$$954.644 = 120\bar{T}_{\text{c}} - 2\,352 + 9.99\bar{T}_{\text{c}} - 257.742 + 37.962\bar{T}_{\text{c}} - 744.055$$

$$954.644 + 23552 + 257.742 + 744.055 = 120\bar{T}_{\text{c}} + 9.99\bar{T}_{\text{c}} + 37.962\bar{T}_{\text{c}}$$

$$4\,308.441 = 167.952\,\bar{T}_{\text{c}}$$

$$\bar{T}_{\text{c}} = \frac{4\,308.441}{167.952} = 25.653$$

The swing in solar heat gain at the environmental node will be:

$$\tilde{Q}_{\text{solar, e}} = \tilde{S}_{\text{e}} A_{\text{g}}\left(I_{\text{s}} - \bar{I}_{\text{s}}\right)$$

where $\tilde{S}_{\text{e}} = \tilde{S}_{\text{el}}$ if $f_{\text{r}} \leqslant 4$ or $\tilde{S}_{\text{e}} = \tilde{S}_{\text{el}}$ otherwise; I_{s} is the total incident solar radiation at the time of interest (W/m^2).

If there are blinds, a swing in solar heat gain at the air node is also applicable:

$$\tilde{Q}_{\text{solar, a}} = \tilde{S}_{\text{a}} A_{\text{g}}\left(I_{\text{s}} - \bar{I}_{\text{s}}\right)$$

$$\tilde{Q}_{\text{solar, e}} = \tilde{S}_{\text{e}} A_{\text{g}}\left(I_{\text{s}} - \bar{I}_{\text{s}}\right)$$

Stage 5: Swing in room solar heat gain.

$A_{\text{g}} = 6$ m^2

$\tilde{S}_{\text{el}} = 0.66$

For light weight, quick-response structure, $\tilde{S}_{\text{el}}$ is used.

Mean July irradiance on a south facing vertical surface = 89+70 = 159 W/m^2 (Table 2.30 (Table 2.13 (g)) in CIBSE Guide A)

Maximum July irradiance on a south facing vertical surface = 361+186 = 547 W/m^2 (Table 2.30 (Table 2.13 (g)) in CIBSE Guide A, values in bold)

$$\tilde{Q}_{\text{solar, e}} = \tilde{S}_{\text{e}} A_{\text{g}}\left(I_{\text{s}} - \bar{I}_{\text{s}}\right) = 0.66\times 6\times(547 - 159) = 1\,536.48\ \text{W}$$

(Compare that to the mean, which was 725 W)

The swing in fabric heat gain is due to the swing in sol-air and external air temperatures and is given by:

$$\tilde{Q}_{\text{fabric, e}} = \left\{\sum(AU)_{\text{opaque}}\right\} f\tilde{T}_{\text{sa}} + \left\{\sum(AU)_{\text{glazing}}\right\}\tilde{T}_{\text{ao}}$$

where $\tilde{T}_{\text{sa}}$ is the swing in sol-air temperature for each opaque element; $\tilde{T}_{\text{sa}}$ is the swing in external air temperature.

Swing in fabric heat gain through solid walls is dependent on the decrement factor, f, the swing in sol-air temperature at time of peak radiation minus time lag, φ, and the sum of the AU products for all external solid opaque walls.

Swing in fabric heat gain through windows is dependent on the swing in air temperature at time of peak radiation, and the sum of the AU products for all external glazing (Figure 5-228).

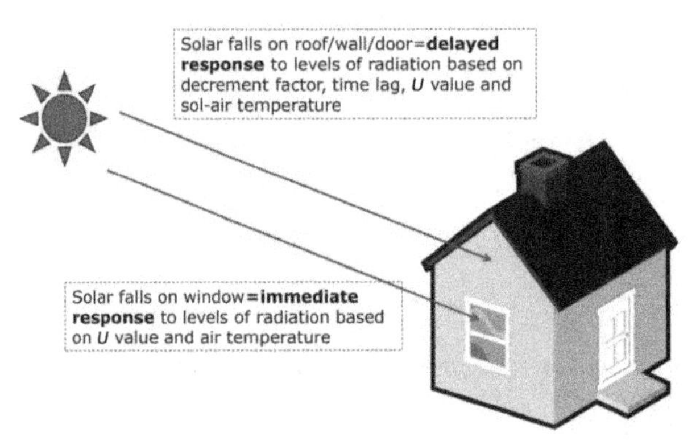

Figure 5-228 Solar falls on fabric

Stage 6: Swing in fabric heat gain.

We want to calculate the swing in fabric heat gain for our example room.

We know from the original question that:

$(AU)_{opaque} = 9\ m^2$

$(AU)_{glazing} = 34.2\ m^2$

f=0.4

Φ=6 h

We need to look up the temperature values.

Maximum solar radiation was taken at 12: 30 (Table 2.30 (Table 2.13 (g)) in CIBSE Guide A), so we need to look up maximum air temperature at this time. Using Table 2.34 (Table 2.14) in CIBSE Guide A, and taking a value which is the average of the 12: 00 and 13: 00 values, $T_{ao,max}$=23.8 ℃ and therefore the swing in air temperature is:

$$\tilde{T}_{ao} = T_{ao,max} - \bar{T}_{ao} = 23.8 - 19.6 = 4.2\ ℃$$

Maximum solar radiation was taken at 12: 30.

The opaque fabric is exhibiting a behavior which is delayed, by a time lag φ of 6 hours. Therefore we need to look up the sol-air temperature for 12:30−6 hours, 6:30. Using Table 2.34 (Table 2.14) in CIBSE Guide A, and taking a value for a dark south facing surface, which is the average of the 6:00 and 7:00 values, $T_{sa,max}$ = 16.6 ℃ and therefore the swing in sol-air temperature is:

$$\tilde{T}_{sa} = T_{sa,max} - \bar{T}_{sa} = 16.6 - 25.8 = -9.2\ ℃$$

We now have all the values we need:

at 06:30 (6 hours lag) at 12:30 peak

$$\tilde{Q}_{\text{fabric, e}} = \left\{\sum (AU)_{\text{opaque}}\right\} f\tilde{T}_{\text{sa}} + \left\{\sum (AU)_{\text{glazing}}\right\} \tilde{T}_{\text{ao}}$$
$$= 9 \times 0.4 \times (-9.2) + 34.2 \times 4.2$$
$$= 110.52 \text{ W}$$

A swing in ventilation heat gain needs to be calculated:

$$\tilde{Q}_{\text{v, a}} = C_{\text{v}} \tilde{T}_{\text{ao}}$$

The swing in internal heat gains is given by:

$$\tilde{Q}_{\text{internal, e}} = Q_{\text{internal, e}} - \bar{Q}_{\text{internal, e}}$$

where $Q_{\text{internal, e}}$ is the maximum casual heat gain at the time of interest.

Stage 7: Swing in ventilation gain.

$$\tilde{Q}_{\text{v, a}} = C_{\text{v}} \tilde{T}_{\text{ao}} = 120 \times 4.2 = 504 \text{ W}$$

Stage 8: Swing in internal gains.

The maximum casual heat gain at 12:30 is the gain from all 4 occupants:

$$\tilde{Q}_{\text{internal, e}} = Q_{\text{internal, e}} - \bar{Q}_{\text{internal, e}} = 90 \times 4 - 135 = 225 \text{ W}$$

The swing in heat gain to the room produces a swing (i.e. deviation from mean) in operative temperature, T_{c}, which can be obtained from the following heat balance:

$$\tilde{Q}_{\text{solar, a}} + \tilde{Q}_{\text{v, a}} + F_{\text{cy}}(\tilde{Q}_{\text{solar, e}} + \tilde{Q}_{\text{fabric, e}} + \tilde{Q}_{\text{internal, e}}) = (F_{\text{cy}} \sum AY + C_{\text{v}}) \tilde{T}_{\text{c}}$$

And thus, the peak operative temperature will be:

$$T_{\text{c}} = \bar{T}_{\text{c}} + \tilde{T}_{\text{c}}$$

F_{cy} and AY: when we explore averages and steady state we consider the thermal TRANSMITTANCE or U value of the building components. But when we are exploring changes in values and swings in temperatures we consider the thermal ADMITTANCE or Y value of the building components.

The thermal admittance (Y) is defined as the flux entering a surface of unit area per unit deviation in ambient temperature. It is therefore a non-steady (or transient) term and thus has a time dependency term associated with it (which is actually a time lead, $\varnothing$). The thermal admittance value can be thought of as a transient counterpart to the element U value but, unlike the U value, the value of thermal admittance will be different when viewed from opposite sides of a multiple-layer element made up of differing materials.

Generally, the Y value is close to the U value of the surface layer of a building element.

Swing heat balance:

$$\tilde{Q}_{\text{solar, a}} + \tilde{Q}_{\text{v, a}} + F_{\text{cy}}(\tilde{Q}_{\text{solar, e}} + \tilde{Q}_{\text{fabric, e}} + \tilde{Q}_{\text{internal, e}}) = (F_{\text{cy}} \sum AY + C_{\text{v}}) \tilde{T}_{\text{c}}$$

Mean heat balance:

$$\bar{Q}_{\text{solar, a}}+F_{\text{cu}}(\bar{Q}_{\text{solar, e}}+\bar{Q}_{\text{internal, e}})=C_{\text{v}}(\bar{T}_{\text{c}}-\bar{T}_{\text{ao}})+...+F_{\text{cu}}\{\Sigma(AU)_{\text{opaque}}(\bar{T}_{\text{c}}-\bar{T}_{\text{sa}})+\Sigma(AU)_{\text{glazing}}(\bar{T}_{\text{c}}-\bar{T}_{\text{ao}})\}$$

For calculation of the mean, we only consider AU for external components, assuming internal mean temperatures are equal and the mean heat loss is only through external building components. But for calculation of the swing, we need to consider AY for all components, assuming that on a transient basis the adjacent spaces could be at different temperatures and therefore heat loss could be through all building components.

Stage 9: Swing in internal operative temperature.

$$\tilde{Q}_{\text{solar, a}}+\tilde{Q}_{\text{v, a}}+F_{\text{cy}}(\tilde{Q}_{\text{solar, e}}+\tilde{Q}_{\text{fabric, e}}+\tilde{Q}_{\text{internal, e}})=(F_{\text{cy}}\Sigma AY+C_{\text{v}})\tilde{T}_{\text{c}}$$

We already know most of these values, we are trying to calculate $\tilde{T}_{\text{c}}$ and we also need to first calculate F_{cy}.

$$F_{\text{cy}}=\frac{3(C_{\text{v}}+6\Sigma A)}{\Sigma AY+18\Sigma A}=\frac{3\times(120+6\times158)}{459.6+18\times158}=\frac{3\,204}{3\,303.6}=0.970$$

$$\tilde{Q}_{\text{internal,e}}=225\text{ W}$$
$$\tilde{Q}_{\text{v,a}}=504\text{ W}$$
$$\tilde{Q}_{\text{fabric, e}}=110.52\text{ W}$$
$$\tilde{Q}_{\text{solar, e}}=1\,536.48\text{ W}$$
$$\tilde{Q}_{\text{solar, a}}=0$$
$$C_{\text{v}}=120$$
$$\Sigma AY=459.6$$

$$0+504\text{ W}+0.970\times(1\,536.48\text{ W}+110.52\text{ W}+225\text{ W})=(0.970\times459.6+120)\tilde{T}_{\text{c}}$$

$$2\,319.84\text{ W}=565.812\tilde{T}_{\text{c}}$$

$$\tilde{T}_{\text{c}}=\frac{2\,319.84}{565.812}=4.10$$

Stage 10: Peak operative temperature.

$$T_{\text{c}}=\bar{T}_{\text{c}}+\tilde{T}_{\text{c}}=25.653+4.10=29.753\ ℃$$

Interpretation of results:

• Evidently, the peak temperature is strongly influenced by the natural ventilation rate i.e. infiltration.

• The natural ventilation rate is partly influenced by buoyancy-driven forces which are in turn dependent on internal temperatures—thus the two variables are partly inter-dependent.

• As to "acceptable" peak temperatures, the criteria are application dependent (e.g. criteria applicable to offices are unlikely to be practical in, for example, a bakery).

• Refer to Section 1: Environmental Criteria for Design (CIBSE Guide Book A) for methods of assessing ranges of acceptable comfort temperatures.

• One criterion for offices that is often used is the agreed standard for UK government offices of the civil estate which sets a mean temperature of 24 ℃ with a maximum range of 4 ℃.

5.3.4 Commercial Buildings and Their Systems

Generic design process defined in CIBSE indicates chapter and publication references where information can be found (Figure 5-229), which enables engineers to design comfortable, environmentally sustainable, energy efficient buildings that are a pleasure to live, work and spend leisure time in.

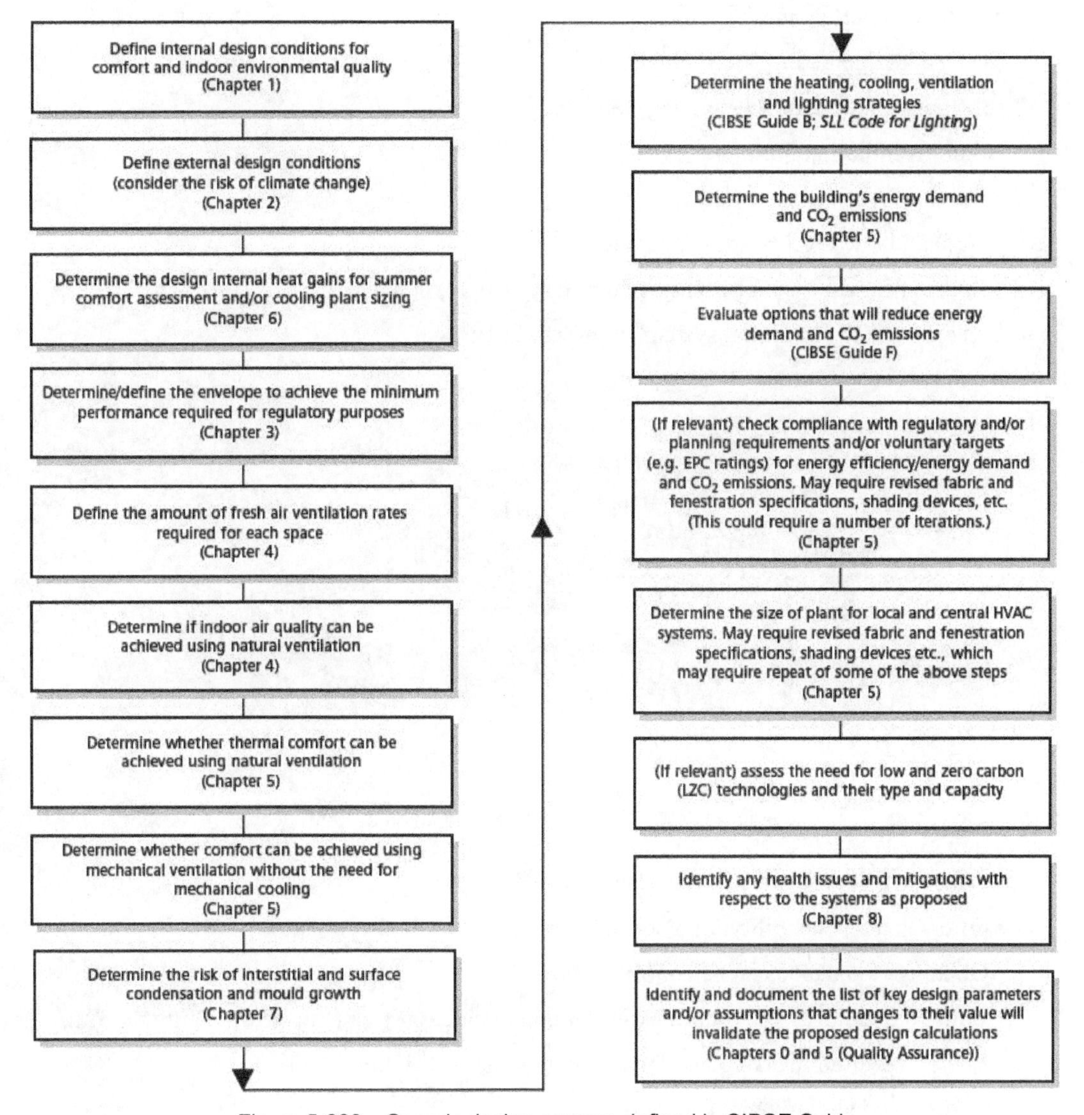

Figure 5-229 Generic design process defined in CIBSE Guide

5.3.4.1 System Approaches Ranked by Priority

1. Natural Ventilation (Highest)

(1) Requires shallow floor plates for effective ventilation.

(2) Direct openings to outside, such as windows, chimneys, dampers.

(3) Limited cooling load achievable, inside drive directly related to outside conditions.

(4) Effectiveness improved by

• Limiting solar gain;

• Exposed building mass.

(5) Provides generous floor heights.

(6) Energy required for heating systems only.

Natural ventilation limitations:

• Depth of room from window;

• High and low level window openings;

• Excess heat gains, solar energy, equipment, people, lights.

Typical natural ventilation system (Figure 5-230):

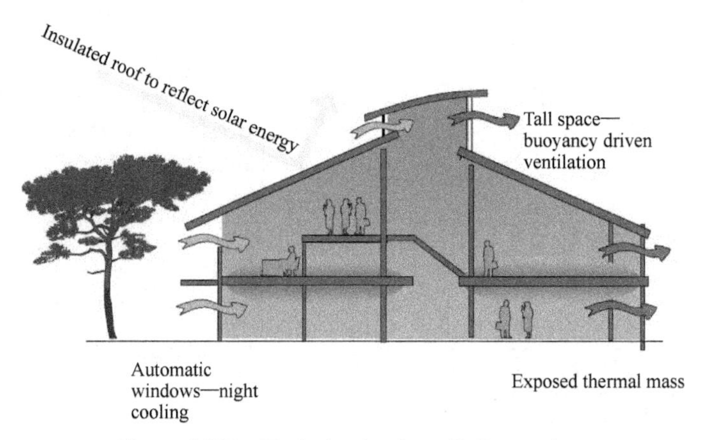

Figure 5-230 Typical natural ventilation system

2. Mechanical Ventilation (Figure 5-231)

• Offers greater flexibility for floor plan design (controlled ventilation).

• Openings to outside can be controlled.

• Limited cooling as with natural ventilation but can assist with air flow (Figure 5-232).

• Energy required for fan power and heating systems (Figure 5-233).

• Added services equipment costs.

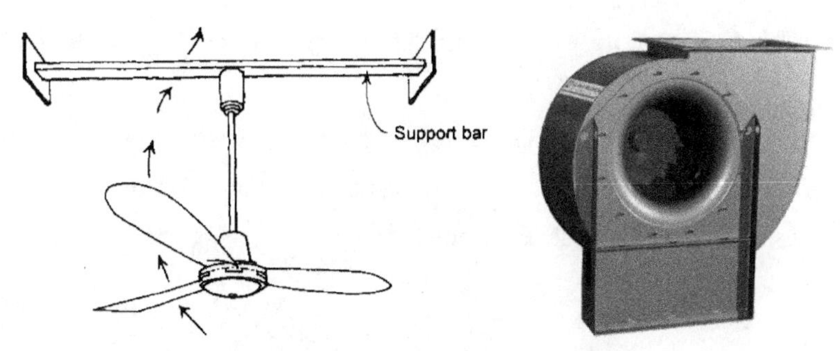

Figure 5-231 Mechanical ventilation

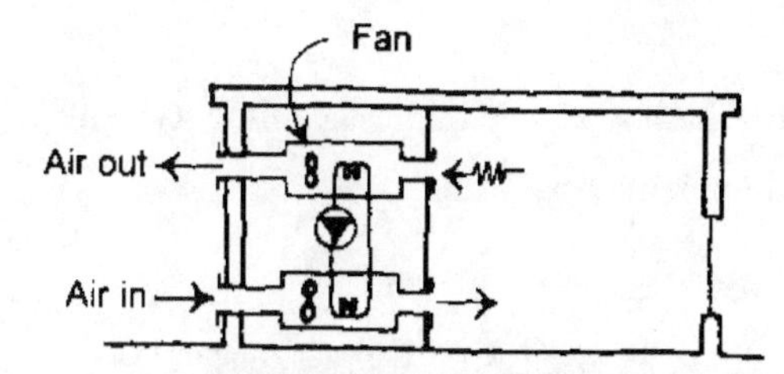

Figure 5-232 Typical mechanical ventilation system with heat recovery

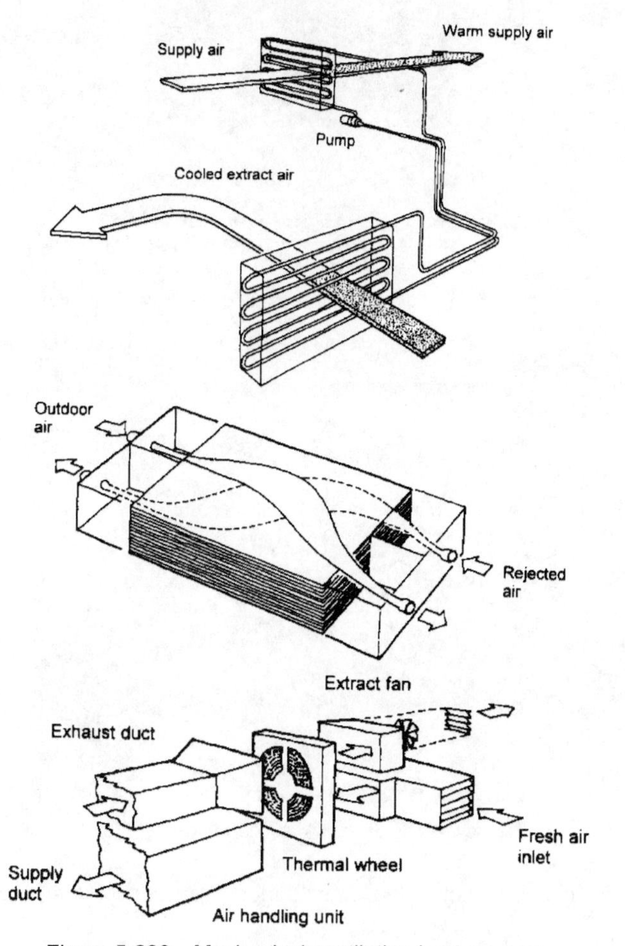

Figure 5-233 Mechanical ventilation heat recovery

3. Mechanical Cooling or Air Conditioning (Lowest) (Figure 5-234, Figure 5-235)

(1) Flexibility in building layout, fabric design, glazing provision.

(2) Can offset high heat gains from equipment, people, solar radiation.

(3) Provides control over internal environment:

• Temperature & humidity;

• Air filtration;

• Air flow;

• Noise levels.

(4) Energy required for fan power, heating and cooling, typically by refrigeration.

(5) Added services equipment costs.

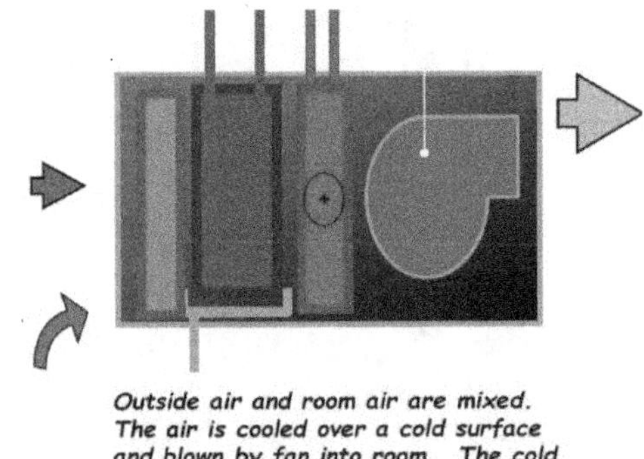

Figure 5-234 Mechanical cooling or air conditioning

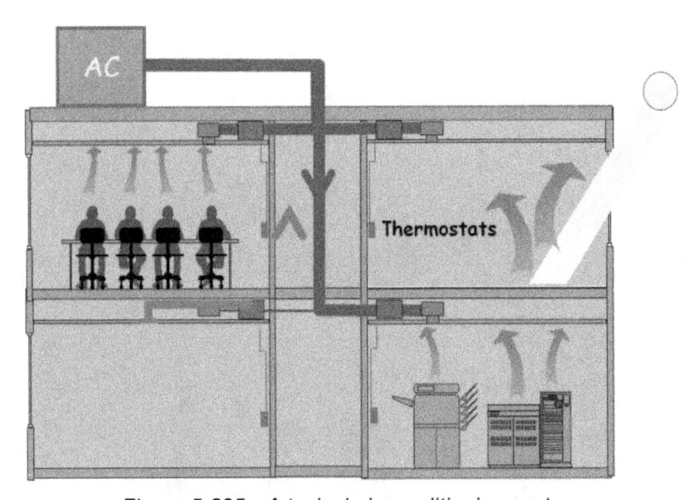

Figure 5-235 A typical air conditioning system

Comparison of natural ventilation, mechanical ventilation and mechanical cooling is shown in Table 5-48.

Table 5-48 Comparison of natural ventilation, mechanical ventilation and mechanical cooling

Item	Natural ventilation	Mechanical ventilation	Mechanical cooling
Facade performance	High	High	Minimum building regulations allow
Effect on building from	Significant	Less significant	Least significant
Energy use	Low	Low-medium	High
Services equipment	Heating only	Heating, fans and ductwork	Heating, refrigeration, fans and ductwork
Effect on floor plan	High	Minimum	Minimum
Cooling capacity	Limited	Limited	As required
Mixed mode	Yes	Yes	Yes

Facade & air conditioning size:

• In an office, heat gain through the facade can be 40% of the total gains.

• Lighting, equipment and people contribute to heat in the building.

• Ventilation load is the capacity required to cool the incoming fresh air allowance (Figure 5-236).

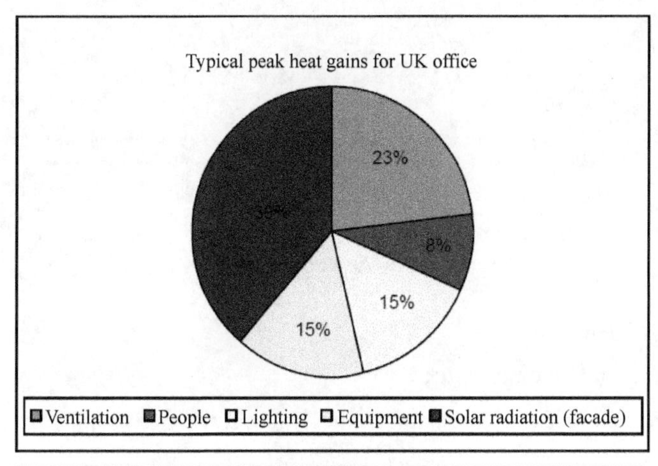

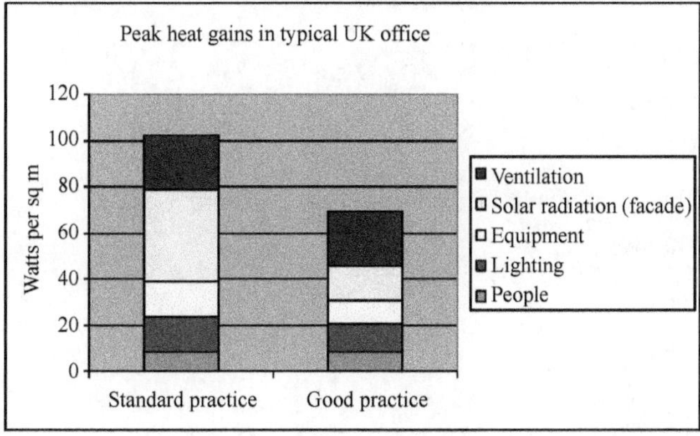

Figure 5-236 Peak heat gains

• Heat gains in a typical office from fresh air ventilation, lighting, equipment, people and windows.

• The effect of improving the facade thermal performance has the most significant impact on air conditioning size.

• Strong relation between facade and air conditioning costs.

• Low performing facade requires a high cost air conditioning installation.

• Limited air conditioning requires a high facade performance.

• Typical facade cost is 20% to 40% of construction cost.

• Typical air conditioning cost is 20% to 40% of construction cost.

5.3.4.2 Basic Air Conditioning System

(1) All air systems (Figure 5-237):

• Central cooling and heating of air;

• Air in ducted to the room;

• Requires large plant rooms and ductwork.

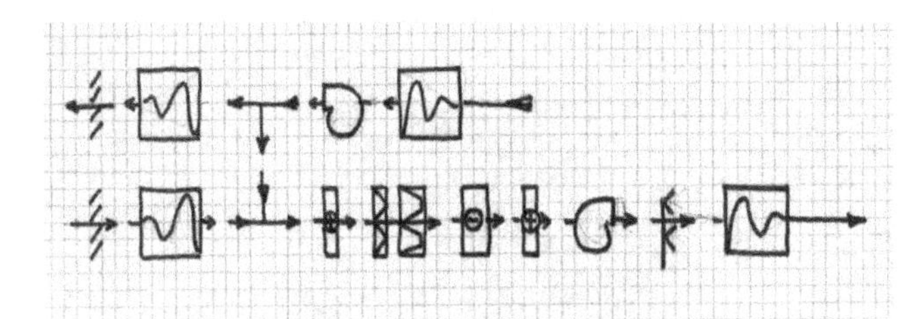

Figure 5-237 Air conditioning system

(2) Minimum outside air:

• Room cooling and heating of air;

• Only minimum outside air is treated centrally and ducted to room (primary air system).

(3) Common types of all air systems:

• Constant air volume (CAV);

• Variable air volume (VAV);

• Displacement.

(4) Air volume based on maximum cooling loads (Figure 5-238):

• 1 000 sqm floor area;

• 2.5 sqm riser space;

• 35 sqm plant space.

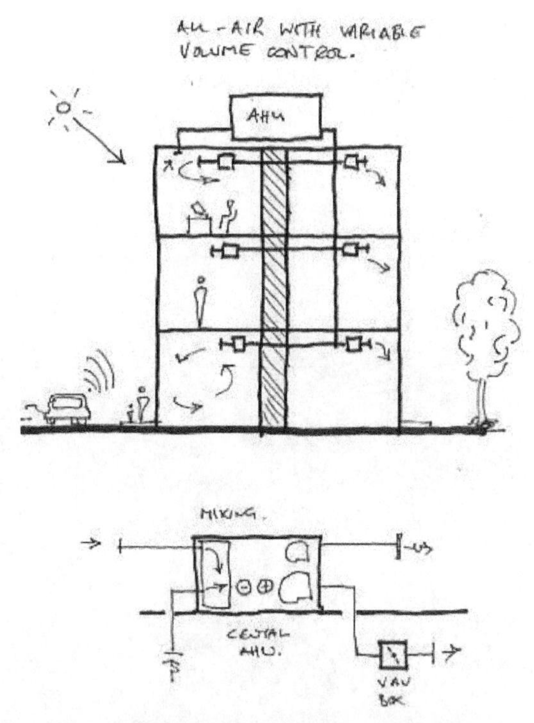

Figure 5-238 Air with variable volume control

1. Constant Air Volume (CAV) (Figure 5-239)

• Air is cooled centrally by an AHU located in the plant room.

• No VAV unit to adjust the supply air volume, i.e. constant volume of air supplied to the room.

• Dealing with various load conditions through controlling the hot and chilled water to the heating coil and the cooling coil, commonly through a three-port valve.

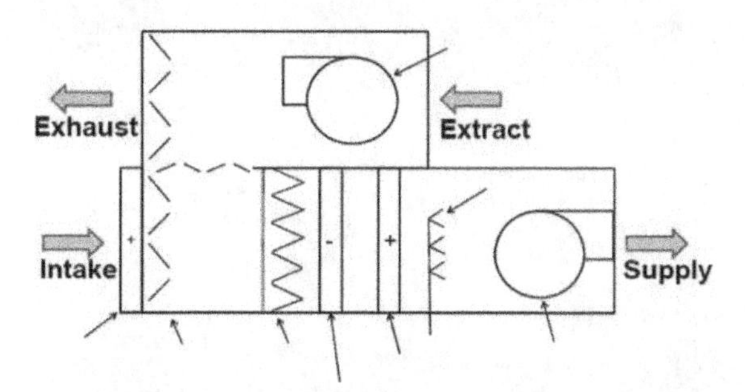

Figure 5-239 Constant volume AHU with mixing box

2. Variable Air Volume (VAV) (Figure 5-240)

• Air is cooled centrally and controlled by a adjustable damper in the VAV unit.

• Air supplied from the ceiling around 14 ℃ (8-11 ℃ difference from the indoor design temperature).

• All air is fully mixed within the room.

- High cooling capacities typically up to 120 W/m^2.
- Not normally considered as a low energy approach.

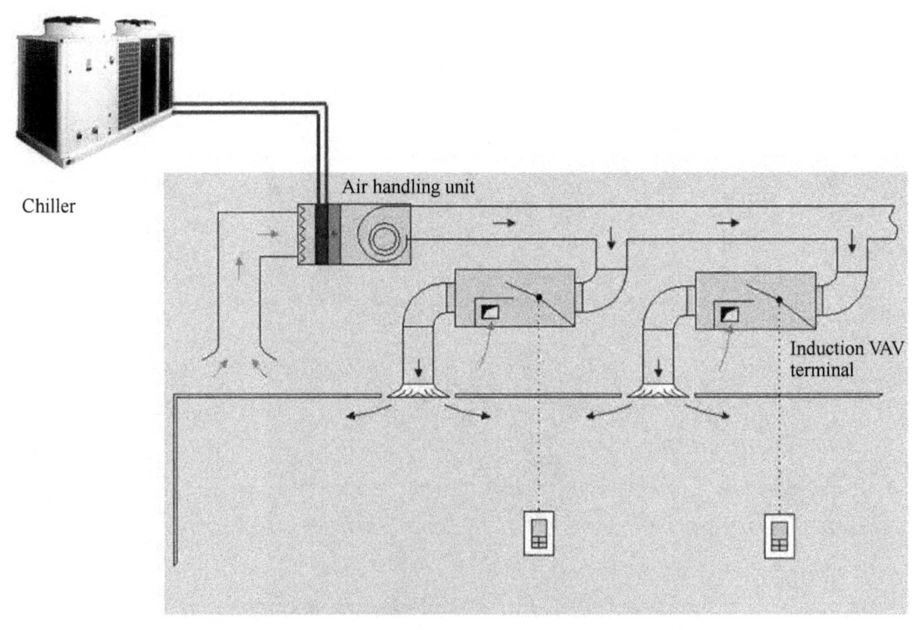

Figure 5-240 Induction variable air volume system

3. Displacement Air Supply (Figure 5-241)

- Air is cooled centrally and required minimal control by occupants.
- Air supplied through a raised floor at 18 ℃.
- Air is allowed to stratify and extracted at 28 ℃ above occupied areas.

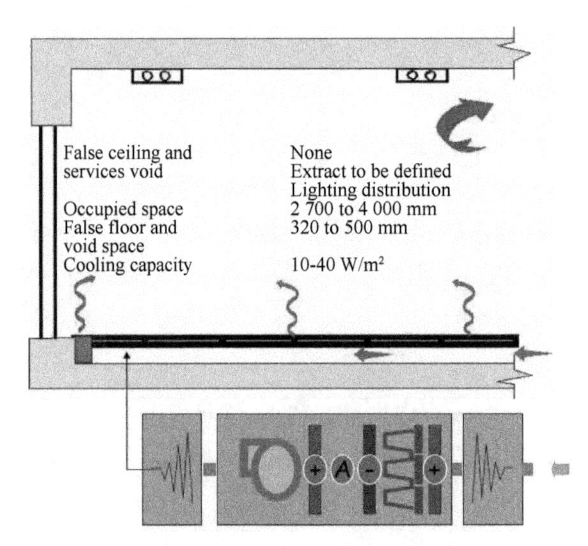

Figure 5-241 Displacement air supply

• Cooling capacity 10-40 W/m², can be increased to 60-70 W/m² by mixing the air at near the floor.

• Considered as a low energy system.

Common types of minimum outside air systems (Figure 5-242):

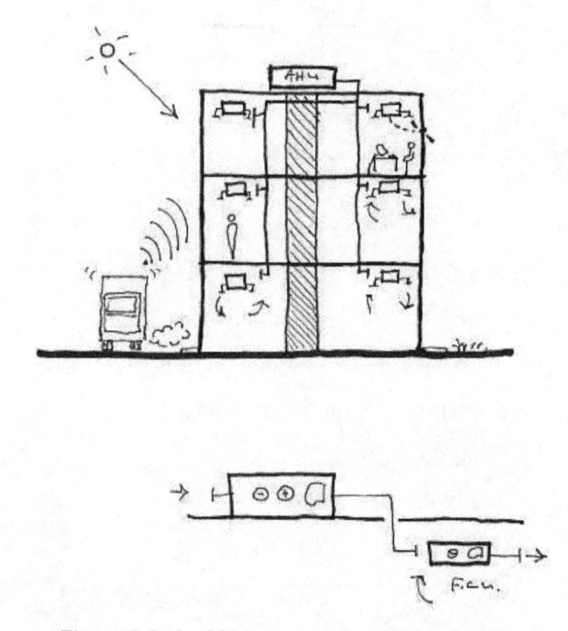

Figure 5-242 Minimum outdoor air load cooling

• Fan coil unit (FCU);

• Chilled beams;

• Chilled ceilings.

Air volume based on maximum cooling loads:

• 1 000 sqm floor area;

• 1.0 sqm riser space;

• 15 sqm plant space.

1) Fan Coil Unit (FCU) (Figure 5-243)

• Air is cooled in the space by the fan coil unit in the ceiling.

• Air is re-circulated by the fan coil unit within the ceiling void.

• The minimum required fresh air is supplied separately.

• Air supplied from the ceiling at around 14 ℃.

• All air is mixed within the room.

• High cooling capacities typically up to 120 W/m².

• Typically not considered as a low energy approach.

Figure 5-243 Fan coil unit (FCU)

2) Chilled ceiling (Figure 5-244)

• Air is cooled in the space by the cooling coils in the ceiling.

• Energy is exchanged by both radiation and convection.

• The minimum required fresh air is supplied separately.

• All air is mixed within the room.

• Cooling capacities typically up to 60 W/m^2.

• Can be considered as a low energy system.

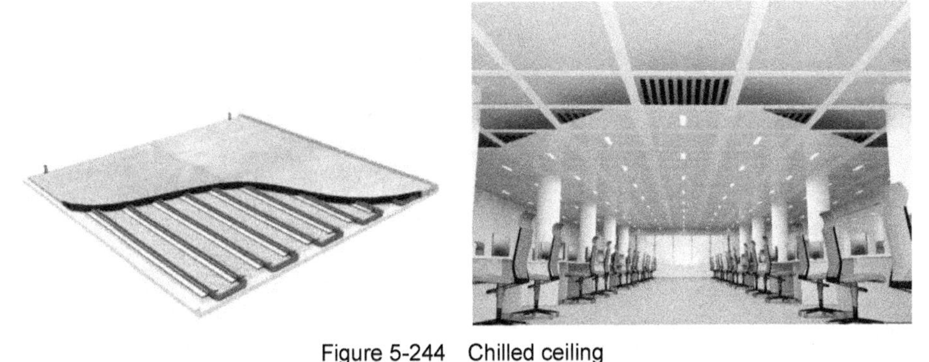

Figure 5-244 Chilled ceiling

(https://www.frenger.co.uk/pdfs/chilled-ceiling-v1.3.pdf)

3) Chilled beam (Figure 5-245)

• Air is cooled in the space by the chilled beam in the ceiling.

• Beam works by using natural convection, i.e. no fans.

• Can be classified as either passive (convection) or active (convection and radiation) chilled beams.

• The minimum required fresh air is supplied separately.

• Air supplied from the ceiling at 14 ℃.

• Cooling capacities typically up to 90 W/m^2.

• More efficient than the chilled ceiling system.

• Can be considered as a low energy system.

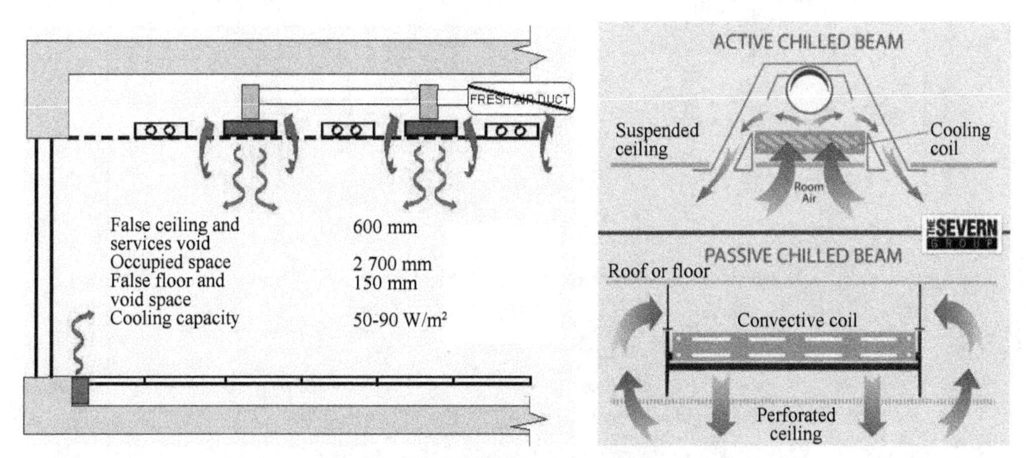

Figure 5-245 Chilled beam

(https://www.theseverngroup.com/chilled-beam-vs-chilled-ceiling/)

5.3.4.3 System Selection and Cooling Loads

Summary of options for HVAC systems (Table 5-49 to Table 5-51, Figure 246):

• Natural ventilation;

• Mechanical ventilation;

• Central air conditioning (all air);

• CAV/VAV air conditioning system;

• Displacement air conditioning system;

• Room based air conditioning (minimum outside air);

• Fan coil unit system;

• Minimum outside air;

• Chilled beams;

• Chilled ceilings.

Table 5-49 Comparison of air conditioning systems

SYSTEM \ FUNCTION	SENSE OF COMFORT	SPEED OF CONSTRUCTION	SPEED OF COMMISSIONING	SUITABILITY FOR OPEN-PLAN	SUITABILITY FOR CELLURISATION	FLEXIBILITY IN DESIGN	ON-FLOOR MAINTENANCE	TYPICAL FLOOR HEIGHTS	LEVEL OF OPEN-PLAN CONTROL	LEVEL OF CELLULAR CONTROL	CENTRAL PLANT SPACE REQUIREMENT	RISER SPACE REQUIREMENT	SUPPLEMENTARY COOLING CAPABILITIES	PERIMETER HEATING REQUIREMENT	CONDENESATE REQUIREMENTS
DISPLACEMENT UNITS	✓	✓	✓	✓	●	✓	✓	●	✓	✗	●	●	✗	✗	✓
DISPLACEMENT AND CHILLED CEILINGS	✓	●	●	✓	●	✓	●	●	✓	●	●	●	✗	✗	✓
ACTIVE CHILLED BEAMS	✓	●	●	✓	●	●	●	●	✓	●	✓	✓	✗	✗	✓
FAN COIL UNITS	✓	●	●	✓	✓	✓	✗	●	✓	✓	✓	✓	✓	✓	✗
VARIABLE AIR VOLUME	●	●	●	✓	✓	●	●	✗	●	✓	✗	✗	✓	✗	✓

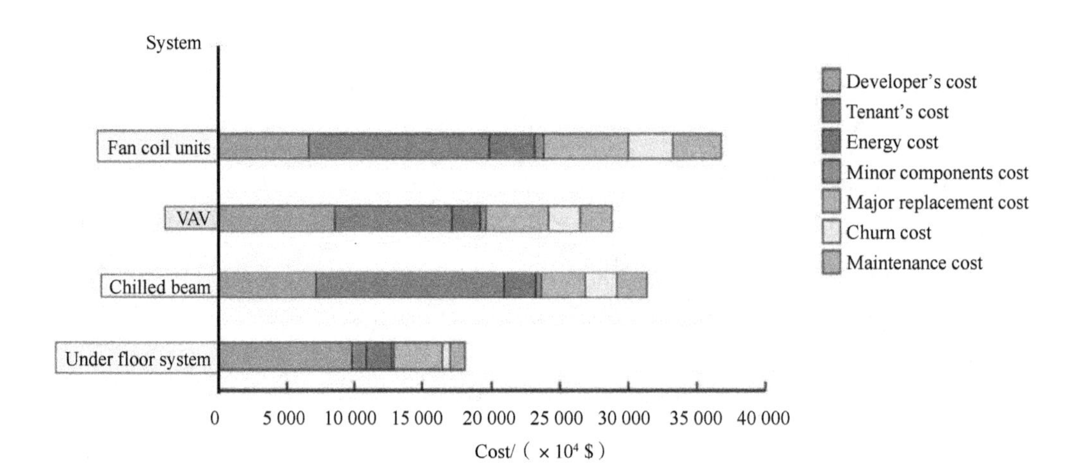

Figure 5-246 Example cost over the life of a building of various air conditioning systems

Table 5-50 Comparison of systems

System	Key points	Effect on building form and facade	Capital costs	Running costs	Space taken	Energy use	Typical applications
Natural ventilation	Limited to low energy	Highest	Low	Low	Low	Low	Residential buildings, offices, schools
Mechanical ventilation	Special extract systems	High	Low	Low	Low	Low	Kitchens, libraries
VAV air conditioning	High capacity	Low	High	High	High	High	Laboratories, dealer offices, hospitals

Continued

System	Key points	Effect on building form and facade	Capital costs	Running costs	Space taken	Energy use	Typical applications
Displacement system	Low energy	Medium	Medium	Medium	High	Medium	Offices, libraries, galleries, auditoria
Fan coil unit system	Flexibility	Low	High	High	Medium	High	Hotels, offices, residential buildings
Chilled beam system	Low energy	Medium	High	Medium	Medium	Medium	Offices, classrooms

Table 5-51 Selecting the right system

Criteria	System suitability	
Building function criteria, occupancy, temperature, noise concerns? Lighting levels?	Natural ventilation	What is the effect of opening windows? What is cooling demand required?
Location, out of town, city center, etc.	Mechanical ventilation	What is cooling demand required? What are opportunities for heat recovery
Flexibility: How often will plan change	Fan coil unit cooling	What is cooling load? Can units be accessed in rooms?
Building—installation and operating cost concerns	Chilled beam cooling	What is cooling requirement? What is fresh air requirement?
Space requirements and maintenance of equipment	Variable air volume (all air)	Cooling load, cost of installation, space available
Life span of building	Displacement system	Cooling load, facade performance, comfort

Selection Criteria:

- Location: out of town or city center.
- Outdoor climate, seasonal temperatures, humidity.
- Site conditions, prevailing winds, noise, air quality.
- Maximum usable or rentable floor area.
- Occupancy patterns, hours of operation.
- Speculative or owner occupied.
- Programme and cost—minimize construction cost.
- Energy use and efficiency.
- Aesthetic quality, image, architectural objective.

Common building types:

- Offices;
- Theaters;

- Concert halls;
- Colleges;
- Schools;
- Airports;
- Shopping malls;
- Hospitals;
- Research laboratories;
- Museums;
- Libraries;
- Hotels;
- Supermarkets.

Summary:

- There are three basic system approaches. Within each there is a wide range of options.
- System selection based on design criteria, building function and location.
- Each system type can be applied to a variety of building uses.
- System choice and building facade are closely linked.
- Selection involves all the project team, not just the engineers.

Chapter 6

Low Carbon Building Design Strategy

6.1 Building Energy Usage and Carbon Emission

6.1.1 Building Energy Usage

6.1.1.1 Energy Crisis

An energy crisis occurs when the economy is affected by a shortage of energy supplies or an increase in prices (Figure 6-1). This usually involves a shortage of oil, electricity or other natural resources. Energy crises usually result in economic downturns. From the consumer's point of view, the increase in the price of petroleum products used in cars or other transportation increases their expenses.

Energy security has been an issue since the industrial revolution, and several oil crises have elevated the issue of energy security to national prominence. The energy crisis will not only cause a big impact on the economy, but also change the structure of the energy market, especially in nowadays' era of rapid economic development, where energy strategy has been the core strategy of countries.

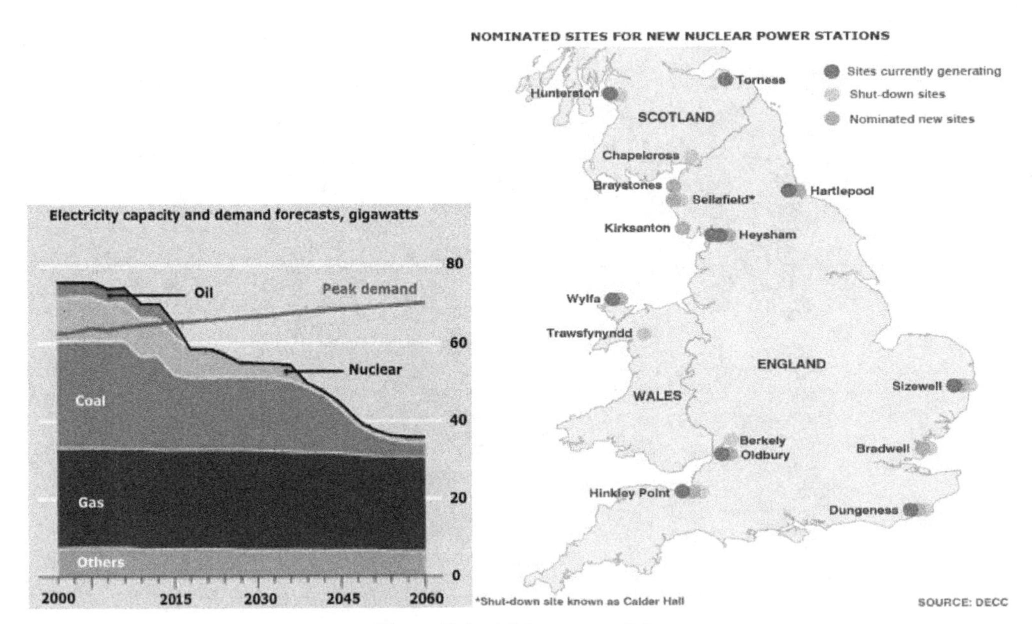

Figure 6-1 UK energy crisis

Although fossil energy has supported the development of industrial civilization, but also brought about environmental pollution, climate change and other real problems affecting human survival and development, so the energy production and consumption methods based on fossil energy are in urgent need of transformation (Figure 6-2). At the same time, the world's wind, solar and other clean energy power generation generally in the accelerated development stage, in terms of technological innovation, equipment development, engineering applications, system safety and economy still faces greater challenges.

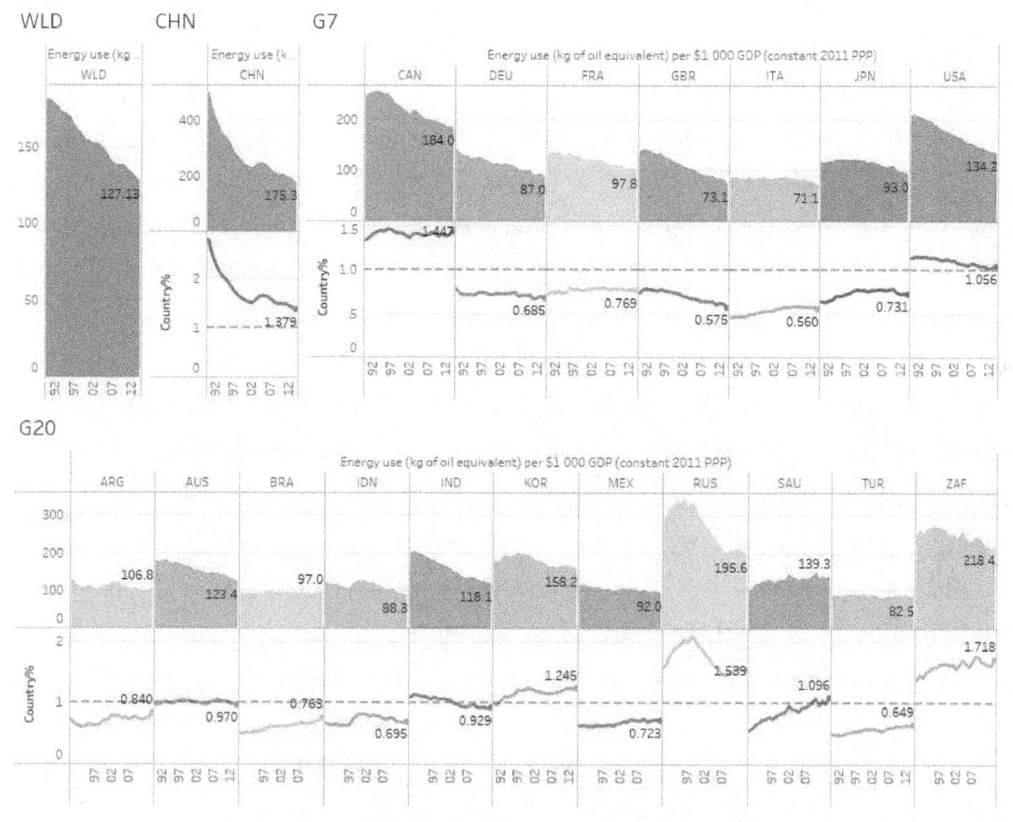

Figure 6-2 World energy use

At present, China has become the biggest energy producer and consumer in the world. China's energy consumption mainly involves high carbon fossil energy, and in the future we will be under increasing international pressure to address the issue of climate change. This not only seriously restricts the total amount of fossil energy used in China, but also poses a challenge to the existing way of using fossil energy. The ecological degradation brought by large-scale fossil energy development and utilization, as well as the pollution emissions caused by energy consumption, is increasingly exceeding the capacity of the environment, mainly in terms of the impact on the atmospheric environment, water resources and ecosystems, etc. The use of fossil energy emits a large amount of SO_2, NO_x, soot and other pollutants.

At present, urban transportation and thermal power have become the main sources of fine particulate matter ($PM_{2.5}$), and particulate matter emissions from thermal power, transportation and other industries will continue to increase. The widespread and high-intensity haze weather is prompting energy transformation.

There is a global consensus on climate change caused by greenhouse gases, and a binding intergovernmental agreement on CO_2 emission reduction has been reached. The Chinese government has pledged to peak carbon emissions by 2030 and reduce CO_2 emissions per unit of GDP by 60%-65% from 2005 levels.

6.1.1.2 Building Energy Consumption

There are two ways to define building energy consumption. Broadly speaking, building energy consumption refers to the energy consumption of the whole process from the manufacturing of building materials, building construction, to the use of the building. The narrow sense of building energy consumption, i.e. the energy consumption of building operation, is the energy consumption of people's daily use, such as heating, air conditioning, lighting, cooking, laundry, etc. , which is the dominant part of building energy consumption. With the growth of economic income and improvement of life quality, the focus of building energy consumption will shift from the consumption of decorative and durable goods to the consumption of function and environmental quality, so the energy consumption required to guarantee indoor air quality (air conditioning, ventilation, heating and hot water supply) will rise rapidly.

There are five main factors affecting building energy consumption.

1. Influence of Outdoor Thermal Environment

The outdoor thermal environment of buildings, i.e. various climatic factors, affects the indoor climatic conditions through the building envelope, exterior doors or windows and various openings. The climate factors closely related to the building are solar radiation, air temperature, air humidity, wind and precipitation, etc.

2. Heating Region and the Number of Days in the Heating Period

The heating region is the area where the average daily temperature is stable below 5 ℃ for more than 90 days in a year. The boundary between the heating region and the non-heating region is roughly south of the eastern and central section of the Longhai Line, extending west to the vicinity of Xi'an and then extending southwest. Heating period degree days is the temperature difference between the indoor benchmark temperature of 18 ℃ and the average outdoor temperature of the heating period, multiplied by the number of days in the heating period, and the unit is ℃ · d.

3. Solar Radiation Intensity

There are many sunny days in winter, with long hours of sunshine, low solar incident angle and solar radiation intensity as well as great depth of sunlight entering the room from south-facing windows, which can achieve the effect of improving indoor temperature and saving heating energy.

4. Heat Insulation and Airtightness of the Building

The thermal insulation of the building envelope and the airtightness of the doors and windows are the main intrinsic factors affecting the energy consumption of the building. The heat loss from heat transfer of building envelope accounts for about 70%-80%; the heat loss from air infiltration of door and window gaps accounts for about 20%-30%.

Strengthening the heat insulation of the envelope, especially strengthening the heat insulation and air tightness of windows, including balcony doors, is a key link to reduce heating energy consumption.

5. Thermal Efficiency of Heating System

The heating system is a system consisting of heat source, heat network and heat users. The thermal efficiency of heating system includes boiler operating efficiency and pipe network delivery efficiency. Boiler operating efficiency is the ratio of the heat generated by the boiler that can be effectively utilized to the heat contained in the coal it burns. Under different conditions, it can be further divided into boiler rated efficiency and boiler operating efficiency. The outdoor pipe network conveying efficiency is the ratio of the total heat output of the pipe network to the total heat input of the pipe network, and the boiler can generally only convert 55%-70% of the heat contained in the fuel into the available effective heat during the operation process, that is, the boiler operating efficiency is 55%-70%. The outdoor pipe network has a transmission efficiency of 85%-90%, that is, the effective heat input to the pipe network from the boiler is lost 10%-15% along the way, and the remaining 47%-63% of the heat is supplied to the building and becomes the heating heat.

6.1.1.3 Building Energy Issues in China

With the continuous development of China's construction industry in recent years, the building area of urban and rural houses has increased significantly, and while the per capita living area is growing, the use of various household appliances is also increasing, so the building energy consumption is rising year by year. From the perspective of energy demand, with the improvement of living standards, the expansion of heating range, the increase of air-conditioned buildings, a comfortable building thermal environment has become a necessity of life. China's climate characteristics determine the energy consumption for winter heating and summer air conditioning is higher than that of the same latitude, the northern heating areas are still the key areas of building energy consumption. Hot-summer and cold-winter regions are also chronically cold in winter, and there is an urgent need to consume large amounts of energy for improving thermal comfort in winter.

According to relevant information, only in 2015, China's building energy consumption has accounted for 43% of the total social energy consumption (Figure 6-3). With the accelerated urbanization process, urban construction will always maintain a high-speed development trend, and people's living standards will also be continuously improved; drawing on the development experience of developed countries, the proportion of building energy consumption in total social energy consumption will rise to about 35% in the future.

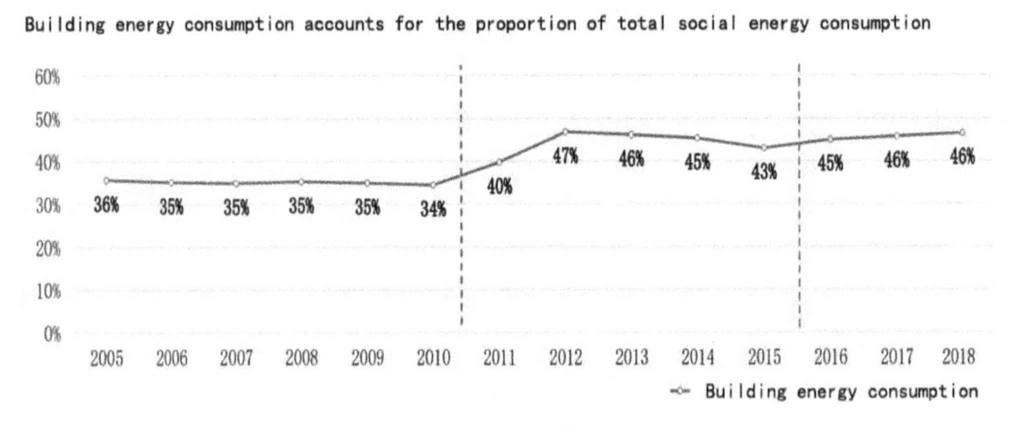

Figure 6-3 Building energy use growth in China

Low energy use efficiency is one of the most prominent problems in China's current building energy use. Compared with Western Europe or North America (with equivalent climate), Chinese residential units need to consume 1-2 times more energy for heating area, and the comfort is also poor. For example, after the implementation of the new energy-saving standard in 1995, the energy consumption of buildings in Beijing has been significantly reduced, but it is 1 time higher than that of countries such as Denmark and Sweden, while the number of heating days in Shanghai is much lower than that of the Nordic countries after the implementation of the building energy-saving standard, but the heat consumption per unit area is higher than that of the Nordic countries. Although the energy consumption per unit of China's buildings is higher than that of developed countries, the comfort level of China's buildings is far lower than that of developed countries.

6.1.1.4 Building Energy Saving

Energy efficiency in buildings means minimizing energy consumption while meeting the requirements for the production of building materials, construction and building use.

There are several ways to reduce energy demand.

1. Architectural Planning and Design

In the planning and design of buildings, according to the influence of the wider climatic conditions, the specific environmental climate characteristics of the building itself are addressed, and attention is paid to the use of the natural environment to create a good indoor microclimate of the building in order to minimize the reliance on building equipment. The specific measures can be summarized in the following three aspects: reasonable selection of the building's location, reasonable design of the external environment (placement of trees, vegetation, water, etc. around the building); reasonable design of the building form (including the determination of the overall building volume and building orientation) to improve the existing microclimate; reasonable design of the building form is a key part of making full use of the building's outdoor microenvironment to improve the building's indoor microenvironment. This is achieved mainly through the structural design of the building components and the rational separation of the internal spaces of the building. At the same time, the design can be optimized with the help of relevant softwares, such as the use of building shadow simulation in Tianzheng Architecture to assist in the design of building orientation, roads and greenery in residential areas and the use of CFD software to analyse the smooth flow of indoor and outdoor air.

2. Building Envelope

The design of the components of the building envelope has a fundamental impact on the building's energy consumption, environmental performance, indoor air quality and the visual and thermal comfort environment in which the user lives. Generally the cost of increasing the envelope is only 3% to 6% of the total investment, while energy savings can be 20% to 40%. By improving the thermal performance of the building envelope, the transfer of heat from outdoors to indoors can be reduced in summer and the loss of heat from indoors can be reduced in winter, so that the thermal environment of the building can be improved, thus reducing the cold and heat consumption of the building. Firstly, the thermal performance of each component of the envelope needs to be improved, and then, according to the local climate, the geographical location and orientation of the building, the design method for optimizing the envelope combination needs to be selected.

3. Improving Total Energy Use Efficiency

Energy losses are significant in the process of conversion from primary energy to end-use energy used in building equipment systems. Therefore, the whole process (including extraction, treatment, transmission, storage, distribution and end use) should be evaluated in order to fully reflect the energy use efficiency and the impact of energy on the environment. Energy-consuming equipment in buildings, such as air conditioners, water heaters, and washing machines, should be supplied with efficient energy. For example, as a fuel, natural gas has a higher total energy efficiency than electricity. The use of second-generation energy systems can make full use of different grades of thermal energy to maximize energy efficiency, such as combined heat and power (CHP) and combined cooling, heating and power (CCHP).

Besides, it is essential to use new energy sources. In terms of energy conservation and environmental protection, the use of new energy plays a vital role. New energy sources usually refer to non-conventional renewable energy sources, including solar energy, geothermal energy, wind energy, biomass energy, etc.

6.1.2 Building Carbon Emission

6.1.2.1 Definition of Building Carbon Emission

Building carbon emissions are the sum of greenhouse gas emissions from buildings during the production and transportation of building materials associated with them, construction and demolition, and operation phases, expressed in carbon dioxide equivalent (Figure 6-4).

Figure 6-4 Sources of carbon emissions

6.1.2.2 Measurement of Carbon Emissions

What is 1 tonne of CO_2? It equals 730 kg oxygen + 270 kg of carbon. 1 tonne of carbon corresponds to 3.7 tonnes of CO_2. This is a balloon over 15.5 m in diameter!

6.1.2.3 Sources of Carbon Emissions

The main source of carbon emissions is fossil fuels. Fossil fuels are a mixture of hydrocarbons or derivatives of hydrocarbons, including natural resources such as coal, oil, natural gas, oil shale, oil sands, and combustible ice under the sea. During the combustion of fossil fuels, carbon is transformed into carbon dioxide and enters the atmosphere, which increases the emission of greenhouse gases. Coal, oil, natural gas and other fossil fuels are widely used and are the main sources of carbon dioxide. Among them, coal is very rich in carbon, and its combustion emits a large amount of carbon dioxide (Figure 6-5).

Any human activity may cause carbon emission, various fuel oil, gas, paraffin, coal, and natural gas all produce large amounts of carbon dioxide during the process of use, and a large quantity of carbon dioxide is also emitted from city operation, people's daily life, and transportation (airplane, train, car, etc.) (Figure 6-6).

CO_2 is released into the environment by the combustion of fossil fuels & cellulose (wood):

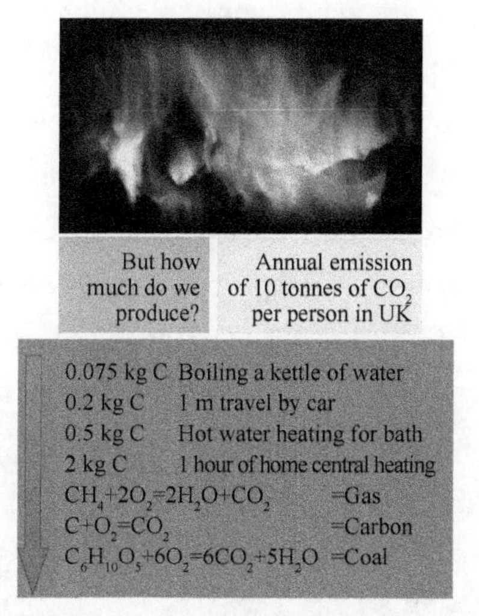

Figure 6-5 CO_2 production and emissions

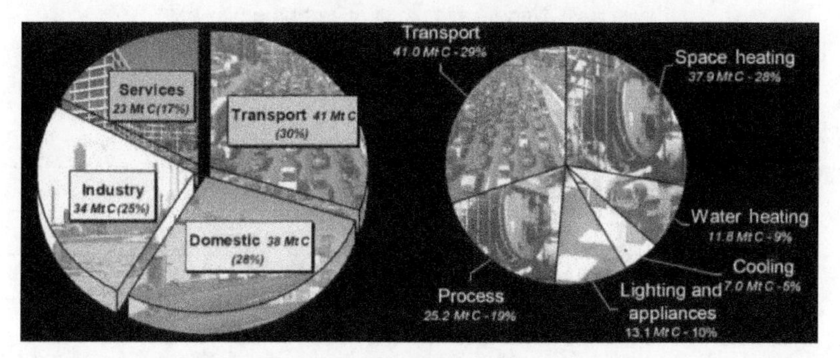

Figure 6-6 Annual carbon emissions in UK

6.1.2.4 Consequences of Carbon Emissions

Almost 70 million tonnes of carbon (over 50% of total emissions) are emitted from environmental treatment of buildings in the UK each year (Figure 6-7)!

Equivalent to 260 million tonnes of CO_2 or a balloon almost 6.5 km in diameter!

A typical tree can absorb up to 10 kg of CO_2 per year → 26 000 000 000 typical trees required!

Figure 6-7 Carbon dioxide absorbed by trees

Carbon emissions is a general term for greenhouse gas emissions, which can absorb large amounts of energy from heat radiation in the environment, producing the well-known greenhouse effect and causing further warming of the atmosphere. This can also produce a heat island effect, causing temperatures in cities to be significantly higher than those in the outer suburbs (Figure 6-8). Due to the high temperature near the ground in the central area of the heat island, the atmosphere does upward movement, forming a difference in air pressure with the surrounding areas, and the atmosphere near the ground in the surrounding areas converges to the central area, thus forming a low-pressure vortex in the central area of the city, so that the air pollutants formed by the burning of fossil fuels in people's life, industrial production and transportation will gather inevitably in the heat island center, endangering people's health and even life.

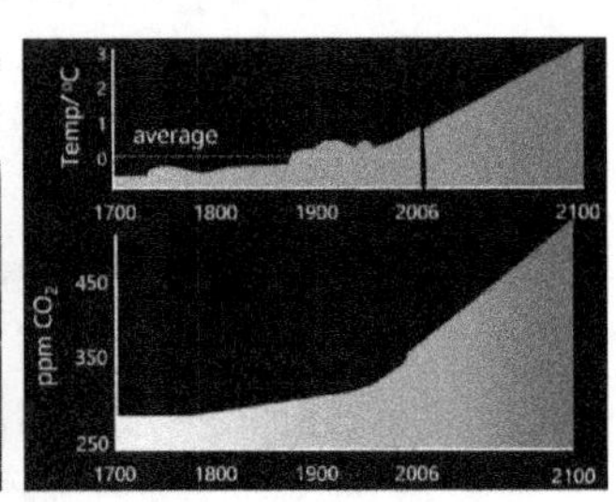

Figure 6-8 Negative impacts of carbon emissions

• CO_2 has increased by 0.007% in the past 50 years (Table 6-1).

Table 6-1 Atmospheric composition

Composition	Molecular formula	Content
Nitrogen	N_2	78.08%
Oxygen	O_2	20.95%
Argon	Ar	0.93%
Carbon dioxide	CO_2	0.039%
Others	Ne, He, CH_4, H_2, N_2O, O_3	0.01%

• Increasing by about 1.7 ppm per year.

The greenhouse effect (Figure 6-9):

• Water vapor (36%-70%);

• Carbon dioxide (9%-26%);

• Methane (4%-9%);

• Ozone (3%-7%).

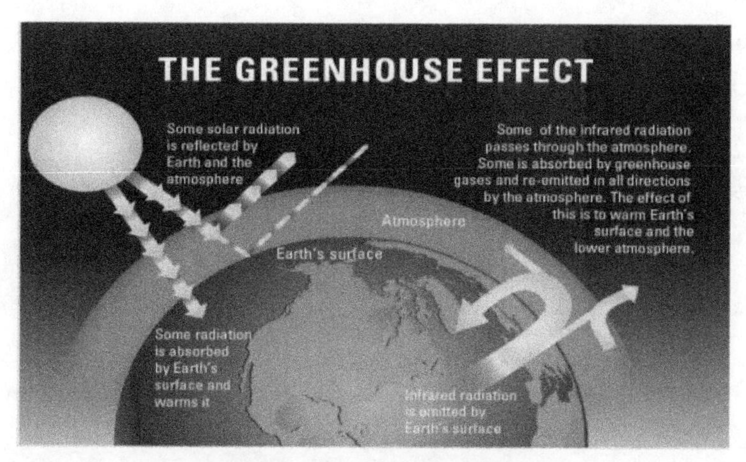

Figure 6-9 The greenhouse effect

(https://energyeducation.ca/encyclopedia/Greenhouse_effect)

The increase in the concentration of greenhouse gases leads to an increase in the ability of the atmosphere to be opaque to infrared radiation, which causes the emission of effective radiation into space from lower temperatures and higher altitudes. This results in a radiative forcing, an imbalance that can only be compensated for by an increase in the temperature of the ground troposphere system. If the greenhouse effect continues to increase, the global temperature will also continue to rise year by year (Figure 6-10).

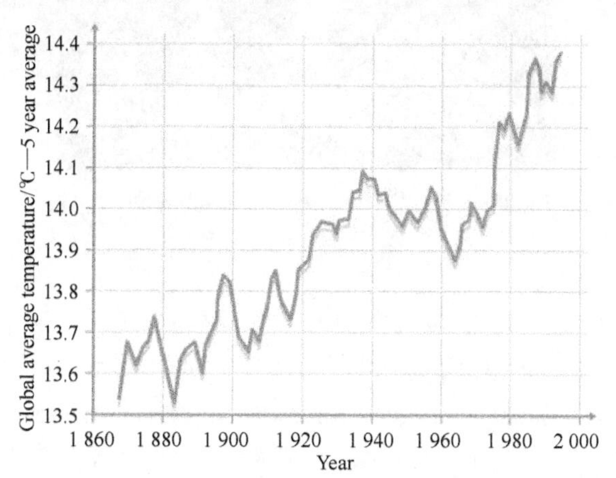

Figure 6-10 Leading to a significant climate change

(http://www.bbc.co.uk/schools/gcsebitesize/science/ocr_gateway/energy_resources/global_warmingrev2. shtml)

Effects of climate change (Figure 6-11):

• Sea temperatures in Antarctica have risen 5 times global average;

• Ice coverage in the Alps 50% is less than a century ago;

• Ice coverage in Greenland is reducing by 1 m per year;

• Rise in sea levels 2 mm/yr currently;

• Increased water cycle/rain fall;

• Dilution of sea water salt content;

• Impact on building performance has become a hot research topic.

Figure 6-11 Effects of climate change

6.1.2.5 Strategies to Deal with Emissions

Energy saving and emission reduction in buildings can save a lot of fossil fuels and make up for the shortage of energy resources in China (Figure 6-12). The low-carbon route is an internationally recognized effective model for reducing CO_2 emissions in recent years, and is the main path to achieve sustainable development and an effective way to cope with climate change.

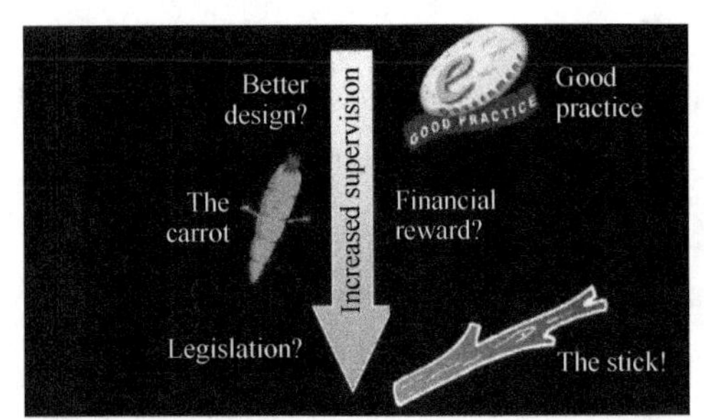

Figure 6-12 Strategies to deal with emissions

1. Building Design

• Better thermal insulation (Is this good for both summer and winter?) (Figure 6-13);

• More air-tight buildings (Is this good for both summer and winter?);

• Efficient shading design (Is this good for both summer and winter?);

• Renewable energy;

• Energy efficient systems;

• Energy efficient occupants.

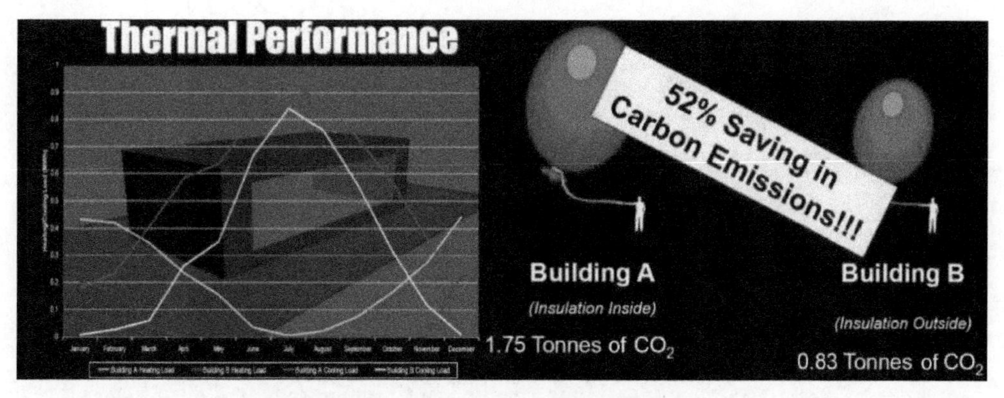

Figure 6-13 Inside and outside insulation

2. Financial Incentives

1) Climate Change Levy (CCL)

As one of the active promoters of the development of low carbon economy, the birthplace of industrial revolution and the pioneer of the existing high carbon economic mechanism, the UK is deeply aware of its historical responsibility in the process of climate change, so it has taken the lead in the world to raise the banner of developing low carbon economy and become the earliest advocate and practitioner of developing low carbon economy. Since April 2001, the UK has introduced a climate change tax to save energy and protect the environment (Figure 6-14).

CCL is an energy tax in the UK. It aims to encourage the efficient use of energy and promote the renewable energy, thereby helping the UK to achieve domestic and international targets for greenhouse gas emission reduction. CCL is one of the key initiatives in the UK government's strategy to shift the pressure of taxation from "positive" factors such as boosting employment to "negative" factors such as controlling environmental pollution.

The tax is based on the amount of coal, natural gas and electricity used, but the use of combined heat and power, renewable energy, etc. is exempt from the tax. The purpose of the tax is not to broaden the tax base and raise financial resources, but to improve energy efficiency and promote energy saving investment, which is a core part of the UK's overall climate change strategy.

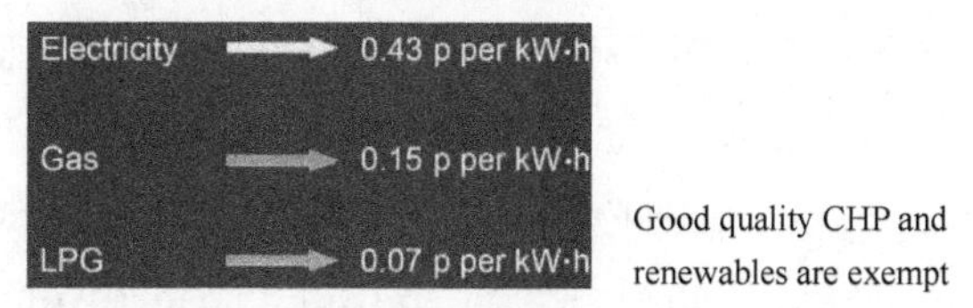

Figure 6-14 Climate change levy

2) Renewable incentives

· Subsidies for installation of new wind farms;

· Subsidies for feedback tariffs for photovoltaic power;

· Subsidies for enhancing insulation of domestic properties—the "Green Deal";

· Very low take up of "Green Deal";

· Average cost to consumers of the above—£130 annually.

3. Domestic Legislation

"Responding to Climate Change: China's Policies and Actions" , which is the first official law on the subject of "combating climate change" in China, aiming at controlling and reducing greenhouse gas emissions and promoting sustainable development (Figure 6-15).

中国应对气候变化的政策与行动
Responding to Climate Change:
China' s Policies and Actions

（2021年10月）
中华人民共和国
国务院新闻办公室
The State Council Information Office of
the People' s Republic of China
October 2021

Figure 6-15 Responding to Climate Change: China's Policies and Actions

Since the 18th National Congress of the Communist Party of China (CPC) convened in 2012, guided by Xi Jinping thought on eco-civilization and committed to the new development philosophy, China has made combating climate change a top priority of national governance. It has steadily reduced the intensity of its carbon emissions, reinforced the effort to achieve its Nationally Determined Contributions (NDCs), and maximized its drive to mitigate climate change. It has adopted green and low-carbon approaches in its economic and social development, and worked to build a modernized country, in which humanity and nature coexist in harmony.

At the general debate of the 75th Session of the United Nations General Assembly on September 22, 2020, President Xi Jinping announced that China would scale up its NDCs by adopting more vigorous policies and measures, strive to peak CO_2 emissions before 2030, and achieve carbon neutrality before 2060. China is taking pragmatic actions towards these goals.

As a responsible country, China is committed to building a global climate governance system that is fair, rational, cooperative and beneficial to all, and makes its due contribution to tackling climate change using its greatest strengths and most effective solutions. Confronted by the challenges of climate change, China is willing to work together with the international community to ensure the Paris Agreement delivers steady and lasting results, and make greater contribution to the global response.

The Chinese government is publishing this white paper to document its progress in mitigating climate change, and to share its experience and approaches with the rest of the international community.

4. Overseas Related Legislation

The European Union introduced the Energy Performance of Buildings Directive (EPBD) in 2003. The UK Government followed with the Energy Performance of Buildings (Certificates and Inspections) Regulation in 2007. The introduction of Display Energy Certificate (DEC) for public buildings is part of a series of legislation to dramatically improve the energy efficiency of the UK's building stock, both new and existing.

1) Energy Performance Certificate (EPC)

An EPC is required whenever a domestic property is:

• Built;

• Sold;

• Rented.

Along with information on the nature and extent of a property, Energy Performance Certificates provide information on the energy efficiency of a property, along with estimates on its running costs (Figure 6-16). An EPC will go on to make recommendations on how to improve the energy efficiency of the premises and reducing the running costs.

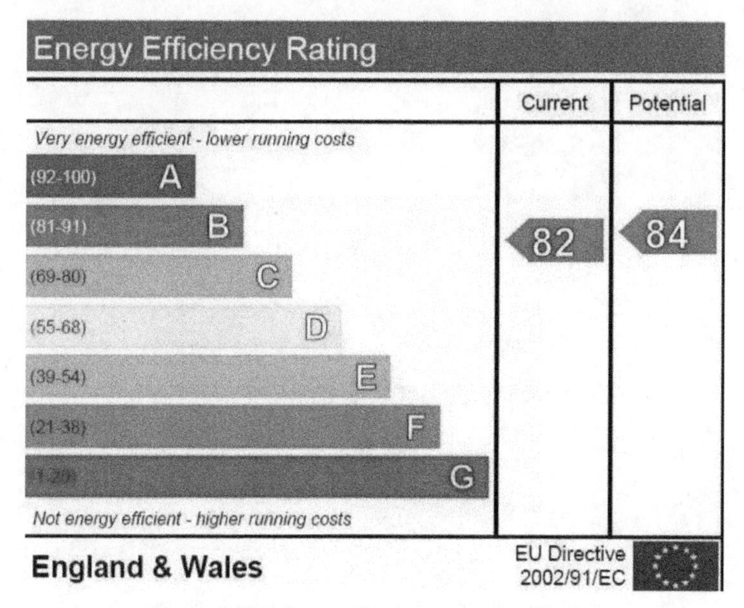

Figure 6-16 Energy Performance Certificates

Valid for 10 years, Energy Performance Certificates in China provide an overall efficiency score for domestic dwellings. Scores are out of 100 and the higher the rating the more energy efficient a home will be. The overall score out of 100 will determine which band a property falls into, with A being the most efficient and G being the least efficient.

2) Display Energy Certificate (DEC)

DEC aims to encourage the owners of public buildings to adopt energy efficiency measures by displaying their energy performance (Figure 6-17). A DEC shows the energy performance of a building based on actual energy consumption. The rating is an indicator of the annual CO_2 emissions from operating the building. This rating is shown on a scale from A to G, where A is the lowest CO_2 emissions and represents the best performing buildings. Also shown on the certificate are the ratings for the previous two years which provides information on whether the energy performance of the building is improving. The rating is based on the amount of energy consumed over a period of 12 months and is taken from meter readings.

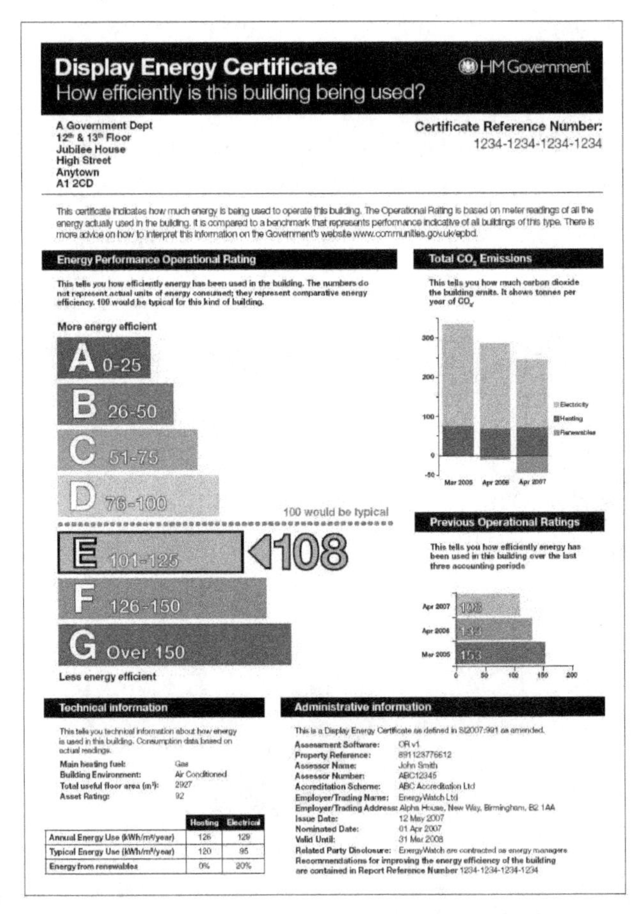

Display Energy Certificate
How efficiently is this building being used?

HM Government

A Government Dept
12th & 13th Floor
Jubilee House
High Street
Anytown
A1 2CD

Certificate Reference Number:
1234-1234-1234-1234

This certificate indicates how much energy is being used to operate this building. The Operational Rating is based on meter readings of all the energy actually used in the building. It is compared to a benchmark that represents performance indicative of all buildings of this type. There is more advice on how to interpret this information on the Government's website www.communities.gov.uk/epbd.

Energy Performance Operational Rating

This tells you how efficiently energy has been used in the building. The numbers do not represent actual units of energy consumed; they represent comparative energy efficiency. 100 would be typical for this kind of building.

More energy efficient
A 0-25
B 26-50
C 51-75
D 76-100
100 would be typical
E 101-125 ◁108
F 126-150
G Over 150
Less energy efficient

Total CO_2 Emissions

This tells you how much carbon dioxide the building emits. It shows tonnes per year of CO_2

Previous Operational Ratings

This tells you how efficiently energy has been used in this building over the last three accounting periods

Technical information

This tells you technical information about how energy is used in this building. Consumption data based on actual readings.

Main heating fuel: Gas
Building Environment: Air Conditioned
Total useful floor area (m^2): 2927
Asset Rating: 92

	Heating	Electrical
Annual Energy Use (kWh/m²/year)	126	129
Typical Energy Use (kWh/m²/year)	120	95
Energy from renewables	0%	20%

Administrative information

This is a Display Energy Certificate as defined in SI2007:991 as amended.

Assessment Software: OR v1
Property Reference: 891123776612
Assessor Name: John Smith
Assessor Number: ABC12345
Accreditation Scheme: ABC Accreditation Ltd
Employer/Trading Name: EnergyWatch Ltd
Employer/Trading Address: Alpha House, New Way, Birmingham, B2 1AA
Issue Date: 12 May 2007
Nominated Date: 01 Apr 2007
Valid Until: 31 Mar 2008
Related Party Disclosure: EnergyWatch are contracted as energy managers
Recommendations for improving the energy efficiency of the building are contained in Report Reference Number 1234-1234-1234-1234

Figure 6-17 Display Energy Certificate (DEC)

(1) All non-domestic buildings must display their DEC if:

· It is at least partially occupied by a public authority (e.g. council, leisure center, college,

NHS trust);

·· Total floor area > 250 m²;

· Frequently visited by the public.

(2) DECs last for 1 year for buildings with a total useful floor area more than 1 000 square meters.

(3) DECs last for 10 years for buildings between 250 and 1 000 square meters.

6.2 Sun and Building

6.2.1 Building Sunshine

6.2.1.1 Definition of Building Sunshine

Building sunshine is an important topic in architectural optics based on the principle of direct sunlight and sunshine standards to study the relationship between sunshine with architecture and the application of sunshine in architecture. The purpose of studying building sunshine is to make full use of sunlight in order to meet the requirements of indoor light environment and hygiene, while preventing overheating in the room. Sunlight can meet the needs of building lighting; in kindergartens, nursing homes, hospital wards and residences, sufficient direct sunlight also has the role of sterilization and promoting human health, etc. , and can raise the indoor temperature in winter. In some places, solar energy is also available as an energy source. Sunlight also has negative aspects, for example, direct sunlight in summer can cause high indoor temperatures; under direct sunlight, it can cause glare. In addition, direct sunlight can also accelerate the aging, fading and deterioration of some items, and even cause some explosive materials to explode. Therefore, measures should be taken to prevent direct sunlight in museums, exhibition halls, galleries, book stores, textile workshops, precision instrument workshops and dangerous goods warehouses.

6.2.1.2 Factors Related to Building Sunshine

Building location and form, layout and orientation, lighting, shading, heat storage and heat collection are all closely related to sunlight.

Solar radiation conditions vary from region to region, in the absence of meteorological effects, the possible hours of sunlight from sunrise to sunset is the possible sunshine duration, i.e. the maximum hours of sunlight at a given location. The possible sunshine duration can be calculated depending on the local latitude and season. In the Northern Hemisphere, it generally increases with latitude in summer and vice versa in winter.

Sunshine duration is the actual hours of light that the sun shines directly on the ground from sunrise to sunset. It is measured in a sunshine recorder. Sunshine duration varies from place to place depending on weather, latitude, and topography. At the same place, sunshine is affected by weather and season. South of the Southern Ridge of China, western Sichuan and eastern Guizhou have only about 1 250-1 500 hours per year; the Huaihe and Hanshui basins have about 2 000 hours, northern China about 2 500 hours, and northwest China about 3 000 hours.

The ratio of sunshine duration to possible sunshine duration is called the percentage of sunshine rate. Buildings in different climatic regions have different sunshine requirements. Cold-winter regions should maximize the acquisition, storage and utilization of solar radiation energy, hot summer regions should focus on shading, while hot-summer and cold-winter regions should take both into account.

6.2.1.3 Sunshine Spacing

With the reform of China's housing system, residential quality has become a hot topic of social concern. People are more and more concerned about the requirements of building orientation, ventilation and sunlight. When planning the building spacing, not only the requirements of ventilation, orientation and fire prevention should be considered, but also sunlight is a key factor in determining the building spacing.

Sunshine spacing refers to the minimum interval distance maintained between the front and rear rows of south-facing houses to ensure that the bottom floor of the rear row gets not less than two hours of full window sunshine on the winter solstice (or the cold day). The distance between houses is determined according to the sunshine standard. In other words, a certain distance is maintained between the front and back of the house in order to ensure that the rooms on the ground floor in the back row receive the required amount of sunlight for a specified period of time.

Taking the flat land as an example, the formula for calculating the daylight spacing between two buildings arranged north and south is (Figure 6-18):

$$D = (H - H_1) / \tan h$$

where h is the solar altitude angle at 12 noon on the winter solstice; H is the height of the front row of buildings; H_1 is the height of the south sill on the ground floor of the rear building; D is the minimum daylight spacing between the front and back rows of buildings.

If the sunshine duration required for a room increases, its spacing will increase accordingly, or when the building is not facing south, its spacing will also change. When the houses are arranged on the slope, the spacing of sunlight will change due to the different slope and direction of the terrain under the same sunlight requirement.

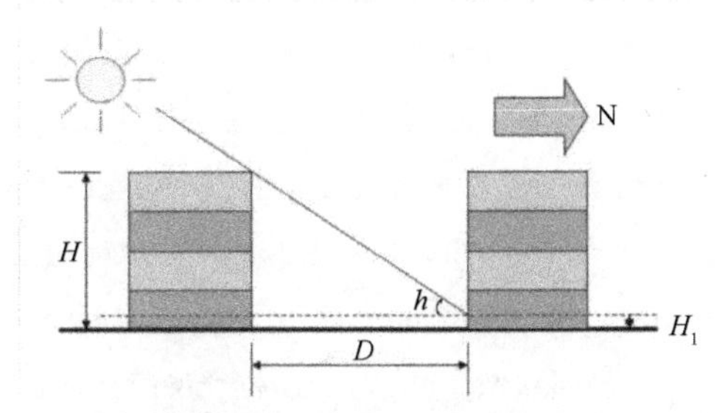

Figure 6-18 Sunshine spacing

When the building is arranged parallel to the contour line and facing the sunny slope, the steeper the slope, the smaller the sunshine spacing can be; vice versa, the larger. Sometimes, in order to fight for daylight and reduce the building spacing, the buildings can be arranged diagonally or perpendicular to the contour line.

6.2.1.4 Solar Building

The sun shines on the earth, and there is no geographical limitation, no matter land or sea, no matter mountain or island; it is available everywhere, can be directly developed and utilized, easy to collect, and doesn't need to mine and transport. Since the birth of life on the earth, it has been living mainly on the thermal radiation energy provided by the sun. In the case of decreasing fossil fuels, solar energy has become an important part of human use of energy. Solar energy is utilized in two ways—photothermal conversion and photoelectric conversion, and solar power is an emerging renewable energy source.

Solar buildings refer to buildings that use direct solar energy as a preferential energy source for heating and cooling. The purpose of its application is to use solar energy to meet the energy needs of buildings, including energy supply for heating, air conditioning, domestic hot water, lighting, household appliances, etc. Solar buildings can save a lot of electricity, coal and other energy, while not polluting the environment. The use of solar energy for heating and cooling is especially beneficial in areas with long annual sunshine hours, clean air, adequate sunlight and lack of other energy sources.

6.2.2 Solar Shading in Buildings

6.2.2.1 Definition of Solar Shading in Buildings

Architectural shading is a necessary measure to avoid direct sunlight into the interior, to prevent the building envelope from being overheated by sunlight, thus preventing local overheating and glare, and to protect various indoor items. Its reasonable design is an important factor to improve indoor thermal comfort in summer and reduce energy consumption of buildings.

Whether they are transparent window components or other opaque building envelopes such as roofs, facades, etc., most of the solar radiation can enter the interior through them, and the room temperature rises rapidly due to the greenhouse effect, which is one of the main reasons for the high indoor temperature in summer. Modern buildings due to the extensive use of large glass on the facade, coupled with the widespread use of industrialized lightweight structures, has exacerbated the deterioration of the indoor thermophysical environment, which exists even in cold regions, so the control of solar radiation heat for this part is very important.

6.2.2.2 Four Methods for Reducing Solar Heat Gain

1. Orientation & Window Size

According to the variation regulation of solar altitude and azimuth, the south-facing windows of the building can reduce solar radiation heat gain in summer and increase solar radiation heat gain in winter, which is the most favorable orientation of the building.

In order to avoid excessive window openings affecting building energy consumption, the relevant standards for building energy efficiency require the window openings of buildings, i.e., window-to-wall ratio (WWR) (Figure 6-19). It is the ratio of the total area of the transparent part of the windows and balcony doors on the exterior wall of the whole building to the total area of the exterior wall of the whole building. Here the area of the window emphasizes the transparent part of the area, that is, the area of the window with light function; if the back of the open window is with a wall or blocking, it should not be included in the window area.

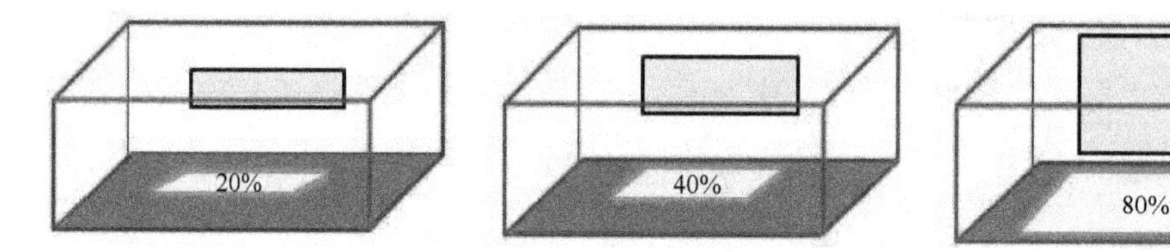

Figure 6-19 Window-to-wall ratio

Usually the heat transfer thermal resistance of the window is much smaller than that of the wall, therefore, the cold and heat consumption of the building increases with the increase of the window-to-wall ratio, on the contrary, the smaller the window-to-wall ratio design, the smaller the heat loss, the better the energy saving effect. Generally speaking, the solar radiation intensity and insolation rate of different orientations are different, and the solar radiation heat obtained by the windows is not the same, the solar radiation intensity and insolation rate of south and north orientations are high, and the solar radiation heat obtained by the windows is more.

The window-to-wall ratio is the measure of the percentage area determined by dividing the building's total glazed area by its exterior envelope wall area (Table 6-2).

Table 6-2 Orientation in relation to windows

Orientation	WWR/%
N	50
NE/NW	40
E/SE/W/SW	32
S	40
Horizontal	12

2. Internal Blinds/Curtains

Shading devices installed within the transparent envelope of a building are internal shading products (Figure 6-20). Due to the relatively simple structure and convenient construction, they are widely used in residential buildings. Internal shading also has its shortcomings. When internal shading is used, solar radiation passes through the glass, causing the internal shading itself to heat up. This part of the heat has actually entered the room, and a large part of it will increase the temperature of the room by convection and radiation.

- Shading devices may be controlled automatically or manually (Figure 6-21);
- For the latter case, occupant behavior plays an important role.

Figure 6-20 Internal blinds/curtains (Gunay et al., 2014)

Figure 6-21 Automatic or manual control

3. Self-shading of Special Glasses

1) Window properties

- Reflectance (ρ): the ratio of luminous flux reflected from a surface to the luminous flux incident on it.

• Transmittance (τ): the ratio of luminous flux transmitted by a surface to the luminous flux incident on it.

• Absorptance (α): the ratio of luminous flux absorbed by a surface to the luminous flux incident on it.

$$\rho + \tau + \alpha = 1$$

2) Solar films for windows

Glass with solar film can effectively insulate and block UV rays, especially in summer when it is difficult to withstand the heat of direct sunlight. With glass with solar film applied, we can not only block heat radiation outside, but also create a comfortable space with a constant temperature range inside (Figure 6-22).

3) Low-E (emissivity) windows

Low-E coatings have been developed to minimize the amount of ultraviolet and infrared light that can pass through glass without compromising the amount of visible light that is transmitted (Figure 6-23).

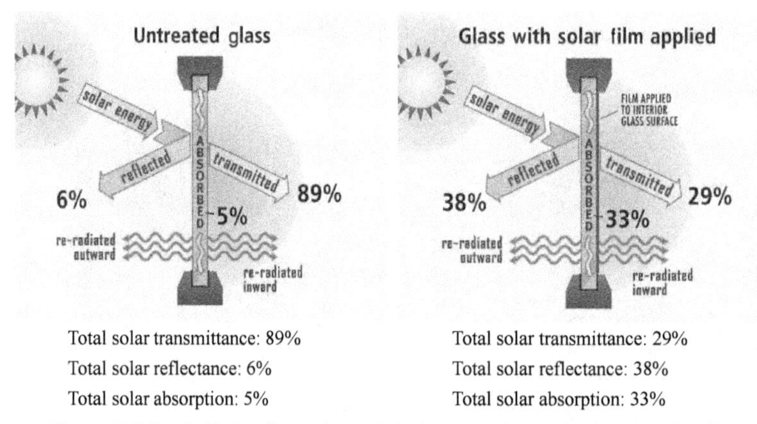

Figure 6-22 Self-shading of special glasses with and without solar film

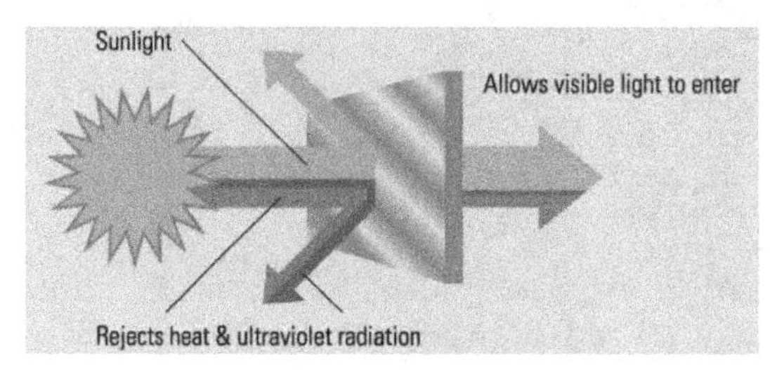

Figure 6-23 Low-E Window

4. External Shading Devices

Among various shading systems, external shading system is the most effective in reducing indoor heat gain in buildings and regulating indoor light and heat environment (Figure 6-24).

Because solar radiation, after being blocked by shading facilities, does not reach the building surface directly, but is reflected and absorbed by shading facilities, and only a small part of it reaches the building surface through shading facilities.

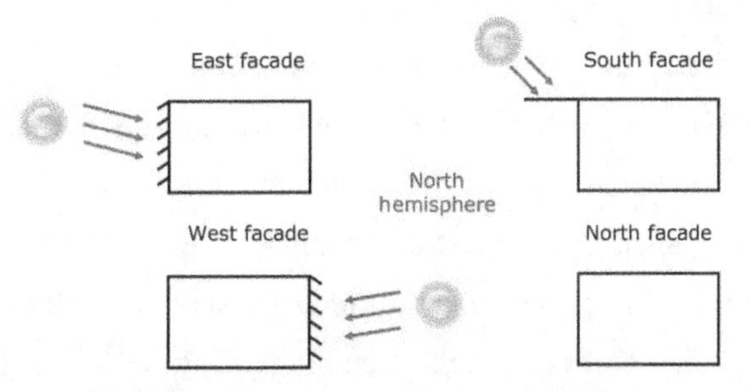

Figure 6-24 Facades in different orientations

On the one hand, the external shading facilities can transfer the direct radiation heat from the sun to the surrounding environment through reflection, reducing the radiation heat gain of the building from the sun; on the other hand, after the external shading facilities absorb the heat radiated from the sun, the temperature rises and they can release heat to the surrounding environment by infrared longwave radiation, only a small part of which is radiated to the building surface.Therefore, the energy saving effect of external shading is very significant and is considered to be the most ideal passive energy saving measure for window openings.

• External shading, i.e. overhang, louvre, shutters, fin and awning (Figure 6-25);

• The selection may be orientation-dependent or application-dependent.

Figure 6-25 External shading

6.2.2.3 Shading Performance

1. Cooling Effect of Shading

Shading has a positive effect on preventing the rise of indoor air temperature. Even in the case of open windows, the difference in indoor air temperature with and without shading is significant. In the hot summer, it has a certain significance for the improvement of indoor thermal environment. In the case of closed windows, the role of shading to prevent the rise of room temperature is more obvious, and because it is set up so that the room temperature fluctuation is smaller, the appearance of the highest indoor temperature will be further delayed, which is meaningful to reduce the cold load in air-conditioned rooms while also allowing a more uniform distribution of the temperature field in the room.

2. Energy-saving Effect of Shading

Architectural shading plays an important role in building energy saving in two aspects: on the one hand, shading measures can effectively block a large amount of solar radiation from entering the room, reducing the air-conditioning and cooling load of the building in summer; on the other hand, shading panels can convert direct sunlight into soft diffuse light, improving the quality of the indoor light environment and thus reducing daytime artificial lighting energy consumption. The energy-saving effect of architectural shading design is generally described by shading coefficient. The shading coefficient is the ratio of the solar radiation penetrating into the window with shading and the solar radiation penetrating into the window without shading during the irradiation time. The smaller the coefficient, the smaller the solar radiation heat through the window, the better the heat protection effect. The effect of shading facilities to block solar radiation heat depends on not only the form of shading, but also the construction of shading facilities, installation location, materials and colors, and other factors.

3. Dimming Effect of Shading

Architectural shading has two effects on natural lighting: on the one hand, architectural shading can block the entry of direct sunlight or convert it into softer diffuse light, avoiding glare directly onto the working surface, thus meeting people's requirements for lighting quality and reducing the energy consumption of daytime artificial lighting; on the other hand, architectural shading measures do reduce the level of indoor illumination. It is observed that the general indoor illuminance is reduced by about 53% to 73%, but the distribution of indoor illuminance is more even.

4. Impact of Shading on Ventilation

The application of shading facilities has a two-fold impact on the natural ventilation of the room. On the one hand, sunshade will have a certain blocking effect on the room ventilation, so that the indoor wind speed is reduced. Experimental data show that in the room with sunshade, the indoor wind speed is reduced by about 22%-47%, depending on the construction method of sunshade; on the other hand, sunshade will play a certain guiding role on the ventilation of the building.

6.3 Occupant Behavior vs. Building

6.3.1 Occupant Behavior vs. Building Energy

6.3.1.1 Performance Gap

The total energy use of 10 identical homes varied, even though they had the same floor area (102 m^2), on the same street, built in same year and with similar efficiencies (Figure 6-26). This variation is even larger at the energy end use level (e.g. up to 10.6 times in space heating energy use).

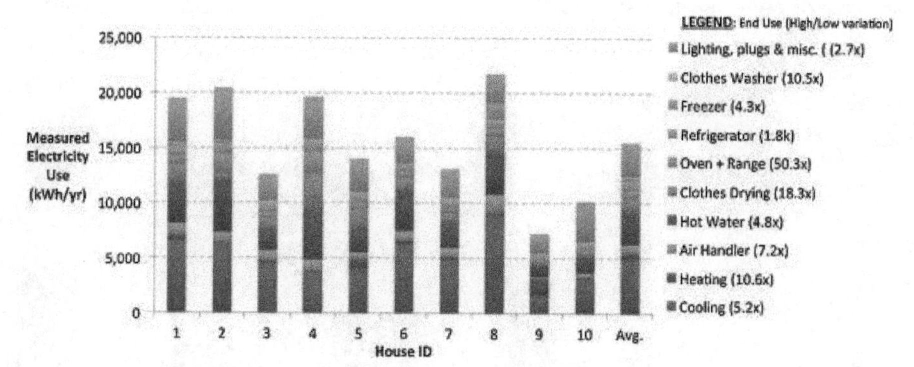

Figure 6-26 The measured electricity use for ten nearly identical homes (Hong et al., 2016)

6.3.1.2 Contents

- What do we mean by occupant behavior (OB) in buildings?
- Importance of OB in buildings.
- Better understanding OB in buildings.
- Changing OB in buildings for promoting building performance.
- Recommended readings.

6.3.1.3 Occupant Adaptive Behavior

In buildings, "if a change occurs such as to produce discomfort, people react in ways that tend to restore their comfort".

Adaptive behavior in buildings mainly includes:

- Opening/closing windows;
- Opening/closing fans;
- Opening/closing internal shading devices, e.g. blinds/curtains;
- Adjusting thermostatic settings of air-conditioning (AC) systems;

• Adjusting clothing insulation;

• Adjusting activity levels;

• Drinking hot/cold water;

• etc.

Significance of occupant behavior is as follows:

• Big impact on both building energy consumption and occupant comfort;

• Big contribution to the performance gap between building design and operation;

• Huge potential on saving energy for buildings by changing behavior.

6.3.1.4 Impact of Occupant Behavior

"Occupant behavioral aspects and socio-economic aspects are critical, partly directly and also significantly indirectly through their influence on choices and decisions about the physical characteristics of buildings and systems"

—Professor Koen Steemers at Cambridge University, UK (Figure 6-27)

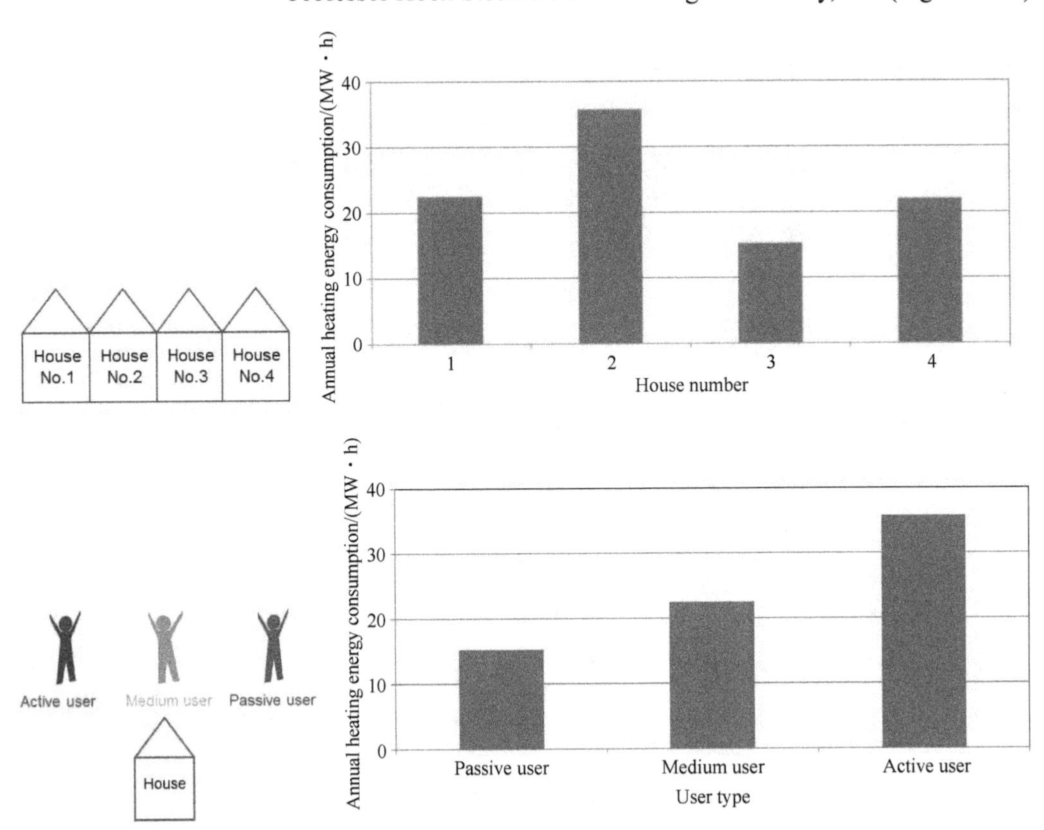

Figure 6-27 The influence of occupant behavior on building energy use

6.3.1.5 Occupant Behavior vs. Building Energy Retrofit

• Retrofit measures (Figure 6-28);

• Improving airtightness;

- Upgrading external wall insulation;
- Upgrading roof insulation;
- Increasing window layers.

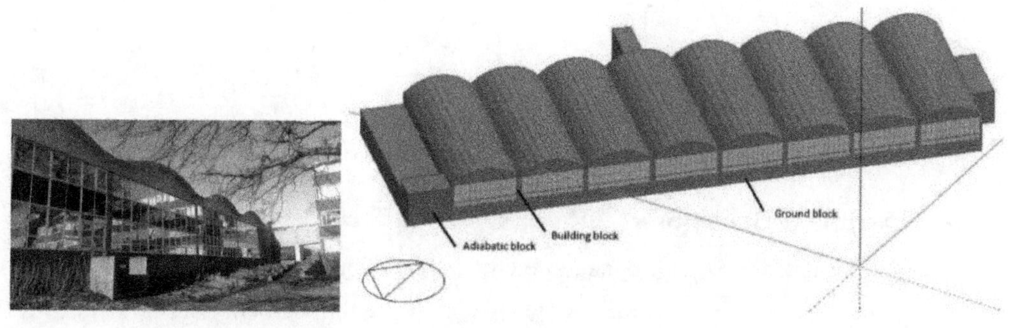

Figure 6-28 Retrofit measures

$$E_{\text{pre-retro}} - E_{\text{post-retro}} = E_{\text{saving}}$$

(1) Heating behavior (Figure 6-29).

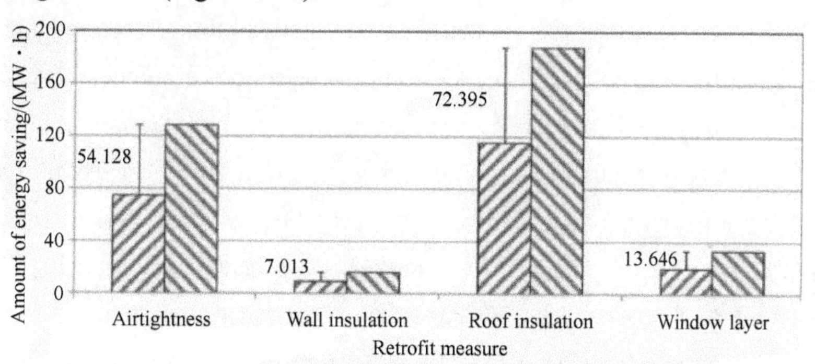

Figure 6-29 Comparison of energy-saving potential of different retrofit measures for various heating conditions

(2) Window behavior (Figure 6-30, Figure 6-31).

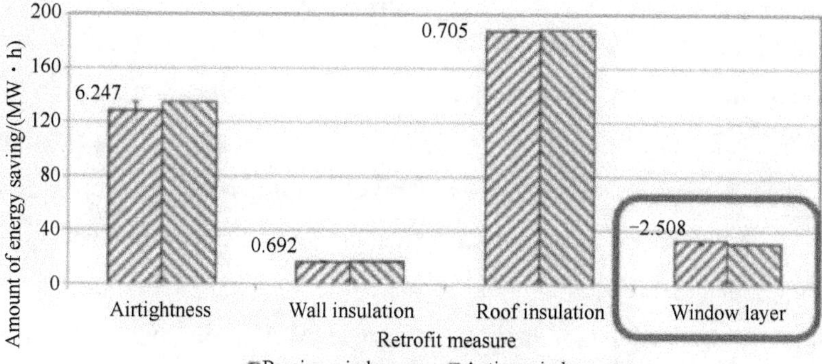

Figure 6-30 Comparison of energy-saving potential of different retrofit measures for various window opening conditions

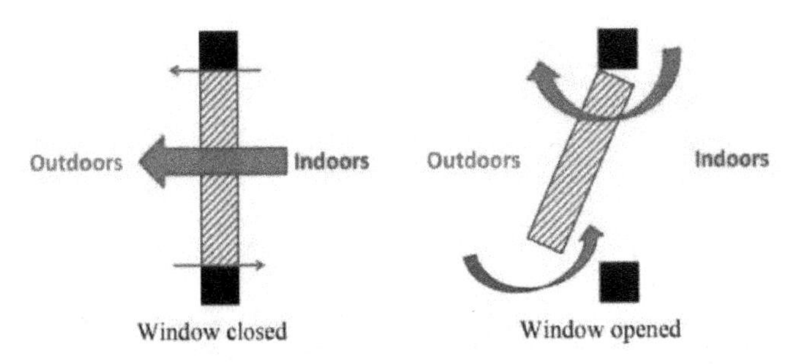

Figure 6-31 Heat loss for different states of windows

6.3.1.6 Occupant Behavior vs. Indoor Thermal Environment

Research presented here uses dynamic thermal simulation to model the effect of passive heat wave mitigating interventions for UK dwellings. The main bedroom (Bed 1) is located at the rear (and therefore south-facing) and the kitchen and bathroom are located within an extension at the rear (Figure 6-32) .

Figure 6-33 show the effect of the interventions on overheating exposure measured by the number of degree hours over 26 ℃ for the main bedrooms in the end terrace and mid terrace houses. The lighter portion of each bar shows the increase in the number of degree hours of overheating under the elderly occupancy profile.The building orientation was found to have a substantial impact on overheating exposure, varying by almost 100% between different orientations. Coating the walls with a high performance solar reflective paint (light walls) was found to be the most effective intervention for the end terrace house in all cases, reducing the number of degree hours over 26 ℃ by between 50% and 60%. External wall insulation, added to reduce heat loss through the walls in the heating season, also shields the outer brick surface from direct solar radiation, whilst leaving the internal plastered walls exposed to provide radiative cooling. Internal wall insulation was found to be less effective than external wall insulation, this is thought to be due to the exclusion of coolth stored in the thermal mass resulting in a loss of radiative cooling from the internal faces of the external walls and also because occupant heat gains are being retained within the rooms. In some cases, the addition of internal wall insulation is seen to increase the number of degree hours compared to the base case by up to 14% for elderly occupancy. External shutters were found to reduce the number of degree hours by up to 39% and, though more expensive, are more effective than internal blinds (up to 20% reduction) and curtains (15% reduction). Internal window shading devices cannot prevent some of the solar radiation being trapped inside the room and converted to long wave radiation. The window rules ventilation strategy was not found to have a great effect on bedroom overheating because during bedroom occupied hours the outside air temperature is rarely higher than the bedroom temperature. The night ventilation intervention, though applied to ground floor windows, still had a benefit for bedroom overheating by cooling the building as a whole. Increasing the level of loft insulation does reduce annual space heating energy use and also has a small benefit (around 4%) in reducing overheating in bedrooms.

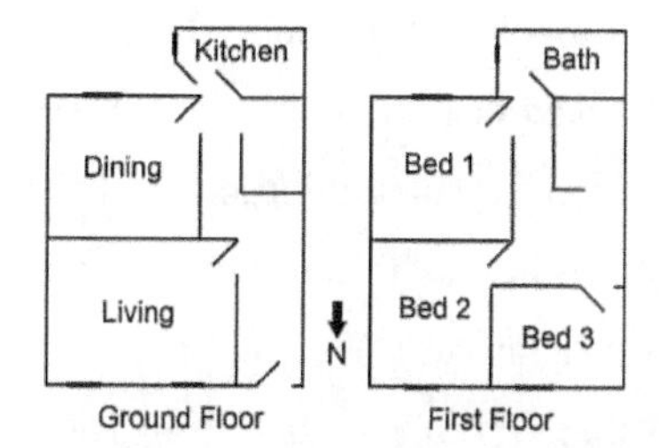

Figure 6-32 Terraced house floor plans of ground floor and first floor

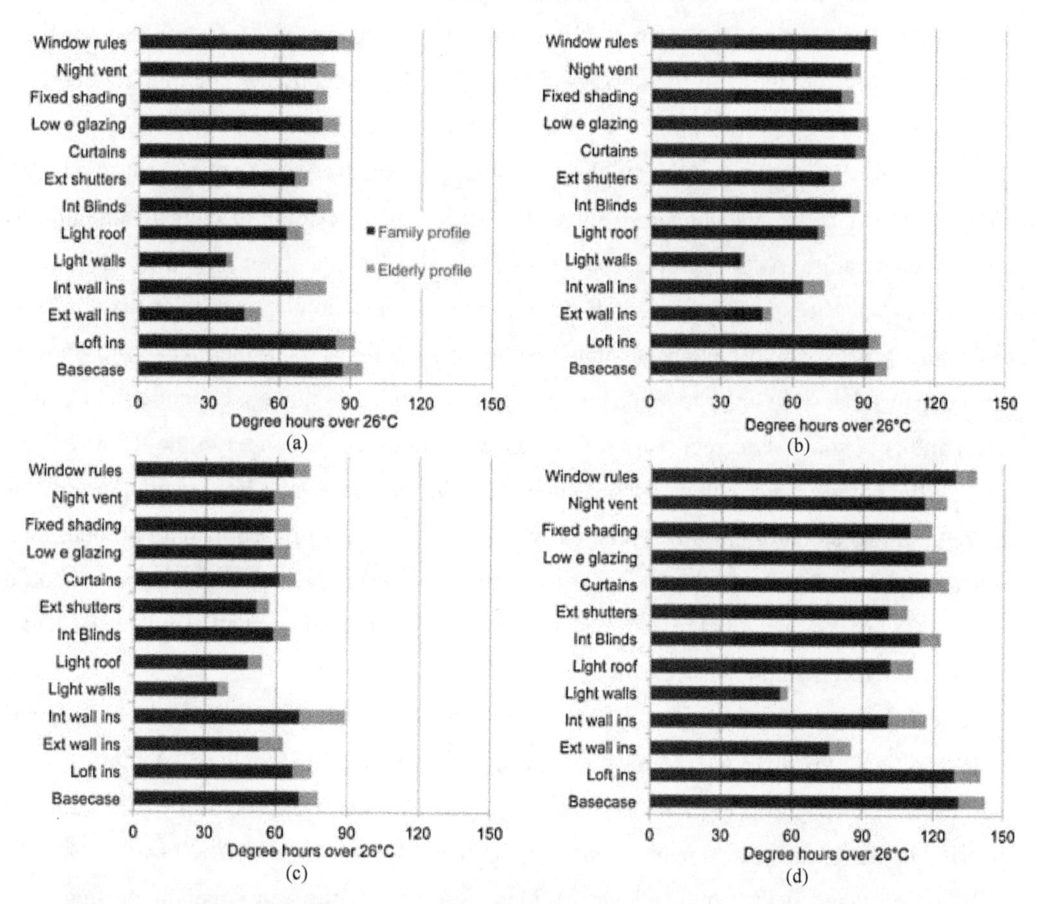

Figure 6-33 End terrace bedroom 1 single interventions for family and elderly profiles (Porritt et al., 2012)

(a) End terrace bedroom 1 facing north (b) End terrace bedroom 1 facing south (b) End terrace bedroom 1 facing east (c) End terrace bedroom 1 facing west

6.3.1.7 Occupant Behavior vs. Occupant Comfort

In the mid-1980s, ASHRAE began funding a series of field studies of thermal comfort in office buildings spread across four different climate zones. Since that time numerous other thermal comfort researchers independently adopted the same procedures for collecting both physical and subjective thermal comfort data in their own field studies. In 1995, ASHRAE RP-884 began by collecting raw field data from various projects around the world that had followed this standardized (or a similar) protocol, and/or where the data met strict requirements regarding measurement techniques, type of data collected, and database structure. Standardized data processing techniques, such as methods for calculating clo and various comfort indices, were then applied consistently across the entire database. This enabled RP-884 to assemble a vast, high-quality, internally consistent database of thermal comfort field studies. The RP-884 database contains approximately 21 000 sets of raw data from 160 different office buildings located on four continents, and covering a broad spectrum of climate zones.

Figure 6-33(a) and (b) shows some of the most compelling findings from separate analysis of HVAC and NV buildings, in the upper and lower panels, respectively. The graphs present a regression of indoor comfort temperature1 for each building against mean outdoor air temperature recorded for the duration of the building study in question. Regressions were based only on buildings that reached statistical significance (P=0.05) in the derivation of their own neutral or preferred temperature. As a result, 20 buildings in the RP-884 database had to be eliminated from this analysis because of their small sample sizes or very homogeneous indoor climates. Use of the 20 cases would have necessitated certain assumptions about thermal sensitivities of their occupants that were unsustainable on the empirical evidence at hand. Each graph in Figure 6-34(a) and (b) , shows two regressions, one based on observed responses in the RP-884 database, and the other on predictions using Fanger's PMV. The latter take into account clo, metabolic rate, air speed and humidity averaged within the building in question. The original data points through which the adaptive model regression line was fitted are also shown, but it should be noted that this is a weighted regression so that outliers representing small sample sizes had a relatively smaller effect on the slope of the model.

Note that the observed comfort temperatures have been corrected for the effects of semantics. In addition, the adaptive model fitted by weighted regression to observed comfort temperatures in Figure 6-34(a) has an R^2=53% (P=0.0001). The adaptive model fitted by weighted regression to observed comfort temperatures in Figure 6-34(a) has an R^2=70% (P=0.0001).

When provided with a higher level of adaptive opportunities (e.g. opening/closing windows, adjusting blind/shading positions and turning up/down thermostatic settings) to rebalance comfort, occupants displayed greater comfort acceptance.

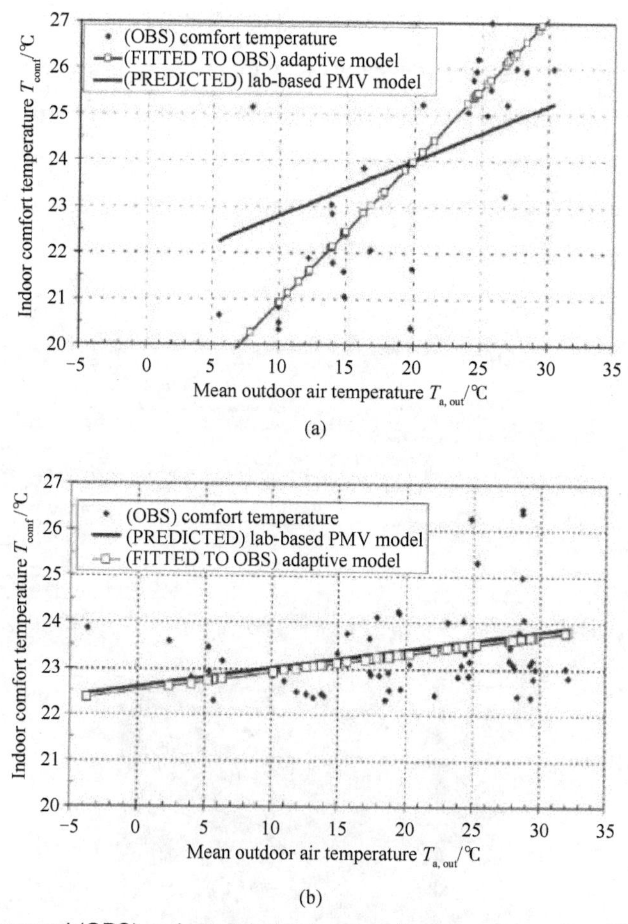

Figure 6-34 Observed (OBS) and predicted indoor comfort temperature from RP-884 database for HVAC buildings and naturally ventilated buildings

(a)HVAC buildings (b)Naturally ventilated buildings

6.3.2 Better Understanding Occupant Behavior

6.3.2.1 Capturing Occupant Behavior

The combination of all collected information has to generate a space for possible interpretation of the occupants' preferences, although collecting information of occupants' behaviors is a challenge. So there is significant room for improvement of predicting the actual energy consumption in the building, currently used four methods for sensing human presence and actions can grant access to the partial information regarding the occupant's behavior.

Method 1: Self-recording by building occupants (Figure 6-35).

Method 2: Recording by electronic measuring devices.

In observational studies, the occupants' behavior and presence and indoor environmental variables are passively monitored. Because occupants rarely interact with some of the building components (e.g. blinds) and these interactions may exhibit seasonal variations (e.g. windows) (Figure 6-36), the monitoring periods to gather data to develop and verify an occupant model typically extend over at least numerous months. Also, the selection and placement of sensors to monitor occupant interacting building components and presence can play an unprecedented role over the way occupants exhibit behavioral patterns.

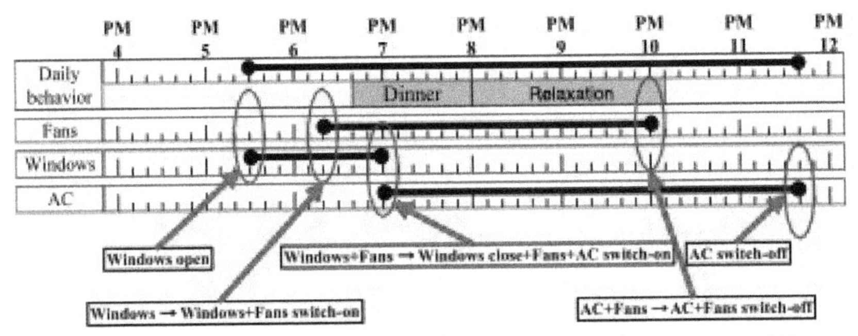

Figure 6-35 Self-recording by building occupants (Nakaya et al., 2008)

Figure 6-36 Contact sensors for monitoring window opening and closing behaviors (Anderson et al., 2013)

Method 3: Observing by surveyors.

Manual observation methods (e.g., photographic or manual recording) posed a limitation on sample size—both number of occupants and duration—and did not lend themselves to developing statistical models that are suitable for building simulation. They were nevertheless valuable for identifying key behaviors and motivators (Table 6-3).

Method 4: Self-estimating by building occupants.

The survey was conducted aiming to present the opening window status of the residents, ascertain the features of indoor thermal sensation, house area, storey number, time at home, household number, feeling of indoor air quality and health consciousness and reflections of these features on the opening window habit (Table 6-4).

Table 6-3 Manual recording (Johnson and Long, 2005)

Visit	Time	Number of open windows per wall	Number of open doors (omit garage)	Floor location of openings (circle one)	Status of car door of attached garage	Likeli-hood of AC operation (circle one)	Likeli-hood of occupancy (circle one)	Evidence supporting occupancy rating	Precip. during last hour	Special conditions (write in)
A	____ am pm	39 Front_ 40 Right_ 41 Left_ 42 Back_ 43 Total_	44 Front_ 45 Right_ 46 Left_ 47 Back_ 48 Total_	49 None 50 Ground 51 Upper 52 Both	53 Closed 54 Open w/vehicle 55 Open w/o vehicle	56 100% 57>50% 58<50% 59 0% 60 Uncertain	61 100% 62>50% 63<50% 64 0% 65 Uncertain	Write in:	66 Yes 67 No 68 Uncertain	
B	____ am pm	69 Front_ 70 Right_ 71 Left_ 72 Back_ 73 Total_	74 Front_ 75 Right_ 76 Left_ 77 Back_ 78 Total_	79 None 80 Ground 81 Upper 82 Both	83 Closed 84 Open w/vehicle 85 Open w/o vehicle	86 100% 87>50% 88<50% 89 0% 90 Uncertain	91 100% 92>50% 93<50% 94 0% 95 Uncertain	Write in:	66 Yes 67 No 68 Uncertain	

Table 6-4 Survey information of opening windows (Huang et al., 2014)

Category	Questions	Answers
Part Ⅰ: Window opening habit	1. What's the frequency in the heating days? 2. What's the frequency in a day? 3. What's the continuous time each time? 4. What's opening size? 5. Single side ventilation or cross ventilation?	A. Scarcely B. Not often C. Often D. Nearly every day A. 1 B. 2 C. 3 D. More A. 0-5 min B. 6-10 min C. 11-12 min D. More A. Small B. Moderate C. Big D. Fully open A. Single side ventilation B. Cross ventilation
Part Ⅱ: Potential related factors of opening window	1. What's the indoor thermal sensation? 2. What's the house area? 3. What's the storey number? 4. What's the total time at home in a day? 5. What's the household number? 6. What's the feeling of IAQ? 7. How much do you care about personal health?	A. Cold B. A little cold C. Moderate D. A little hot E. Hot A. 20-60 m^2 B. 61-100 m^2 C. 101-140 m^2 D. >140 m^2 A. ≤8 B. 9-16 C. more A. 8-11 h B. 12-16 h C. 17-19 h D. 20-24 h A. 1 B. 2-3 C. 4-6 D. More than 6 A. Good B. No feeling C. Bad A. Care a lot B. Care a little C. Don't care

Advantages and disadvantages of the above mentioned four methods are listed in Table 6-5.

Table 6-5 Advantages and disadvantages of four methods

Method name	Advantages	Disadvantages
Method 1	1. Easy to use; 2. Large sample size	1. Survey-fatigue; 2. Accuracy
Method 2	1. Accurate; 2. Continuous measurement; 3. No human factors	1. Limited sample size (cost); 2. Malfunction of devices

Continued

Method name	Advantages	Disadvantages
Method 3	1. Easy to use; 2. Large sample size; 3. Accurate	1. Time involvement; 2. Limited visit per object
Method 4	1. Easy to use; 2. Large sample size	1. No real-time data; 2. Do people do what they said?

6.3.2.2 Factors Influencing Occupant Behavior

Factors influencing occupant behavior, both external and individual, that could be named with the general term "drivers" , are the reasons leading to a reaction in the building occupant and suggesting him or her to act (they namely "drive" the occupant to an action). There is a distinction to be made within the factors influencing the occupant behavior in relation to the natural ventilation. In Table 6-6, the major parameters found in literature driving the occupant behavior aimed at controlling the indoor environment in relation to natural ventilation are split into five categories of influencing factors for residential and office buildings.

In Table 6-7, Columns 2 and 3 provide a series of numbers that establish the current research findings on each factor. Column 2 provides the number of papers that report a correlation between the factor and space-heating behavior, and Column 3 indicates the number of papers that report no correlation between the factor and space-heating behavior.

Within these factors, some have been studied more frequently than others, and the verdict on the individual factors is quite varied. Although the number of existing studies on each factor varies, the following factors can be said to be unambiguously assumed to be influential on space-heating behavior in residential buildings: outdoor climate, dwelling type, room type, house insulation, type of temperature control, occupant age, time of day and occupancy. The remaining factors can be classified into three categories: ① the influence of that factor has been confirmed in a small number of existing studies and no papers reject its influence; this category includes indoor relative humidity, type of heating system, occupant gender, occupant culture/race, social grade, previous dwelling type, perceived IAQ and noise, and health; ② the influence of that factor has been rejected in a small number of existing studies and no papers confirm the influence; this category only includes dwelling size; and ③ the influence of that factor has both been confirmed and rejected in nearly equal numbers of existing studies; this category includes dwelling age, type of heating fuel, heating price, energy use awareness, occupant education level, household size, family income, house ownership, thermal sensation and time of week. The factors belonging to these three categories still need further investigations to fully establish their influence.

Table 6-6 Driving forces for energy-related behavior with respect to ventilation/window operation in residential buildings (Fabi et al., 2012)

Physiological	Psychological	Social	Physical environmental	Contextual
Age	Perceived illumination	Smoking behavior	Outdoor temperature	Dwelling type
Gender	Preference in terms of temperature	Presence at home	Indoor temperature	Room type
			Solar radiation	Room orientation
			Wind speed	Ventilation type
			CO_2 concentrations	Heating system
				Season
				Time of day

Table 6-7 Overview of literatures evaluating the influencing factors of occupant space-heating behavior

Potential drivers	Does the driver influence occupant space-heating behavior?	
	No. of papers reporting a correlation	No. of papers reporting no correlation
01. Outdoor climate	7	0
02. Indoor relative humidity	1	0
03. Dwelling type	7	0
04. Dwelling age	3	1
05. Dwelling size	0	1
06. Room type	8	0
07. House insulation	5	0
08. Type of heating system	3	0
09. Type of temperature control	9	0
10. Type of heating fuel	1	1
11. Occupant age	14	2
12. Occupant gender	2	0
13. Occupant culture/race	2	0
14. Occupant education level	1	1
15. Social grade	1	0
16. Household size	4	2
17. Family income	5	4
18. Previous dwelling type	1	0
19. House ownership	3	1
20. Thermal sensation	2	1
21. Perceived IAQ and noise	1	0

Continued

22. Health	1	0
23. Time of day	9	0
24. Time of week	1	1
25. Occupancy	8	0
26. Heating price	2	1
27. Energy use awareness	3	1

(WEI S, JONES R, DE WILDE P. Driving factors for occupant-controlled space heating in residential buildings[J]. Energy and Buildings, 2014, 70:36-44)

6.3.2.3 Modeling Occupant Behavior (Deterministic Modeling)

• Conventional behavioral modeling approach in most dynamic building performance simulation tools, e.g. EnergyPlus, IES VE and DesignBuilder;

• Easy and straightforward;

• Generating the same result among simulations;

• Question: Will people perform an adaptive behavior, e.g. opening the window, at a certain time or certain condition? (Table 6-8)

Table 6-8 Modeling Occupant Behavior

	Time	Value	
1	00:00	0.00	Time-based profile
2	09:00	0.00	
3	09:00	1.00	
4	17:00	1.00	
5	17:00	0.00	
6	24:00	0.00	

	Time	Value	
1	00:00	0.00	Condition-based profile
2	09:00	0.00	
3	09:00	(ta>23)	
4	17:00	(ta>23)	
5	17:00	0.00	
6	24:00	0.00	

6.3.2.4 Modeling Occupant Behavior (Probabilistic/Stochastic Modeling)

• Occupant behavior should be presented as "an algorithm for the likelihood of a control being used rather than a simple on/off condition" (Figure 6-37).

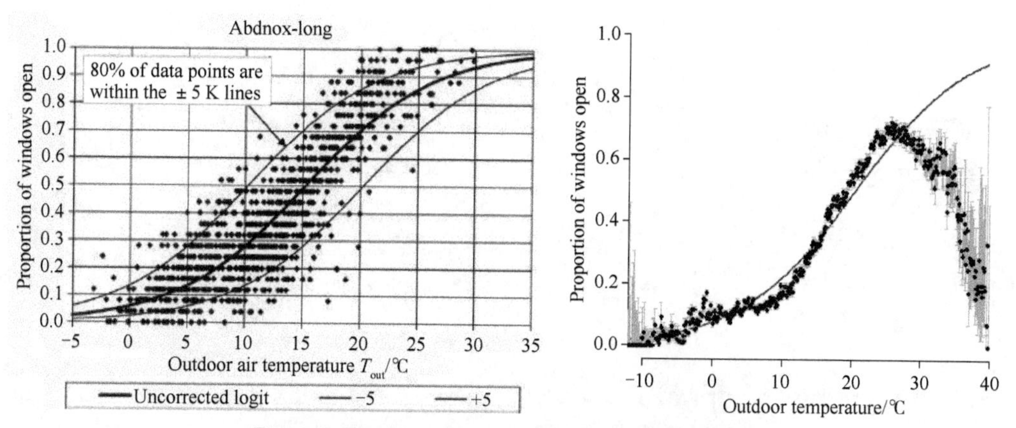

Figure 6-37 The effect of behavior on architecture

(Rijal et al., 2007; Haldi and Robinson, 2009)

- Logistic regression analysis (Figure 6-38):

$$Logit(p_i) = \ln\frac{p_i}{1-p_i} = A + B_1x_{1,i} + \ldots + B_kx_{k,i}$$

$$p_i = e^{A+B_1x_{1,i}+\ldots+B_kx_{k,i}} / \left(1 + e^{A+B_1x_{1,i}+\ldots+B_kx_{k,i}}\right)$$

- Bernoulli process/Markov chain:

$$pr\left(X_{n+1} = x \middle| X_1 = x_1, X_2 = x_2, \ldots, X_n = x_n\right) = pr\left(X_{n+1} = x \middle| X_n = x_n\right)$$

$$pr\left(X_{n+1} = x \middle| X_1 = x_1, X_2 = x_2, \ldots, X_n = x_n\right) = pr\left(X_{n+1} = x\right)$$

- Inverse function method (Figure 6-39):

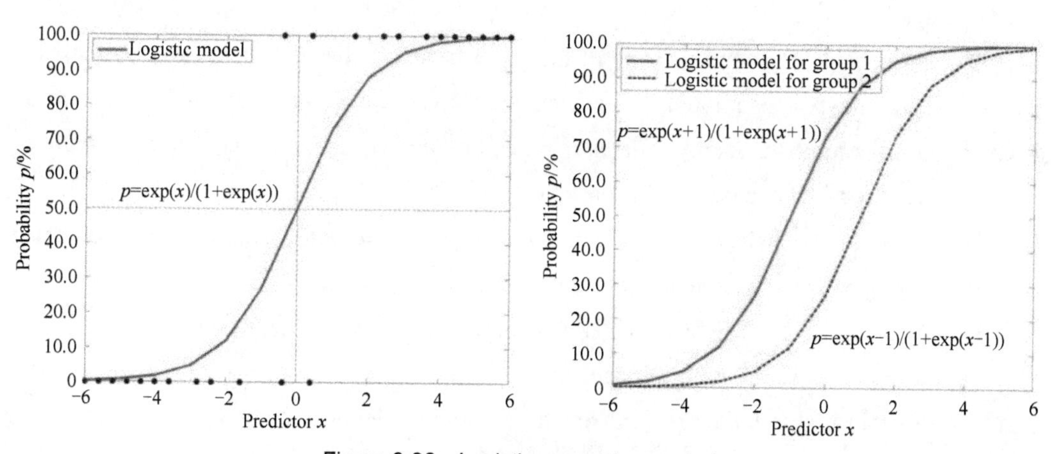

Figure 6-38 Logistic regression analysis

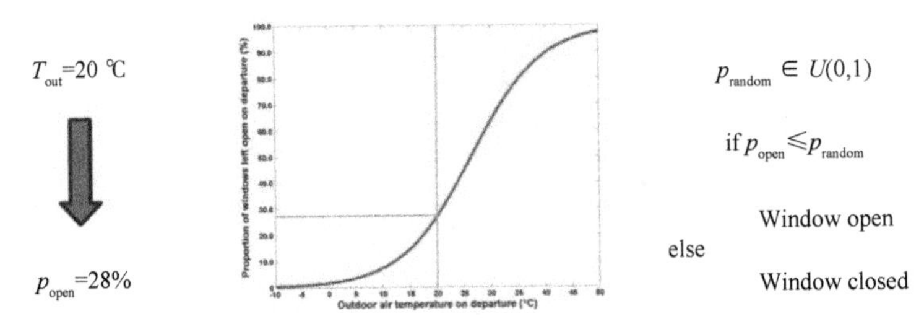

Figure 6-39 Inverse function method

6.3.3 Changing Occupant Behavior

6.3.3.1 Changing Occupant Behavior for Promoting Building Performance

In order to limit the increasing energy demands of the building sector, the impact of occupant behavior has become a growing research topic. Numerous studies have investigated the impact of occupants on the energy consumption in buildings in order to qualitatively and quantitatively interpret occupant behavior, foster energy efficiency, and reduce the gap between the predicted and real energy consumption. Behavioral impacts on energy savings, simulations of building users' adaptive behaviors, control systems for energy, and comfort management of buildings were the subjects of research conducted in the literature. A growing awareness about how all these topics are highly interconnected seems necessary in order to significantly influence building energy use.

A major conclusion from this state of the art overview is that user behavior has a significant impact on energy consumption in residential and commercial buildings. Behavioral change is crucial in order to foster energy efficiency in the building sector, and users can be influenced to use less energy when exposed to feedback.

The impact of user behavior is difficult to quantify for methodological reasons. The human decision-making process is complex and multifactorial; thus, factors influencing behavior are also numerous and varied. The inner dynamic nature of occupant's energy behavior represents a challenge and multi-disciplinary approaches are needed to provide new insights into the domain. In order to assess building occupant behaviors, a scientific study which describes the dominant factors that are involved in energy behaviors has to be conducted with the occupants. Since occupants do not always make rational decisions, the manner of presenting the choice itself becomes determinant in adopting energy-efficient behaviors. Introducing energy conservation measures without taking into account user satisfaction can often be counter-productive because users are likely to try to adapt their environment to obtain satisfying conditions. Emphasizing behavior change would better achieve energy efficiency.

6.3.3.2 Related Tools

(1) Decision support tools:

· Retrofit advice tool by De Montfort University, UK;

· Energy Efficiency Educator by Plymouth University, UK.

(2) Retrofit advice tool:

· Developed by De Montfort University, UK;

· CREW (Community Resilience to Extreme Weather) project;

· Developed to assist when choosing retrofit adaptations to reduce dwelling overheating during heat wave periods;

· Based on dynamic building performance simulation tool, IES VE;

· http://www.iesd.dmu.ac.uk/crew/.

6.3.3.3 Energy Efficiency Educator

· Developed by Plymouth University, UK;

· eViz (Energy Visualization for Carbon Reduction) project;

· Developed to help building occupants make decision on changing behavior for reducing residential energy demand;

· Based on dynamic building performance simulation tool, EnergyPlus;

· Been tested with 14 potential tool users for its usability;

· Figure 6-40 shows the main window of the EEE. The EEE consists of three main steps to give householders advice on behavioral changes (e.g. turning down the thermostatic setting or reducing the daily use of the heating system). In Step 1, users (usually building occupants) are asked to provide some basic information about their building and building systems (Figure 6-41). In Step 2, they are required to input information about how they currently use the heating systems in their buildings and where do they want to change this behavior to. In Step 3, the EEE will estimate the effectiveness of all behavioral changes defined by the householders in Step 2 (Figure 6-42). This is done by predicting the building energy consumption using EnergyPlus for both the current behavioral circumstance and the ones they would like to change to, and then calculating the difference between them. The predicted financial saving for each behavioral change intention will be shown in Pound Sterling based on the information imported by the users in Step 1 and Step 2 and the simulation results obtained in Step 3.

Figure 6-41 depicts the sub-window where users input information about their building and building systems according to their real living conditions. If the user does not know some essential information such as the main orientation of the house and the energy efficiency rate of the boiler, further help can be obtained online by clicking the buttons below the question.This tailored information helps to increase the accuracy of the later predicted building performance by

EnergyPlus.

Home Energy Efficiency Simulator

File Steps Help

Energy Efficiency Educator

Step 1: Please input some information about your building and systems — Building and Systems

Step 2: Please input some information about your use of the heating system — Heating Behaviour

Step 3: Estimate the financial savings per month — Estimate

Effect of Change No.1: 33 Pounds 31 Pennies

Effect of Change No.2: 0 Pounds 0 Pennies

Effect of Change No.3: 0 Pounds 0 Pennies

eViz Energy Visualisation for Carbon Reduction

Copyright © Shen Wei, Building Performance Analysis Group, Plymouth University, 2014

Figure 6-40 Main window of energy efficiency educator

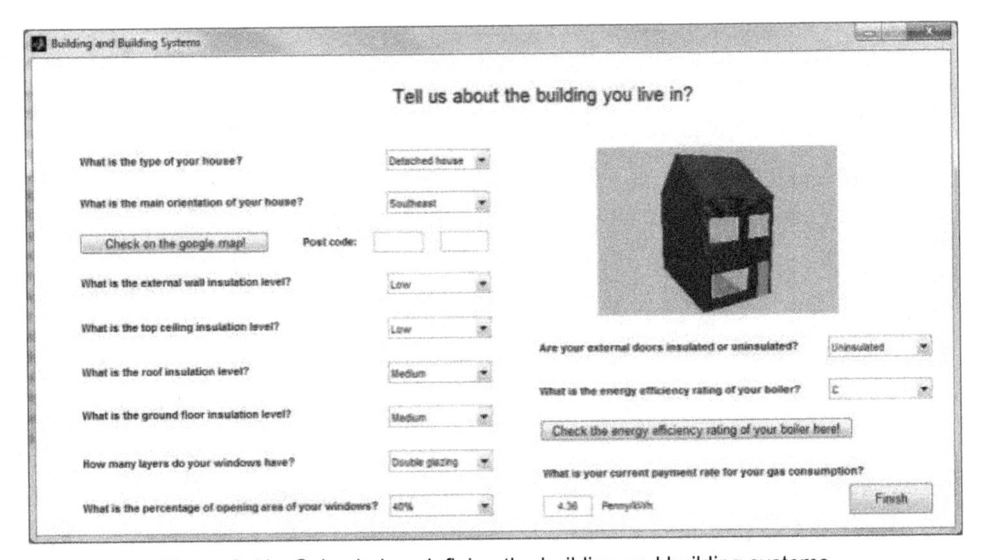

Figure 6-41 Sub-window defining the building and building systems

Figure 6-42 depicts the sub-window where users define their heating use behavior, for both the current state of behaviors and the one(s) they would like to change to (e.g. defined as behavior 01, 02 and 03). This section starts with a question about the user's normal occupancy schedule, which will be used to control the heating behavior defined later. The second group of questions asks for information about the user's preferred indoor air temperature, at the occupied time, unoccupied time and sleeping time respectively. The last group of questions is about the use of the heating boiler for both unoccupied and sleeping times. The first column in these two groups is used to define the user's current heating behavior and they have been provided with up to three options to change their behavior in columns 01, 02 and 03. This information will then be used to drive the dynamic building performance simulation shown in Figure 6-40, for an evaluation of the impact of various heating operations on the building energy consumption.

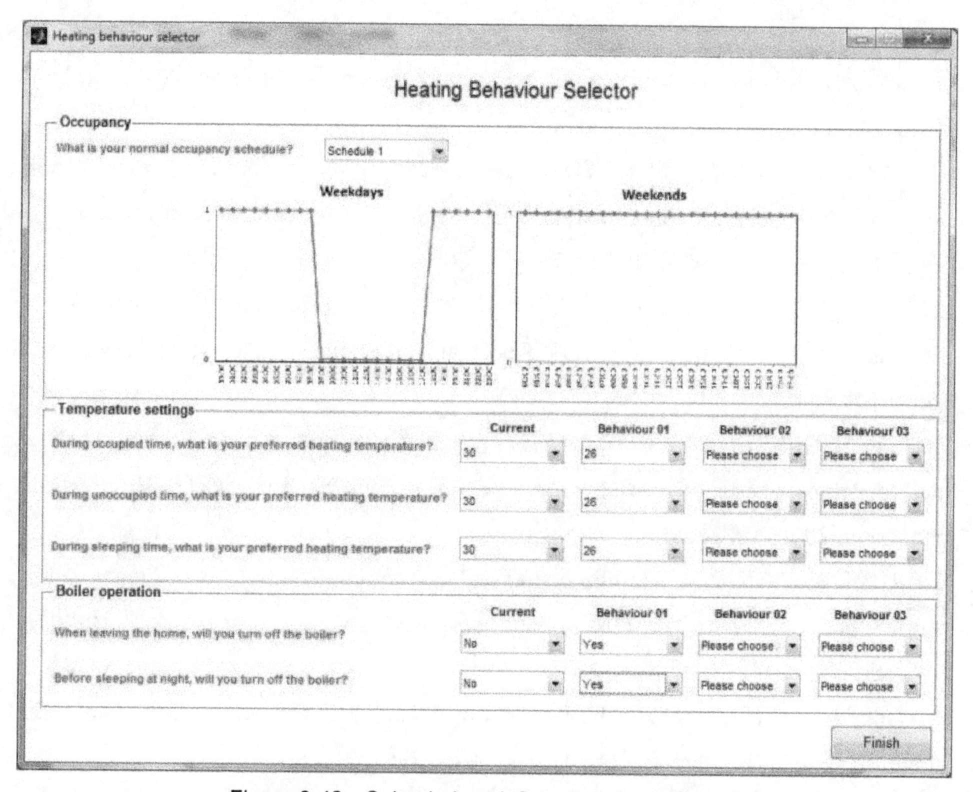

Figure 6-42 Sub-window defining heating behavior

References

[1] WEI S, JONES R, DE WILDE P. Driving factors for occupant-controlled space heating in residential buildings[J]. Energy and buildings, 2014, 70: 36-44.

[2] ROETZEL A, TSANGRASSOULIS A, DIETRICH U, et al. A review of occupant control on natural ventilation[J]. Renewable and sustainable energy reviews, 2010, 14(3): 1001-1013.

[3] GUNAY H B, O'BRIEN W, BEAUSOLEIL-MORRISON I. A critical review of observation studies, modeling, and simulation of adaptive occupant behaviors in offices[J]. Building and environment, 2013, 70: 31-47.

[4] LOPES M A R, ANTUNES C H, MARTINS N. Energy behaviors as promoters of energy efficiency: a 21st century review[J]. Renewable and sustainable energy reviews, 2012, 16(6): 4095-4104.

[5] FABI V, ANDERSEN R K, CORGNATI S P, et al. Occupants' window opening behavior: a literature review of factors influencing occupant behavior and models[J]. Building and environment, 2012, 58: 188-198.

[6] O'BRIEN W, GUNAY H B. The contextual factors contributing to occupants' adaptive comfort behaviors in offices: a review and proposed modeling framework[J]. Building and environment, 2014, 77: 77-87.

[7] VAN DEN WYMELENBERG K. Patterns of occupant interaction with window blinds: a literature review[J]. Energy and buildings, 2012, 51: 165-176.

[8] O'BRIEN W, KAPSIS K, ATHIENITIS A K. Manually-operated window shade patterns in office buildings: a critical review[J]. Building and environment, 2013, 60: 319-338.

[9] PEFFER T, PRITONI M, MEIER A, et al. How people use thermostats in homes: a review[J]. Building and environment, 2011, 46(12): 2529-2541.